Introduction to Nuclear Science

Third Edition

Introduction to Nuclear Science

Third Edition

Jeff C. Bryan
University of Wisconsin
La Crosse, Wisconsin

CRC Press
Taylor & Francis Group
Boca Raton London New York

CRC Press is an imprint of the
Taylor & Francis Group, an **informa** business

CRC Press
Taylor & Francis Group
6000 Broken Sound Parkway NW, Suite 300
Boca Raton, FL 33487-2742

© 2018 by Taylor & Francis Group, LLC
CRC Press is an imprint of Taylor & Francis Group, an Informa business

No claim to original U.S. Government works

Printed on acid-free paper

International Standard Book Number-13: 978-1-138-06815-5 (Hardback)

Visit the Taylor & Francis Web site at
http://www.taylorandfrancis.com

and the CRC Press Web site at
http://www.crcpress.com

For Chris

Contents

Preface to the Third Edition xi
Preface to the Second Edition.. xiii
Preface to the First Edition xv
Author xvii

Chapter 1   Introduction 1

    1.1   Radiation.. 1
    1.2   Atomic Structure 3
    1.3   Nuclear Transformations 6
    1.4   Nuclear Stability... 9
    1.5   Ionizing Radiation 12
    1.6   A Biological Threat? 13
    1.7   Natural and Anthropogenic Radiation 15
    1.8   The Chart of the Nuclides 19
    Questions.. 19

Chapter 2   The Mathematics of Radioactive Decay 23

    2.1   Atomic Masses and Average Atomic Masses.. 23
    2.2   The Nature of Decay 24
    2.3   Specific Activity 29
    2.4   Dating 30
    2.5   Branched Decay.. 32
    2.6   Equilibria.. 34
            2.6.1   Secular Equilibrium 35
            2.6.2   Transient Equilibrium 37
            2.6.3   No Equilibrium.. 39
    2.7   Statistics... 41
    Questions.. 46

Chapter 3   Energy and the Nucleus.. 51

    3.1   Binding Energy... 51
    3.2   Total Energy of Decay.. 55
    3.3   Decay Diagrams 57
    Questions.. 62

Chapter 4   Applications of Nuclear Science I: Power and Weapons 65

    4.1   Nuclear Power... 65
            4.1.1   Nuclear Fission 67
            4.1.2   Nuclear Reactors 69

4.1.3   Nuclear Fuel 72
4.1.4   Reactor Safety 75
4.1.5   Nuclear Waste.. 79
4.1.6   Cost of Nuclear Power.. 84
4.1.7   Proliferation of Nuclear Weapons 86
4.2   Nuclear Weapons.. 87
4.2.1   Fission Bombs 87
4.2.2   Fusion Bombs.. 90
4.2.3   Other Bombs 92
4.3   Nuclear Forensics 93
Questions.. 95

Chapter 5   Radioactive Decay: The Gory Details 97

5.1   Alpha Decay 97
5.2   Beta Decay... 102
5.3   Positron Decay.. 106
5.4   Electron Capture.. 108
5.5   Multiple Decay Modes 113
5.6   The Valley of Beta Stability 114
5.7   Isomeric Transitions 119
5.8   Other Decay Modes... 122
5.8.1   Spontaneous Fission 122
5.8.2   Cluster Decay 123
5.8.3   Proton/Neutron Emission.. 124
5.8.4   Delayed Particle Emission... 124
5.8.5   Double Beta Decay... 126
Questions.. 127

Chapter 6   Interactions of Ionizing Radiation with Matter............................... 131

6.1   Ionizing Radiation 131
6.2   Charged Particles.. 131
6.3   Photons 143
6.3.1   Compton Scattering.. 143
6.3.2   The Photoelectric Effect... 145
6.3.3   Pair Production... 145
6.4   Attenuation of Gamma and X-Radiation....................................... 148
Questions.. 151

Chapter 7   Detection of Ionizing Radiation 155

7.1   Gas-Filled Detectors... 155
7.1.1   Ionization Chambers 155
7.1.2   Proportional Counters 156
7.1.3   Geiger-Müller Tubes.. 157

7.2   Scintillation Detectors 159
      7.2.1   Photomultiplier Tubes.................................... 159
      7.2.2   Inorganic Scintillators 160
      7.2.3   Organic Scintillators 162
7.3   Other Detectors 163
      7.3.1   Semiconductor Detectors 163
      7.3.2   Thermoluminescent Dosimeters 164
7.4   Gamma Spectroscopy... 165
Questions 176

Chapter 8   Applications of Nuclear Science II: Medicine and Food 179

8.1   Radiology... 180
8.2   Radiation Therapy 182
8.3   Food Irradiation... 187
8.4   Nuclear Medicine 188
      8.4.1   Radionuclide Production 188
      8.4.2   Radiopharmaceuticals 192
      8.4.3   Gamma Cameras 199
Questions 201

Chapter 9   Nuclear Reactions.. 203

9.1   Energetics 204
9.2   Cross Section 212
9.3   Yield 214
9.4   Accelerators 219
9.5   Cosmogenic Nuclides 222
Questions 224

Chapter 10   Fission and Fusion 227

10.1   Spontaneous Fission 227
10.2   Neutron-Induced Fission 229
10.3   Fusion 237
10.4   Stellar Nucleosynthesis.. 240
10.5   Synthesis of Unknown Elements 241
Questions 243

Chapter 11   Applications of Nuclear Science III: More about Nuclear
Reactors 247

11.1   Reactions in Reactors 247
11.2   Other Reactor Types.. 250
      11.2.1   Pressurized Heavy Water Reactors (PHWR)........... 251
      11.2.2   Gas-Cooled Reactors (GCR) 253

11.2.3   Light Water Graphite Reactors (RBMK) 253
11.2.4   Small Modular Reactors (SMR) 254
11.2.5   Thorium in Reactors 255
11.2.6   Generation IV Reactors 257
11.3   Reactor Safety Systems 259
11.4   Nuclear Power Plant Accidents 262
11.4.1   Three Mile Island 263
11.4.2   Chernobyl 264
11.4.3   Fukushima 266
11.5   Fusion Reactors 269
Questions 270

Chapter 12   Radiation Protection 273

12.1   Terms 273
12.2   Regulations and Recommendations 276
12.3   Risk 280
Questions 288

Chapter 13   X-ray Production 291

13.1   Conventional X-ray Beams 291
13.2   High-Energy X-ray Beams 296
Questions 306

Chapter 14   Dosimetry of Radiation Fields 309

14.1   Percent Depth Dose 309
14.2   Tissue–Air Ratio 320
14.3   Tissue Maximum Ratio 321
14.4   Isodose Curves 323
14.5   Moving Fields 324
14.6   Proton and Electron Beam Dosimetry 325
Questions 331

Appendix A: Nuclide Data 333

Appendix B: Table of Chi-Squared 403

Appendix C: Useful Constants, Conversion Factors, SI Prefixes,
               and Formulas 405

Appendix D: Periodic Table of the Elements 409

Bibliography 411

Index 413

Preface to the Third Edition

Those of you who made the transition from the first to second editions of this book can let out a sigh of relief. The changes this time are much less extensive. This book is only about 5% bigger than the second edition, and the table of contents remains unchanged. My primary goals for the third edition were three-fold: updating contemporary information, using more standard conventions, and adding a few more questions at the end of each chapter.

Bringing this book up to date mostly involved changes in the nuclear power industry. When I wrote the first edition, there was a lot of excitement over what was assumed to be a renaissance for construction of new nuclear reactors. The second edition came out shortly after the Fukushima meltdowns, and there was a great deal of uncertainty. Now (mid-2017), there seems to be some divergence. China, Russia, India, and some Middle Eastern countries are continuing to move aggressively forward with nuclear power plant construction, while most Western democracies are stepping away. I can't help but wonder if the rather pervasive, and largely irrational, fear of anthropogenic sources of ionizing radiation is at least partly to blame for so many countries turning away from nuclear.

When I first wrote this book, I used units and symbols that were familiar to me. I favored units such as curies and rems because I was accustomed to seeing them in my work at the U.S. National Laboratories. I also used k to represent the decay constant instead of λ because many general chemistry textbooks did the same, reinforcing the relationship between radioactive decay and first-order chemical kinetics while avoiding potential confusion with the symbol for wavelength. I did not anticipate this book enjoying any level of popularity outside of the United States. I've been pleasantly surprised. A couple of the anonymous reviewers responded to Taylor & Francis' request for suggestions by encouraging me to use more conventional units. So has my colleague, Shelly Lesher, a faculty member in the physics department at the University of Wisconsin–La Crosse. While I did not abandon U.S. units, I tried to provide more parity to international units. For any other k fans out there, I'm sorry, but this book was rather singular among nuclear science textbooks in eschewing λ as the decay constant.

When adding questions at the end of each chapter, I looked for holes in coverage. Remarkably lacking were vocabulary questions. I try to emphasize the importance of speaking the language of nuclear science to my students, yet I wasn't holding them to it with the end of chapter questions. I hope that more students will now correctly differentiate between "nuclide" and "isotope."

While I corrected a number of the errors present in the second edition, I probably missed a few and likely created some more. I accept full responsibility for all of them. Please feel free to let me know when you find them.

Jeff C. Bryan
University of Wisconsin
La Crosse, Wisconsin

Preface to the Second Edition

I never expected to be writing this. When I wrote the first edition of this textbook, I thought I had really gotten away with something. I wrote this book for students in my nuclear chemistry and radiation physics classes—almost all of whom are nuclear medicine or radiation therapy majors with a limited background in physical sciences and math. My main motivation was to provide them with an accessible textbook that would be a useful tool in their studies. I honestly didn't expect that anyone outside of my students (besides my mom) would be interested in buying it. Much to my surprise, it exceeded the publisher's sales expectations, and a second edition became possible.

I was happy to have this opportunity, as I wanted to add a chapter on dosimetry as well as flesh out a few sections in the existing text. It was also a chance to find out what readers would like to see in a new edition. I asked a student, Lucas Bartlett, to survey teachers who used the text in their classes, as well as those who had examined the text but were not yet using it with their students. I am grateful to all who responded, especially Jason Donev, Thomas Semkow, Brett McCollum, Craig McLachlan, Lawrence Duffy, Pierre St. Raymond, Dot Sulock, and Magdalen Normandeau. I wasn't able to incorporate all of their suggestions, but they will likely see their influence in some additions listed here:

- The mathematics in some of the examples in the first few chapters are explained in more detail.
- An appendix containing nuclear data for all nuclides mentioned in the text (plus some others) is added.
- Discussion of applications are expanded and, generally, moved to earlier chapters.
- Coverage of nuclear reactors is significantly expanded, with a new chapter looking at more reactor types, their safety systems, and recent accidents.
- The number of end-of-chapter problems was increased for every chapter.

Along with the new chapters and appendix, new sections, or subsections, were added on nuclear forensics, radiology, gamma cameras, and decay through proton or neutron emission. In my hands, the manuscript is about 60% larger than the first edition, so it is definitely bigger. Hopefully, you will also find that it is better and more flexible.

In trying to write a textbook that would be more generally useful, I was especially conscious of how much the first edition centered on nuclear science in the United States. I have tried to broaden this perspective somewhat by pointing out how things are sometimes done differently outside of the United States. I was also concerned that those teaching classes more focused on applications of nuclear science be better served by the second edition. By moving the applications chapters earlier in the text, it gives those instructors an opportunity to get their classes into discussions of applications sooner.

As was true with the first edition, I had a lot of help with this one. I'd like to thank all the students in my nuclear chemistry and radiation physics classes over the past few years for their suggestions and attention to detail. I'd especially like to thank those who worked directly on this edition with me: Mikaela Barrett, Travis Warmka, Kelsey Schilawski, Lucas Bartlett, Dan Hanson, and Mike Hoppe. I'd also like to thank my colleagues at the University of Wisconsin–La Crosse for their suggestions and review of portions of the manuscript, especially Shelly Lesher, Sherwin Toribio, Melissa Weege, Nishele Lenards, Bruce Osterby, and Aileen Staffaroni. I would also like to thank Mary Ellen Jafari of Gundersen-Lutheran Hospital for her comments on Chapter 8. Finally, I need to thank Douglas Boreham for getting me to think more critically of the LNT and ALARA (you'll have to read this book if these acronyms are unfamiliar!).

While I was able to correct a number of the errors in the first edition, the extensive nature of the revisions for this edition have undoubtedly created some more. I accept full responsibility for all of them. If you spot one, I'd appreciate hearing about it, as we may be able to correct it before the next printing.

I have enjoyed writing this book. I hope you enjoy reading it.

Jeff C. Bryan
University of Wisconsin
La Crosse, Wisconsin

Preface to the First Edition

Like many textbooks for specialized classes, this one was born out of frustration with the applicability of existing offerings. This text was initially designed for the nuclear chemistry course I have taught for the past five years. This class is almost entirely populated by nuclear medicine technology majors who need a solid background in fundamental nuclear science but have not studied math beyond college algebra and statistics. There are many excellent nuclear chemistry textbooks available to teach science or engineering seniors or graduate students but very few for those lacking a more extensive science and math background. Ehmann and Vance's *Radiochemistry and Nuclear Chemistry* served as my primary text for this class for the past four years. It is very well written and nicely descriptive, but it is really intended for radiochemists, not medical professionals. It is also slipping out of date, with no future edition planned. I also learned that some of my students were passing my class without ever cracking the primary text. Clearly, a more appropriate text was necessary.

Writing a nuclear chemistry textbook for nuclear medicine majors seemed too limiting, so this text also includes material relevant to other medical professionals using ionizing radiation for diagnosis and treatment of diseases. In particular, chapters were added discussing radiation therapy and X-ray production. Additionally, it seemed that an introductory nuclear science text should address important contemporary topics such as nuclear power, weapons, and food/mail irradiation. All too often, those speaking passionately to these issues are inadequately versed and resort to emotional appeals while ignoring some of the science. Therefore, this text is also meant to serve as a general primer in all things nuclear for the general public. Very little science and math background is assumed, only some knowledge of algebra and general chemistry.

The more extensive chapters on fission nuclear reactors are also designed to help those who will soon be entering the nuclear workforce. The nuclear power industry in United States has been static for almost 30 years but now seems on the verge of renaissance. Nuclear power now appears to a more environmentally benign source of electricity, and a significant number of new plants may be constructed in the near future. Workers at these plants will need a fundamental understanding of nuclear science as well as a basic idea of how the plant works. This text can serve has a primer for these workers.

Finally, this text can also be of use to scientists making a career start or move to the growing nuclear industry or to the National Laboratory system. I began 13 years of work at Los Alamos and Oak Ridge National Laboratories in 1988. At that point, I was thoroughly trained as a synthetic inorganic chemist and was largely ignorant of nuclear science. Some of my motivation in writing this text is to provide scientists in similar situations an easier way to better understand their new surroundings.

The term nuclear science is used rather deliberately in the title and throughout the text. It is meant to encompass the physics, mathematics, chemistry, and biology related to nuclear transformations and all forms of ionizing radiation. Fundamentally, nuclear processes are within the realm of physics, but it can also be argued that all the disciplines mentioned above play an important role in understanding the uses of these processes and the resulting ionizing radiation in today's world.

When this text was initially proposed, some of the reviewers rightly questioned whether my background was appropriate. After all, much of it could be characterized as synthetic inorganic chemistry and crystallography—there's not a lot of hard-core nuclear work there. In my defense, I would point out that I did pay attention when my National Laboratory colleagues "went nuclear." After 13 years at Los Alamos and Oak Ridge, some of it rubs off. I would also argue that a book that wishes to bring nuclear science to a broader audience must be written by someone who originally comes from outside the field and can still see it from an external perspective. My successful experiences teaching nuclear science to those who have a limited math and science background suggests that I might be a good choice to write this book. Ultimately, it will be the readers who decide if I have been successful in this effort.

This book draws heavily on those that have proceeded it, especially Ehmann and Vance's *Radiochemistry and Nuclear Chemistry*, and Choppin, Rydberg, and Liljenzin's *Radiochemistry and Nuclear Chemistry*. It is not meant as a replacement to either, as they are both much more detailed; rather it is a contemporary resource for those wishing to learn a little bit of nuclear science. For those who wish to explore nuclear science beyond this text, a complete listing can be found in the bibliography.

I had often considered writing this textbook over the past five years of teaching nuclear chemistry and physics at the University of Wisconsin–La Crosse. It did not become a reality until Lance Wobus of Taylor & Francis asked to meet with me at a meeting of the American Chemical Society about a year and a half ago. I am very grateful to Lance for seeking me out and for his constant support during the writing process.

This is easily the most ambitious writing project I have undertaken, and it would not have been possible without the assistance and encouragement of many others. First, I must acknowledge my predecessor in my current position at the University of Wisconsin–La Crosse, Willie Nieckarz. Based on his background in the nuclear navy, he pioneered our nuclear science programs. My nuclear chemistry course remains strongly based on his materials. I also acknowledge my students who have never been reluctant to question me beyond the limits of my knowledge and for their critical reading of earlier drafts of this text. I would especially like to single out former students Stephanie Rice, Rachel Borgen, and Jim Ironside for their extensive suggestions and help preparing the solutions manual and figures. I must also thank my colleagues in the Departments of Chemistry and Physics at the University of Wisconsin–La Crosse, for without their support and encouragement, this book would not have been written. I would also like to acknowledge my former colleagues and teachers who took their time to introduce me to nuclear science. Without them, I would not have been hired into the wonderful job I now enjoy. Finally, I need to thank my family. Not only have they endured my many absences while preparing this text, but my father (William Bryan Sr., formerly an engineer who worked on the nuclear rocket program NERVA) and son (Lars Bryan, currently a sophomore studying mathematics at the University of Wisconsin—Madison) have read much of the text and provided numerous helpful suggestions. It is rare that three family generations have the opportunity to work together on a project like this.

Jeff C. Bryan
University of Wisconsin
La Crosse, Wisconsin

Author

Jeff C. Bryan was born in Minneapolis in 1959, the second of four children. His arrival coincided with the purchase of a clothes dryer and was, therefore, somewhat joyous—no more freeze-dried diapers during the long Minnesota winters. After two winters of being bundled up and thrown out into the snow to play with his brother, the family moved to Sacramento. Dr. Bryan was raised in Northern California during the 1960s and 1970s on hotdish, Jell-O™* salad, reticence, and hard work. The conflict of the free and open society with the more reserved home life is the cause of many of Dr. Bryan's current personality disorders.

Dr. Bryan's father felt it was important that his children pay for their own way for their post-secondary education, thereby gaining an appreciation for advanced learning as well as being able to budget. Dr. Bryan decided to stay at home for his first two years of college, attending American River College in Sacramento. This was fortuitous, as he began his study of chemistry from excellent teachers in a small classroom and laboratory-intensive environment. Dr. Bryan later earned a bachelor of arts degree in chemistry from the University of California, Berkeley, and his doctorate in inorganic chemistry from the University of Washington, Seattle, under the supervision of Professor Jim Mayer. His thesis presented a new chemical reaction, the oxidative addition of multiple bonds to low-valent tungsten.

After one year of postdoctoral work with Professor Warren Roper at Auckland University, Dr. Bryan spent five years at Los Alamos National Laboratory, initially as a postdoctoral fellow, then as a staff member. At Los Alamos, his mentor, Dr. Al Sattelberger, encouraged him to begin research into the chemistry of technetium, which turned out to be modestly successful. Dr. Bryan then accepted a staff position at Oak Ridge National Laboratory, working with Dr. Bruce Moyer's Chemical Separations Group as the group's crystallographer. During this time, there was a great deal of work done in the group investigating the separation of ^{137}Cs from defense wastes.

In 2002, Dr. Bryan made a long-anticipated move to academia by accepting a faculty position at the University of Wisconsin–La Crosse. He was hired to teach nuclear and general chemistry courses, primarily based on his 13 years of research experience in the National Laboratories. Apparently, the chemistry faculty was in desperate need to fill this position, as this background did not adequately prepare him for such a dramatic change. Despite the odds, this arrangement has worked out well for all concerned. It has gone so well that Dr. Bryan was granted tenure in 2008 and was promoted to full professor in 2011. He also taught radiation physics from 2005 until 2009.

---

* Jell-O is a delicious dessert and a registered trademark of the Kraft Foods Inc., Northfield, Illinois.

Dr. Bryan has more than 100 publications in peer-reviewed journals, including one book chapter and an encyclopedia article. His research has been featured on the covers of one book, two journals, and two professional society newsletters.

Dr. Bryan married the love of his life, Chris, in 1983. They are still happily married and are very proud of their three children, Katrina, Lars, and Tom.

1 Introduction

1.1 RADIATION

Radiation is scary. We can't see, hear, smell, or touch it, and we know that it can do terrible things to us. In large enough doses, it can kill us. Significant doses received over a long period of time can increase our odds of getting cancer. It is easy to fear. Radiation is also well studied, and this knowledge can be used to dispel some of the fear it so easily engenders. This chapter will focus on the fundamentals of physics that can help us answer questions such as "What is radiation?", "Where does it come from?", and "What does it want with us?"

Much of this chapter will be devoted to definitions. In a sense, this chapter is a foreign language tutorial where you'll learn to speak some nuclear science. Let's start with radiation. **Radiation** is energy that moves. More scientifically, it is energy propagated through a medium. There are all kinds of radiation all around us, all the time, such as radio waves, microwaves, and infrared and visible light. Radio stations use radio waves to transmit audio signals to our homes and cars. We also use radio waves to transmit signals to and from our growing number of wireless devices, such as cell phones, networked computers, and remote controls. Remote controls can also use infrared light to transmit signals to televisions, game consoles, and other video or audio equipment. We use microwaves to transmit energy to water molecules in our food. Visible light is also a form of radiation. Green light generated by the sun travels through space and our atmosphere, bounces off a plant, and travels through some more air to our eyes, which tells us the plant is green.

All examples of radiation in the preceding paragraph use photons to transmit their energy. Photons are commonly thought of as the individual packets of visible light, but they are also handy at transmitting all kinds of energy, as shown in Figure 1.1. Collectively, these types of radiation are known as electromagnetic (EM) radiation.

The various forms of EM radiation interact with matter in different ways. Radio waves don't interact much with most matter but produce electrical currents in wires and circuits. By varying aspects of the waves, complex signals such as those containing audio and video can be transmitted from one electronic device to another. Microwave radiation causes molecular agitation. Specifically, a microwave oven causes water molecules to rotate faster, thereby increasing their temperature. This cleverly allows the food to be warmed without directly heating up everything else in the oven (because the microwaves do not interact with plastic, glass, or most dishware). Infrared radiation causes molecular vibration and bending, again increasing the temperature of the matter it hits. Warming lamps in restaurants typically use infrared lamps to keep prepared meals warm while they are waiting to be distributed to customers. Visible light causes some electron excitations within atoms and molecules. This means an electron is temporarily promoted to a higher energy level. Chemical receptors in our eyes use these temporary changes to send electrical signals to our brain, allowing us to see.

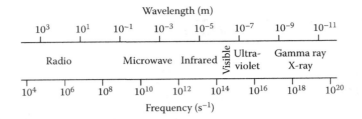

FIGURE 1.1   The electromagnetic spectrum.

Ultraviolet radiation also causes electron excitations, but it does so to a broader number of chemical compounds and is more likely to result in a lasting chemical change. Sufficient exposure to our skin can cause enough chemical change to take place to cause a burn. Gamma rays and X-rays are powerful enough to cause ionizations in any atom or molecule and are, therefore, rather destructive in how they interact with matter. Ionization occurs when an electron is completely separated from its atom or molecule.

Like most things in the subatomic universe, photons are a little strange. Not only do they behave like individual packets of energy, but they also exhibit wave behavior. For example, the type of energy they propagate depends on the wavelength (λ) and frequency (ν) of the photon. **Wavelength** is the distance between two equivalent points on a wave (like two peaks), and **frequency** is how often the same part (like peaks) of the wave goes past. If the waves are traveling the same speed, then frequency is also how many waves are packed into the same distance. Wavelength and frequency for any wave are illustrated in Figure 1.2. The top wave has a longer wavelength (more distance between peaks) but a smaller frequency (fewer waves in the same distance). In fact, there are twice as many waves in the bottom wave than the top wave, meaning its frequency must be two times larger. Notice also that wavelength of the top wave is twice as long as that of the bottom, that is, two bottom waves fit in the same space as one top wave. This pattern is also apparent in Figure 1.1 where there is a regular decrease in wavelength and a regular increase in frequency when moving from left to right.

It would seem that there is a simple mathematical relationship between wavelength and frequency, and there is. If the wavelength of a particular photon is multiplied by its frequency, the result is always the same: the speed of light ($c = 2.998 \times 10^8$ m/s).

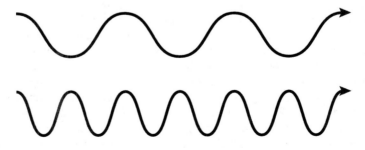

FIGURE 1.2   Two waves.

This relationship is quite handy for calculating wavelength from frequency or frequency from wavelength.

$$c = \lambda \nu \tag{1.1}$$

Example 1.1

Green light has a wavelength of 525 nm. What is its frequency?

Solving Equation 1.1 for frequency and substituting in the wavelength (525 nm) and the speed of light gives the equation and answer that follow. Note that the wavelength units need to be converted to meters to cancel properly. Notice also that frequency can be expressed in units of per second (s^{-1}) or Hertz (Hz), and the two are numerically equivalent:

$$\nu = \frac{c}{\lambda} = \frac{2.998 \times 10^8 \, \frac{m}{s}}{525 \, nm \times \frac{m}{10^9 \, nm}} = 5.71 \times 10^{14} \, s^{-1} \quad \text{or} \quad 5.71 \times 10^{14} \, Hz$$

The amount of energy each photon is packing is directly related to the frequency:

$$E = h\nu \tag{1.2}$$

where h is Planck's constant ($6.626 \times 10^{-34} \, J \cdot s$).

Example 1.2

A therapeutic X-ray machine delivers a beam of photons with a maximum frequency of $9.67 \times 10^{20} \, s^{-1}$. Calculate the energy of one of these photons.

$$E = h\nu = (6.626 \times 10^{-34} \, J \cdot s) \times (9.67 \times 10^{20} \, s^{-1}) = 6.41 \times 10^{-13} \, J$$

Combining Equations 1.1 and 1.2 shows that energy and wavelength are inversely related:

$$E = hc/\lambda$$

Therefore, electromagnetic radiation with high frequency and short wavelength, such as X-rays and gamma rays, are also the highest energy photons known, while radio waves are relatively low energy. This is why Figure 1.1 is called the electromagnetic *spectrum*. It is an arrangement of electromagnetic radiation according to energy (increasing from left to right). We can also begin to appreciate why certain forms (X-rays and gamma rays) of electromagnetic radiation are feared. Being more energetic has its consequences.

1.2 ATOMIC STRUCTURE

Before we get into the more energetic forms of radiation, we need to know a bit about how atoms are put together, since most of the radiation we'll be concerned

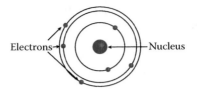

FIGURE 1.3   The Bohr atomic model.

with in this book has its origin in the nuclei of atoms (**radioactivity**). Atoms are made up of a small, positively charged nucleus surrounded by a relatively large space (orbitals) occupied by tiny, fast-moving, negatively charged **electrons**. The nucleus is 10,000 to 100,000 times smaller than the atom, which is defined by the approximate boundaries of the electron orbitals. This means that if the nucleus were the size of a marble, the atom would be as big as a football stadium! Atoms and all the objects that they make up, like this book, bananas, and people, are therefore mostly empty space. Despite its relatively small size, the nucleus contains over 99.9% of the mass of the atom, which means that almost all the mass of an atom occupies very little space within the atom. How could this be? It is because the particles that make up the nucleus are about 2000 times more massive than the electrons.

In this book, we'll use the **Bohr model** to describe the electronic structure of the atom. As shown in Figure 1.3, the Bohr model shows the electrons orbiting the nucleus in circles. Readers should know that this is not terribly accurate. In reality, electrons occupy three-dimensional spaces called orbitals. These spaces are not exactly delineated since there is always a finite (however small) probability of finding the electron just a little further from the nucleus. The Bohr model works just fine for us because the circular orbits nicely represent the average distance between the electron and the nucleus. Note also that the nucleus is not drawn to scale in Figure 1.3—if it were, it would be so small we wouldn't be able to see it.

The nucleus is made up of positively charged **protons** and neutral **neutrons**. Both have a mass of approximately 1 unified atomic mass unit or u (1 u = 1.66×10^{-24} g), while the mass of an electron is a mere 5.5×10^{-4} u.[1] The number of protons in a nucleus is equal to the atom's atomic number and determines what element it is. For example, an atom with eight protons in its nucleus is an oxygen atom, the eighth element listed in the periodic table. It is an oxygen atom regardless of the number of neutrons present. The number of neutrons simply affect the mass of the atom and determine which **isotope** of oxygen an atom is.

If both protons and neutrons vary, we'll need a more general term than isotope. The term **nuclide** refers to any isotope of any element, that is, a nucleus with any number of protons and neutrons. We can now define isotope a bit more efficiently as nuclides with the same number of protons but different numbers of neutrons. Any nuclide can be represented using a nuclide symbol:

$$\ce{^{A}_{Z}X_{N}}$$

---

[1] Some readers may be familiar with the "atomic mass unit" (amu). It is an older, albeit numerically equivalent, way of expressing atomic masses. 1 u = 1 amu.

where X is the elemental symbol from the periodic table, A is the **mass number** (number of protons plus neutrons), Z is the **atomic number** (number of protons), and N is the **neutron number** (number of neutrons). Since the mass number is equal to the number of protons plus the number of neutrons, then:

$$A = Z + N \tag{1.3}$$

It is fairly rare that the neutron number is included on a nuclide symbol. To find the number of neutrons, one only needs to subtract the number of protons from the mass number. Also, the atomic number is sometimes left off a nuclide symbol, since the elemental symbol also gives us that value. When spoken aloud, the nuclide symbol given as "element-A." For example, ^{99}Tc would be said as "technetium-99." Technetium-99 is another way to represent a nuclide symbol in writing. Finally, ^{99}Tc can also be represented as "Tc-99." These other methods of representing a nuclide in print are likely holdovers from days when it was difficult to place superscripts into manuscripts.

If isotopes have a constant number of protons, then isotones have the same number of neutrons. More formally, **isotones** are nuclides with the same number of neutrons but different numbers of protons. Finally, **isobars** are nuclides with the same mass number but differing numbers of protons and neutrons. As examples, consider the following nuclides:

$$^{16}_{8}O_8 \quad ^{18}_{8}O_{10} \quad ^{15}_{7}N_8 \quad ^{18}_{9}F_9$$

Of these, only ^{16}O and ^{18}O are isotopes, ^{16}O and ^{15}N are isotones, and ^{18}O and ^{18}F are isobars. Isotope, isotone, and isobar are all more narrowly defined groups of nuclides, that is, nuclide is the most general term, and the "iso-" terms are more specific.

Collectively, neutrons and protons are known as **nucleons**. Can nucleons be further subdivided into even smaller particles? Yes, as both are partially made up of **quarks**. There are six types, or flavors, of quarks: up, down, top, bottom, charmed, and strange. We will only be concerned with up quarks, which have a $+2/3$ charge, and down quarks, which have a $-1/3$ charge. Protons have two up quarks and one down quark, for a net charge of $+1$. Neutrons are made up of one up quark and two down quarks, for a net charge of 0. It is interesting to note that the difference between a proton and a neutron is a single quark. We can imagine the interconversion between them as simple as flipping a quark, from down to up or up to down. Viewed this way, it would seem relatively easy for protons and neutrons to convert to each other.

What holds the nucleus together? At first glance, it should appear impossible to assemble a number of protons so closely. All those positive charges in a relatively small space should push themselves apart. The force pushing them apart is electrostatic: like charges repel each other, and opposite charges attract. This is also known as **Coulomb repulsion** (or attraction), and it acts over fairly long distances. Since we know that nuclei force protons into close quarters, there must be another force holding them together. This force is known as the nuclear **strong force**, but it has a very short range ($\sim10^{-15}$ m). The average nucleus is $\sim10^{-14}$ m in diameter, so this is a force that has a very limited range, even in a really small space such as a nucleus. While it acts over a very short distance, the strong force is more powerful ($\sim$100$\times$) than the Coulomb repulsion threatening to rip the nucleus apart. It could be said that the ability to destroy a nucleus (through Coulomb repulsion) is insignificant next to the power of the (strong) force. After all, it is the strong force that binds the nucleons together in the nucleus.

1.3 NUCLEAR TRANSFORMATIONS

Now that we have an idea of what nuclei are made of, why they hold together, and how to represent them as symbols, let's start messing with them. Fundamentally, there are two types of nuclear transformations: decay and reaction. This is wonderful simplicity when compared to the world of chemistry where the numbers and types of chemical reactions are constantly proliferating. However, it is an oddity of nuclear science that decay processes are not referred to as "reactions." Nuclear reactions, which we'll study in more detail in Chapter 9, are only those transformations where a particle (such as a neutron or a ^{4}He nucleus) or a photon interacts with a nucleus, resulting in the formation of another nuclide and some other stuff. However, we're getting ahead of ourselves.

The first main transformation type we'll consider is nuclear decay. **Decay** starts with an unstable nuclide that typically spits out a particle and/or a photon while turning itself into something more stable. An example would be ^{222}Rn decaying to ^{218}Po by ejecting two neutrons and two protons all at once. This combination of nucleons is a ^{4}He nucleus and is also known as an **alpha particle**.

$$^{222}_{86}\text{Rn} \rightarrow {}^{218}_{84}\text{Po} + {}^{4}_{2}\text{He}$$

This is an example of **alpha decay**, or it could be said that ^{222}Rn undergoes alpha decay. Notice that the atomic and mass numbers are the same on both sides of the arrow. ^{222}Rn starts out with 86 protons and 222 total nucleons. Together, ^{218}Po and an alpha particle have 86 protons and 222 total nucleons (check this for yourself). This is how most nuclear equations are "balanced," by simply comparing the mass numbers (superscripts) and atomic numbers (subscripts). They should always add up in the same way on either side of the arrow.

There are two other basic types of nuclear decay: **beta** and **gamma**. Additionally, there are three different kinds of beta decay. The first is often referred to simply as **beta decay** but is sometimes called "beta minus" or "negatron" decay. This form of beta decay involves the ejection of an electron from the nucleus—a statement that shouldn't make much sense! Electrons are supposed to spend their time in the space *around* the nucleus, so how can one be emitted *by* the nucleus? Let's look at an example and see if we can understand its origin. ^{14}C decays to ^{14}N by spitting out an electron, also known as a **beta particle**:

$$^{14}_{6}\text{C} \rightarrow {}^{14}_{7}\text{N} + {}^{0}_{-1}\text{e}$$

The beta particle is symbolized here as ${}^{0}_{-1}\text{e}$, but it is sometimes represented as ${}^{0}_{-1}\beta$ or β^-. The mass number for an electron should be equal to zero because it contains no protons or neutrons. The atomic number of –1 simply accounts for its negative charge. It also balances our nuclear decay equation quite nicely.

How does this electron come from the nucleus? Look at what happens to the numbers of protons and neutrons in this decay. The carbon nuclide has six protons and eight neutrons, while the nitrogen nuclide has seven protons and seven neutrons. A neutron has turned itself into a proton in this decay (flipped a quark), as shown in the following equation:

$$_0^1 n \rightarrow {}_1^1 p + {}_{-1}^0 e$$

When that positive charge is created, a negative charge must also be created, for balance. We can, therefore, think of a neutron converting to a proton and an electron within the ^{14}C nucleus. The proton remains a part of the nucleus, while the electron is sent packing. Note that a neutron is represented in nuclear equations as $_0^1 n$. As a nuclide, it has no protons, but it is a single nucleon; therefore, it has a mass number of 1 and an atomic number of 0. Similarly, a proton is a nucleon with an atomic number of 1 and can be symbolized as $_1^1 p$ or as $_1^1 H$ (a hydrogen-1 nucleus) in nuclear equations.

The second kind of beta decay is known as **positron** or "beta plus" decay. In this type of decay, a proton is converted into a neutron and a positron. A positron is a particle with the same mass as an electron but with a positive charge. An example is the decay of ^{18}F to ^{18}O:

$$_9^{18} F \rightarrow {}_8^{18} O + {}_{+1}^0 e$$

Positrons can also be symbolized as $_{+1}^0 \beta$ or β^+. Positrons are a form of antimatter (the same stuff that fuels the starships in *Star Trek*). Every particle that we know about in our universe, such as electrons, has a corresponding antiparticle. In this case, the difference between particle and antiparticle is simply charge.

The third, and last, type of beta decay is **electron capture**. Just like positron decay, electron capture involves the conversion of a proton into a neutron, but it's done a little differently. As its name implies, one of the electrons surrounding the nucleus is drawn into the nucleus, transforming a proton to a neutron. An example is the decay of ^{67}Ga:

$$_{31}^{67} Ga + {}_{-1}^0 e \rightarrow {}_{30}^{67} Zn$$

The words "orbital electron" are often added below the electron in the decay equation to indicate this particle does not originate in the nucleus. In other words, we should not think of it as a beta particle, nor should it ever be represented with the Greek letter beta (β). Electron capture is unique among all forms of nuclear decay in that it has more than the unstable nuclide on the left-hand side of the arrow. We can excuse this poor behavior because the electron is not a nuclear particle.

Notice that this is the first time since atomic structure was described earlier in this chapter that the atom's (orbital) electrons have been mentioned. Generally speaking, this textbook is not concerned with the electrons or even the chemical form of the atom. It is instead focused on the nucleus and nuclear changes. So why all the stuff about the Bohr model a few pages back? Patience, young padawan.

Our final major type of nuclear decay is **gamma decay**. Gamma decay is unique in that no particles[2] are emitted or absorbed, and both the numbers of protons and neutrons remain constant. It is simply the rearrangement of the nucleons in a nucleus. Initially, the nucleons are arranged poorly, in terms of energy. One or more nucleons occupy a higher energy state than they need to, and the nucleus is said to be in an **excited state**. Rearrangement allows the nucleons that are in unusually high-energy states to drop down into lower ones. The lowest energy configuration is called the **ground state**. Rearrangement from a higher energy state to a lower one requires the release of energy, which can be done in the form of a high-energy photon or gamma ray. An example is the decay of ^{99}Tc* to ^{99}Tc:

$$^{99}_{43}\text{Tc}^* \rightarrow \, ^{99}_{43}\text{Tc} + \gamma$$

The asterisk indicates that the nuclide to the left of the arrow is in an excited state. Two nuclides that differ only in energy state (such as ^{99}Tc* and ^{99}Tc) are called **isomers**.[3] The gamma photon can also be represented as $^{0}_{0}\gamma$, but the zeroes are usually left off. We all know that, being a photon; it has no protons, neutrons, or electrical charge.

Some excited (isomeric) states exist for reasonable amounts of time. If it hangs around for more than a second or two, it is often referred to as a **metastable** (or simply "meta") state, and a lowercase m is added after the mass number in the nuclide symbol instead of the asterisk following the element symbol. For example, ^{99}Tc has a metastate, which is symbolized as ^{99m}Tc.

Nuclear reaction is the other main kind of nuclear transformation we'll study. It involves the collision of a particle or photon with a nucleus (or two nuclei crashing into each other) with the formation of a product nuclide and either a particle or a photon. An example would be a neutron colliding with a ^{98}Mo nucleus to form ^{99}Mo and a gamma photon:

$$^{98}_{42}\text{Mo} + \, ^{1}_{0}\text{n} \rightarrow \, ^{99}_{42}\text{Mo} + \gamma$$

In this reaction, the neutron is the **projectile**, ^{98}Mo is the **target**, and ^{99}Mo is the **product**. The easy way to distinguish between decay and nuclear reactions is that reactions always have at least two nuclides or a nuclide with a particle or photons on either side of the arrow. Decay equations generally only have a nuclide on the left side (except electron capture). Like all nuclear equations in this section, this one is balanced, both in terms of atomic and mass numbers. Like decay, there are special kinds of nuclear reactions, but we'll worry about them later.

---

[2] Some would correctly argue that photons *are* particles. This text is trying to keep things simple and is restricting particles to things that have mass.

[3] As noted in Section 5.7, many nuclear science textbooks define isomers as only those excited nuclear states with measurable lifetimes. Because of the arbitrarily limiting nature of this definition, isomers are defined more broadly here.

1.4 NUCLEAR STABILITY

A **radioactive** nuclide is one that spontaneously undergoes nuclear decay. Why do some nuclides decay, but others do not? The question is really one of nuclear stability, and the answer depends on numbers. Perhaps the most important number related to nuclear stability is the neutron-to-proton ratio (N/Z). If we plot the number of protons (Z) versus the number of neutrons (N) and make a mark for every stable nuclide known, the result will look like Figure 1.4. The region where stable nuclides are almost always found is called the **belt of stability**. Notice that for low Z nuclides (near the top of the periodic table, or the bottom left of Figure 1.4), the belt is narrow, and stable nuclides have ratios very close to 1:1. It is difficult to see on this figure, but there are only two stable nuclides with an N/Z ratio less than 1: ^{1}H and ^{3}He. As atomic number increases, the belt widens, and the belt's N/Z ratio also increases.

How can we use this information? If a nuclide's N/Z ratio is too high or too low, it will not lie within the belt, and odds are very good that it will not be stable. Lying outside of the belt means that the nuclide has too many protons or too many neutrons to be stable. Therefore, it will likely undergo some type of beta decay to an isobar. A large number of known (unstable) nuclides lie off the belt, so examination of the N/Z ratio is an important method to determine nuclear instability.

The belt is not completely filled with dots, telling us that there are a number of unstable nuclides with good N/Z ratios. Why? Again, we need to look at some numbers. If we take all the known *stable* nuclides and count how many have an even number of protons (Z) and an even number of neutrons (N), there are about 159. Statistically speaking, there should be equal numbers of stable nuclides with an even Z and odd N, odd Z and even N, or odd Z and odd N. Curiously enough, this is not true. As can be seen in Table 1.1, if either Z or N is odd and the other is even, there are only about 50 stable nuclides. If *both*

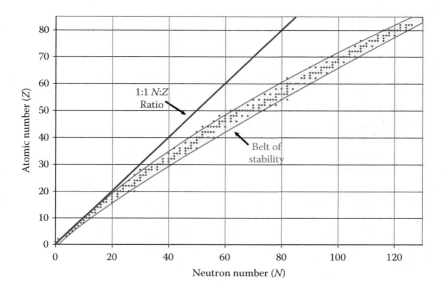

FIGURE 1.4 The belt of stability—dots indicate stable nuclides.

TABLE 1.1
Stable Nuclides

Z	N	#Stable
Even	Even	159
Even	Odd	53
Odd	Even	50
Odd	Odd	4

Z and N are odd, there are only four. These four oddballs are $^2_1H_1$, $^6_3Li_3$, $^{10}_5B_5$, and $^{14}_7N_7$. It seems that odd numbers for both Z and N generally means a nuclide is unstable. Why do these exceptions exist? It's related to the fact that they all have a low number of protons (low Z). A more detailed explanation will have to wait until Chapter 5. For now, we know that even is good and odd is bad, but why? Apparently, protons and neutrons like to pair up with others of their kind, and when they do, they are more stable. Therefore, an odd/odd nuclide can become even/even by converting a proton to a neutron or a neutron to a proton, so some type of beta decay looks likely for odd/odd nuclides.

Why does Figure 1.4 stop at $Z = 82$? Beyond element 82 (Pb), there are no stable nuclides. They are simply too big for the very short-ranged strong force to hold the nucleus together against the tremendous Coulomb repulsion of all those protons. Nuclides this large *tend* to undergo alpha decay, as this lowers both Z and N.

Finally, close examination of Figure 1.4 reveals that some dots seem to lie just outside of the belt of stability. Apparently, there are some *stable* nuclides with *bad* N/Z ratios. All the rationalizations (bad N/Z, odd/odd, too big) we've seen so far are different ways to understand *instability*. How can we rationalize unusual *stability*? We'll need more information first. Are there other instances where unusual stability is observed? Yes, certain elements seem to have an unusually large number of stable isotopes. Most elements only have one, two, or three stable isotopes, but lead has four, nickel has five, calcium has six, and tin has 10. There is something special about certain numbers of protons or neutrons. Detailed analysis reveals that the numbers 2, 8, 20, 28, 50, 82, and 126 seem to impart special stability to nuclides that contain these numbers of protons and/or neutrons. Because of this characteristic, they are sometimes referred to as **magic numbers**.

Far from having supernatural qualities, these numbers have a great analogy in chemistry. Students of chemistry (and some others) know that chemical stability is imparted on those elements occupying the far-right column of the periodic table, the so-called noble gases. This stability comes from exactly filling their valence shells with electrons. In chemistry, these numbers would be 2, 8, 18, and 32. They are not referred to as "magic," but they could be. Applying this analogy to numbers of nucleons, we could say that nuclei with magic numbers of nucleons have complete, or filled, nuclear shells and, therefore, exhibit extra stability. Therefore nucleons, just like electrons, are filling some sort of pattern of energy levels. When a level is filled, the nucleus gives a sigh, as it suddenly feels more stable.

Table 1.2 lists all stable nuclides containing magic numbers of nucleons. Note that a few have magic numbers of both protons *and* neutrons. These nuclides are referred

TABLE 1.2

Stable Nuclides with Magic Numbers of Nucleons

2	8	20	28	50	82	126
Neutrons						
^{4}He	^{15}N	^{36}S	^{48}Ca	^{86}Kr	^{136}Xe	^{208}Pb
	^{16}O	^{37}Cl	^{50}Ti	^{88}Sr	^{138}Ba	
		^{38}Ar	^{51}V	^{89}Y	^{139}La	
		^{39}K	^{52}Cr	^{90}Zr	^{140}Ce	
		^{40}Ca	^{54}Fe	^{92}Mo	^{141}Pr	
					^{142}Nd	
					^{144}Sm	
Protons						
^{3}He	^{16}O	^{40}Ca	^{58}Ni	^{112}Sn	^{204}Pb	
^{4}He	^{17}O	^{42}Ca	^{60}Ni	^{114}Sn	^{206}Pb	
	^{18}O	^{43}Ca	^{61}Ni	^{115}Sn	^{207}Pb	
		^{44}Ca	^{62}Ni	^{116}Sn	^{208}Pb	
		^{46}Ca	^{64}Ni	^{117}Sn		
		^{48}Ca		^{118}Sn		
				^{119}Sn		
				^{120}Sn		
				^{122}Sn		
				^{124}Sn		

to as "doubly magic" and can be expected to be extra stable compared to others with similar numbers of nucleons.

Not all stable nuclides lying outside the belt have magic numbers of protons and/ or neutrons, nor are all nuclides with magic numbers of nucleons stable. In general, if a nuclide is known to be stable, lies outside the belt of stability, *and* has magic numbers, *then* the numbers can be used to rationalize its apparently unusual stability.

Let's look at some examples and use this newfound knowledge about numbers and nuclear stability to help rationalize why specific nuclides are stable or unstable.

Example 1.3

Why is ^{58}Ni stable?

For questions like this, it's best to start by looking at the numbers of protons and neutrons in the nucleus (are they odd or even?) and the N/Z ratio. Nickel is element number 28, so ^{58}Ni has 28 protons and 30 neutrons ($58 - 28 = 30$). It is an even/even nuclide, which increases the odds of stability but does not determine it. Its N/Z ratio is approximately 1.07:

$$\frac{N}{Z} = \frac{30}{28} \cong 1.07$$

TABLE 1.3
Z, N, and N/Z for Specific
Unstable Nuclides

Nuclide	Z	N	N/Z
^{138}Te	52	86	1.65
^{120}Xe	54	66	1.22
^{241}Am	95	146	1.54
^{46}Sc	21	25	1.19

Close examination of Figure 1.4 suggests this ratio is a little low. In fact, it appears that a dot representing ^{58}Ni lies just outside (above) the belt of stability. If this nuclide is outside the belt, how can its stability be rationalized? It is because of the extra stability imparted by having a filled nuclear shell (magic number) of protons ($Z=28$).

Example 1.4

Why are ^{138}Te, ^{120}Xe, ^{241}Am, and ^{46}Sc unstable?

Table 1.3 summarizes the relevant numerical information for these four nuclides. Of these four nuclides, only ^{46}Sc has odd numbers of both protons and neutrons. Since it is not one of the previously mentioned exceptions, we can say that it is unstable because it is odd/odd. Note that its N/Z ratio is actually pretty good, placing it near the center of the belt of stability.

The first two nuclides listed in Table 1.3 are even/even but are unstable because they have poor N/Z ratios. ^{138}Te has a high N/Z ratio and is below the belt of stability, while ^{120}Xe has a low ratio and is above the belt.

We can only guess about the quality of the N/Z ratio for ^{241}Am because Figure 1.4 doesn't go that high. ^{207}Pb is in the middle of the belt in the upper right corner of Figure 1.4 and has an N/Z ratio of 1.52. Since this ratio (within the belt) increases as atomic number (Z) increases, a value of 1.54 for $Z=95$ is probably okay. With all our hard work examining the N/Z ratio for this nuclide wasted, we suddenly realize that this nuclide is unstable simply because it is too big.

1.5 IONIZING RADIATION

The first section of this chapter discussed radiation as energy propagated through a medium but stayed focused on electromagnetic radiation (photons). We have since learned about radioactive decay and know that this happens to achieve greater nuclear stability—therefore, energy must be released as part of all nuclear decay processes.

Ionizing radiation is when a particle or a photon has enough energy to turn an unsuspecting atom into an ion (Figure 1.5). In other words, the radiation needs to have enough energy to knock an electron or a proton loose from an atom. The former process is much more probable, so much so that we'll ignore the latter.

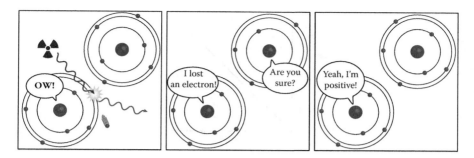

FIGURE 1.5 Ionizing radiation.

As we've already seen, the only types of electromagnetic radiation with sufficient energy to ionize matter are X-rays and gamma rays. All alpha and beta particles produced in nuclear decay also have enough energy to ionize matter. Therefore, X-rays, gamma rays, and alpha and beta particles can all be considered forms of ionizing radiation.

Unfortunately, the more general term "radiation" is commonly, but incorrectly, used to refer to the energetic photons and particles emitted as part of nuclear processes. This text will strive to use the terms "radiation" and "ionizing radiation" properly. The reader should know that the misusage of "radiation" is very common, and even those who know better will sometimes slip up.

Even though the three major forms of decay (alpha, beta, and gamma) can all create ions when their emitted particles or photons interact with matter, their relative abilities to penetrate matter differ greatly. Alpha particles have a high probability of interacting with matter. They carry a 2+ charge and, relative to electrons and photons, are very large. Alpha particles can be effectively stopped by thin pieces of paper, the outer layers of your skin, or by just a few centimeters of air. Beta particles are electrons, so they have a lot less mass and only half the charge of alpha particles, but they still have a reasonably high probability of interacting with matter. They can generally be stopped by a centimeter (or two) of water or plastic. Energetic photons such as X-rays and gamma rays have no charge or mass and, therefore, have a relatively low probability of interacting with matter. The best shielding material for them is high-density matter, such as lead.

1.6 A BIOLOGICAL THREAT?

One of the reasons humans fear ionizing radiation is that we know that, in large enough doses, it can hurt us. From what we've learned so far, it shouldn't be hard to figure out how this might happen. Ionizations inside our bodies and possibly in our DNA can cause cells to die or to replicate poorly. With enough of a dose, we start to observe measurable effects such as lower blood cell counts. As the dose increases, we know that the health effects gradually get worse until death becomes pretty much certain.

What about below the level where measurable adverse health effects are known? Since the health effects cannot be measured, there is some uncertainty as to what

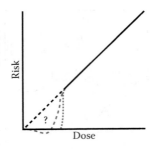

FIGURE 1.6   Dose vs. risk.

effect low doses might have. Regulatory agencies have decided to be prudent and assume that the roughly linear relationship between risk and dose at higher exposure levels should be extrapolated through low doses. This is known as the "linear no-threshold (LNT) model" and is illustrated by the straight, dashed line in Figure 1.6. This means that any additional exposure *could* lead to negative health effects. If this is true, then it is a good idea to minimize dose due to exposure to ionizing radiation when possible. This is the basis of the **ALARA** (*as low as reasonably achievable*) policy followed by those who work with radioactive materials. If there's more than one reasonable way to do something, pick the path that gives a lower dose.

While used by regulators and radiation workers, it should be noted that the LNT is not supported by science. Our bodies can repair or replace many of the cells that are damaged or killed due to exposure to ionizing radiation. We are all exposed to naturally occurring ionizing radiation every minute of every day. Some are exposed to more than others, such as those who live at higher altitudes or in areas with relatively high concentrations of naturally occurring radioactive nuclides. There are no measurable negative health effects due to the extra exposure on the large populations of humans who live in these areas despite the fact that background radiation doses vary by as much as a factor of 100. This suggests that our bodies can safely cope with low-levels of ionizing radiation and that there might be a threshold dose below which no health effects can be anticipated. The *dotted* gray curve in Figure 1.6 represents this threshold model.

A few others have gone a bit further than those that suggest there is a threshold dose. These scientists suggest there is a health benefit to low doses of ionizing radiation (hormesis), and there is some evidence to indicate this might be the case. This is represented by the *dashed* gray curve in Figure 1.6. Hormesis acts like an immunization. A flu shot stimulates the immune system such that, when the body is exposed to the virus, it can fend off the virus. Exposure to ionizing radiation can stimulate cell repair mechanisms, making the body healthier and better able to effectively deal with future exposures.

The bottom line is that none of these hypotheses have enough science behind them to gain widespread acceptance. For the time being, those of us working with ionizing radiation will continue to follow prudence in the form of ALARA. However, we must also recognize that application of ALARA outside of the radiation workplace can have drastic consequences. For example, how extensively must areas contaminated by the Fukushima reactors be cleaned before we allow people to return? ALARA and the LNT suggest that we decontaminate until it is no longer more radioactive than before the accident. This may be ridiculously expensive and unnecessary if we instead assume a threshold dose or hormesis.

1.7   NATURAL AND ANTHROPOGENIC RADIATION

If we are being exposed to ionizing radiation all the time, where is it coming from? **Naturally occurring** ionizing radiation comes from several sources. The universe is a very radioactive place, constantly bombarding the Earth with cosmic radiation. The Earth's magnetic field and atmosphere shield us from most of it, but some makes it all the way to the surface. People living at higher altitudes receive two to three times the dose from cosmic radiation than those living near sea level because there's less shielding from the atmosphere. This also means those who spend a lot of time flying (watch out Supergirl!) are getting a significantly higher dose than those of us who stay on the ground.

What about radioactive materials here on planet Earth? Well, there's a lot, and almost all of it has always been here. There are three main sources of radioactive materials. First, some of the lighter elements present in the upper atmosphere (such as N, O, and H) can undergo nuclear reactions that are a result of cosmic ray interactions. ^3H and ^{14}C are constantly produced in this way. These nuclides find their way back to the planet surface and are incorporated into molecules that make up living things since their chemistries remain the same as other (stable) hydrogen and carbon atoms. Thus, all living things are naturally radioactive.

Another source comes from radioactive nuclides that take a very long time to decay. If that time is longer than the age of the solar system (4.6×10^9 years), then they're still around us. All together, there are about 20 of them. One of the most notable is ^{40}K. Since all isotopes of an element have the same chemical behavior, and because we humans need potassium in our bodies, we all have some ^{40}K in our bodies. Anything that contains potassium (bananas, Gatorade®, your friends) is naturally radioactive.

All but three of these 20 long-lived radioactive nuclides decay to stable nuclides. The three that don't are ^{232}Th, ^{235}U, and ^{238}U. They all eventually decay to stable isotopes of lead, but along the way, they produce a number of other naturally occurring radioactive nuclides, including ^{222}Rn. This is our final source of naturally occurring radioactive nuclides. The paths they take to Pb are called the naturally occurring **decay series** (or **chains**). To get there, several alpha and beta (β^-) decays need to take place, as illustrated in Figure 1.7 for the series originating with ^{238}U. Notice that an alpha decay lowers both the proton and neutron numbers by 2, while a beta decay lowers N by 1 but increases Z by 1. At certain nuclides (such as ^{218}Po), the path splits into two. This means that more than one decay pathway exists for that particular nuclide, that is, it can decay via alpha or beta.

What about all the radioactive materials released by humans (anthropogenic sources)? Despite our best efforts, we've allowed a fair bit of radioactive material to get into the environment. The following are some of the more significant sources:

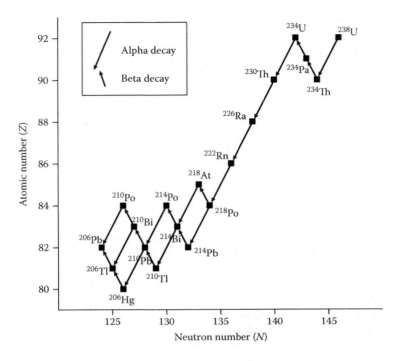

FIGURE 1.7   The ^{238}U decay series.

1. The development, manufacture, testing, and use of nuclear weapons.
2. Accidental releases from nuclear power plants such as Fukushima (2011), Chernobyl (1986), and Three Mile Island (1979).
3. Regulated releases from nuclear power plants.
4. Irresponsible disposal of radioactive waste.
5. The manufacture, use, and disposal of radioactive nuclides for research or medical applications.

It sounds bad, but all five sources combined account for less than 0.1% of the average annual human dose in the United States. They are combined under "Other" in Figure 1.8, which illustrates the major sources for average annual doses from ionizing radiation in the United States. The dose rates are given in millisieverts per year. The millisievert is a commonly used unit for dose and is explained in Chapter 12. Notice also that the unit abbreviation for years is "a," which comes from the Latin word for year, *annum*.

The first four bars shown in Figure 1.8 are from natural sources. By far, we get the most from radon. That's because it is a gas. Once formed in a decay series, it can percolate through the ground and get into the water we drink or into the air we breathe. It gives us a large dose because we ingest it, and it can undergo alpha decay while inside of us. Because of its high probability of interacting with matter, the alpha particle cannot escape our body and ends up doing a lot of local damage. Don't stop breathing the air or drinking the water, as your friends and relatives might miss you! Besides, radon has always been present on the planet, and humans have

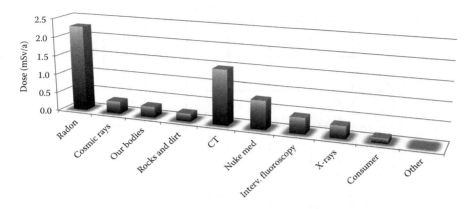

FIGURE 1.8   Average annual radiation doses in the United States. (From NCRP Report #160, Ionizing Radiation Exposure of the Population of the United States, 2009.)

evolved with some ability to repair the damage it causes. At most concentrations we humans normally encounter, there is no adverse health effect; in fact, there seems to be a hormetic effect at moderate concentrations. At very high concentrations, long-term exposure to radon gas can cause lung cancer. Since a good chunk of this dose comes from the air we breathe at home, it is a good idea to test your home for radon, if your home is in an area known to have high radon levels. If it is found in high concentration,[4] there are ways to lower it. As we'll soon learn, the best results come from a longer test—choose one that is in your home for a few months or longer.

You might have expected cosmic rays to account for a larger percentage of our annual dose. Much of the cosmic radiation we're exposed to is in the form of high-energy photons or particles, most of which have a very low probability of interacting with the matter of which we are made. That's the difference between exposure and dose. Our greatest exposure to naturally occurring ionizing radiation is cosmic, but our greatest dose comes from radon because radon is inside us and emits ionizing radiation with a very high probability of interacting with matter, while most cosmic radiation has a relatively low probability of interacting.

The final six sources shown in Figure 1.8 are anthropogenic. Notice that the first four are all medical diagnostic procedures and make up almost the entire U.S. dose from human-produced ionizing radiation. Of course, not everyone has a computed tomography (CT) scan or nuclear medicine procedure every year, so total dose for individuals will vary somewhat. These numbers come from the dose for all sources of ionizing radiation each year divided by the total population. The total average dose also varies from country to country. Adding up all the bars in Figure 1.8 tells us that the average American picks up 6.2 mSv each year, yet the average Canadian

---

[4] How high is too high? Governments typically set a recommended threshold for potential remediation, but they are set much lower than they should be based on scientific evidence. For example, the United States sets its threshold at 4 pCi/L, yet 56 pCi/L is the observed threshold for negative health effects.

Z \ N	0	1	2	3	4	5	6	7	8	9	10	11	12	13	14	15	16	17	18
8					O-12 1.77	O-13 17.77	O-14 5.1439	O-15 2.7542	O-16 99.757	O-17 0.038	O-18 0.205	O-19 4.822	O-20 3.815	O-21 8.11	O-22 6.5	O-23 11.3	O-24 11.5	O-25 15.9	O-26 16.9
7				N-10 2.64	N-11 2.290	N-12 17.338	N-13 2.2205	N-14 99.636	N-15 0.364	N-16 10.421	N-17 8.68	N-18 13.90	N-19 12.53	N-20 18.0	N-21 17.2	N-22 22.8	N-23 23.8	N-24 1.2	
6			C-8 2.142	C-9 16.495	C-10 3.6480	C-11 1.982	C-12 98.93	C-13 1.07	C-14 0.1565	C-15 9.772	C-16 8.010	C-17 13.17	C-18 11.81	C-19 16.6	C-20 15.8		C-22 21		
5			B-7 2.200	B-8 17.980	B-9 0.277	B-10 19.9	B-11 80.1	B-12 13.369	B-13 13.437	B-14 20.64	B-15 19.10	B-16 0.040	B-17 22.7		B-19 26.9				
4			Be-6 1.372	Be-7 0.8619	Be-8 0.092	Be-9 100	Be-10 0.556	Be-11 11.51	Be-12 11.71	Be-13 0.50	Be-14 16.3								
3		Li-4 3.100	Li-5 1.970	Li-6 7.59	Li-7 92.41	Li-8 16.005	Li-9 13.607	Li-10 0.42	Li-11 20.62										
2		He-3 0.0001	He-4 100	He-5 0.890	He-6 3.508	He-7 0.440	He-8 10.65	He-9 1.15	He-10 15.8										
1	H-1 99.989	H-2 0.0115	H-3 0.1859	H-4 2.910	H-5 25	H-6 24.3													

$N \rightarrow$

$\leftarrow Z$

FIGURE 1.9 A chart of the nuclides for the first eight elements.

sees only about 2.6 mSv per year. The difference between the two numbers seems to largely be an American affinity for diagnostic medical procedures.

1.8   THE CHART OF THE NUCLIDES

The chart is a valuable tool for anyone studying nuclear science. We've already seen a couple of extremely condensed versions of it in Figures 1.4 and 1.7. A more conventional, albeit limited, representation is shown in Figure 1.9. Basically, it is a way to provide information on all known nuclides, stable (white) and unstable (gray). It is laid out with neutron number (N) increasing from left to right (the x-axis) and atomic number (Z) increasing from bottom to top (y-axis). Thereby, each square represents a unique N and Z combination, that is, a unique nuclide. Percent natural abundance is given at the bottom of the square for each stable nuclide, and total energy (MeV, see Chapter 3) given off during decay is given for all the unstable nuclides.

Printed versions are prepared by scientists at the Knolls Atomic Power Laboratory (KAPL) and are available for purchase through http://www.nuclidechart.com/. The printed charts also come with extensive text explaining how to properly interpret the information provided for each nuclide. There are also Web-based versions of the chart; just ask your favorite AI assistant. Many of the questions following each chapter in this textbook assume that the reader has access to the kinds of information available in the chart. For the readers' convenience, Appendix A also lists limited information about select nuclides. This appendix should be sufficiently complete to perform the problems posed in this text.

QUESTIONS

1.1   Which of the following statements are true?
   a.   All radiation is harmful.
   b.   All ionizing radiation is anthropogenic.
   c.   Atoms cannot be changed from one element to another.
   d.   You are exposed to ionizing radiation every minute.
   e.   The human body is naturally radioactive.
   f.   The human body is capable of detecting radioactivity at any level.
1.2   Which of the following statements are false?
   a.   An electron is an example of a nucleon.
   b.   There are three different kinds of beta decay.
   c.   A nuclide having a magic number of protons and/or neutrons will always be stable.
   d.   Anthropogenic radiation is more harmful than naturally occurring radiation.
   e.   A nuclide with an N/Z ratio of exactly 1 will always be stable.
   f.   The nucleus accounts for almost all the mass of an atom.
1.3   List some of the apprehensions you have or have had concerning ionizing radiation.
1.4   What is radiation? What is ionizing radiation, and where does it come from?
1.5   What is the wavelength of a photon with a frequency of $3.86 \times 10^{18}\,\text{s}^{-1}$?

1.6 What is the energy of a photon with a wavelength of 6.87×10^{-5} m? What type of radiation is this?

1.7 Millimeter wave imaging systems are used for personnel screening. They emit electromagnetic radiation with wavelengths in the range of 1–10 mm, which penetrate clothing but are reflected by denser matter such as the human body. Calculate the frequency range for these photons. Are they ionizing? The radiation used by these scanners is typically characterized as high-frequency radio waves. Is this correct? What might motivate the use of this moniker?

1.8 What is the wavelength of a photon with an energy of 7.84×10^{-18} J? Is this ionizing radiation? Briefly explain your answer.

1.9 What do gamma rays and UV light have in common? What distinguishes them?

1.10 Is the phrase "nature abhors a vacuum" accurate based on our knowledge of atomic structure?

1.11 If you built a scale model of an atom and used a baseball to represent the nucleus, how big would the atom have to be?

1.12 Define the following: nuclide, nucleon, beta particle, positron decay, the electromagnetic spectrum, isotope, isobar, and ALARA.

1.13 Define the following: belt of stability, metastable state, alpha particle, isotone, electron capture, and isomer.

1.14 Which of the following are isotopes? Isotones? Isobars?

$$^{14}C \quad ^{16}O \quad ^{16}N \quad ^{12}C \quad ^{13}B \quad ^{19}F$$

1.15 How many electrons could fit in the same space occupied by a proton?

1.16 Complete and balance the following. Identify each as nuclear decay or reaction. If it is decay, give the type of decay.

$$^{81}_{37}Rb \rightarrow \,^{81}_{36}Kr + \underline{\quad} \qquad ^{14}_{7}N + \underline{\quad} \rightarrow \,^{1}_{1}H + \,^{14}_{6}C \qquad \underline{\quad} \rightarrow \,^{137}_{56}Ba + \,^{0}_{-1}e$$

$$^{239}_{94}Pu \rightarrow \,^{4}_{2}He + \underline{\quad} \qquad ^{201}_{80}Hg + \,^{2}_{1}H \rightarrow 2\,^{1}_{0}n + \underline{\quad} \qquad ^{208}_{83}Bi + \,^{0}_{-1}e \rightarrow \underline{\quad}$$

1.17 Give three reasons why a nuclide might be stable.

1.18 Why are ^{32}P, ^{64}Ge, ^{20}O, $^{100}Ru^*$, and ^{212}Po unstable nuclides? Write out a possible decay equation for each.

1.19 Are the following nuclides stable or unstable? Why? ^{11}C, ^{190}W, ^{114}In, ^{232}Th, ^{76}Se.

1.20 Complete and balance the following nuclear decays. Is the product nuclide of each of the decays stable? If not, why not?

$$^{225}_{90}Th \rightarrow \,^{221}_{88}Ra + \underline{\quad} \qquad ^{184}_{73}Ta \rightarrow \,^{184}_{74}W + \underline{\quad}$$

$$^{133m}_{58}Ce \rightarrow \,^{133}_{58}Ce + \underline{\quad} \qquad ^{126}_{53}I + \underline{\quad} \rightarrow \,^{126}_{52}Te$$

1.21 ^{40}Ca shouldn't be a stable nuclide, but it is. Explain this apparent contradiction.

1.22 One of the naturally occurring decay series begins with ^{232}Th and ends with ^{208}Pb. What is the minimum number of alpha and beta decays required for this series?

1.23 A fourth naturally occurring radioactive decay series starting with ^{237}Np may have existed in the past. Using a chart of the nuclides, graph this decay series (similar to Figure 1.7), and explain why it does not exist today.

1.24 Using Figure 1.8, estimate the percentage of the average annual U.S. dose due to natural sources.

1.25 List at least two anthropogenic sources of radioactive nuclides now found in the environment. Do a little research outside of this textbook, and explain how these nuclides were (are) released.

1.26 Why are so many more neutron-rich nuclides shown in Figure 1.9 than proton-rich nuclides?

1.27 What is the most common nuclide in air?

1.28 Otto Hahn and Lise Meitner reported the discovery of an isotope of a new element in 1918. They described it as follows:

   a. The previously hypothetical mother substance of actinium has been found and prepared in alkali acids in radioactively pure form. It is a homologue of tantalum.

   b. It emits alpha rays with a penetration of 3.14 cm.

   c. The half-life is a minimum of 1200 years and a maximum of 180,000 years.

   What nuclide did they isolate? Briefly explain.

2 The Mathematics of Radioactive Decay

2.1 ATOMIC MASSES AND AVERAGE ATOMIC MASSES

Before looking at the mathematics of decay, it is important to remember the difference between atomic mass and average atomic mass. **Atomic mass** is the mass of a single atom, and it is numerically equivalent to the mass of 1 mole of atoms. Atomic mass is expressed in unified atomic mass units (u). When we are talking about a mole of atoms, the units are grams per mole (g/mol). For example, the mass of a single ^{12}C atom is 12.00 u, while the mass of a whole mole (6.022×10^{23}) of ^{12}C atoms would be 12.00 g. The unified atomic mass unit is more convenient when referring to the mass of single atoms and is more commonly used in nuclear science. To convert one to the other, one unified atomic mass unit is equal to 1.66054×10^{-24} g.

The masses listed in the periodic table are not always the mass of a single atom of each element; rather, it lists the element's *average* atomic mass. **Average atomic mass** is a weighted average of the masses of each naturally occurring isotope for a particular element. "Weighted" means that if one isotope has a greater natural abundance for a particular element, it contributes more significantly to the average atomic mass. The only time the periodic table gives the mass of a single atom of an element is when there is only one naturally occurring isotope for that element. The best way to see where the average atomic masses in the periodic table come from is through a sample calculation.

Example 2.1

Calculate the average atomic mass of copper.

Copper exists in nature as a mixture of two isotopes, ^{63}Cu and ^{65}Cu. The percent natural abundances are 69.15% and 30.85%, respectively. This means that if we find 100 copper atoms in nature, we'll likely have ~69 atoms of ^{63}Cu and ~31 atoms of ^{65}Cu. Their masses are 62.9296 u and 64.9278 u, respectively. These data are available in Appendix A for all elements.

To determine the average atomic mass, calculate the sum of the products of the atomic masses and the *fractional* abundances. In this way, the average mass of copper atoms is weighted by the amount of each isotopic mass present:

$$62.9296 \text{ u} \times 0.6915) + (64.9278 \text{ u} \times 0.3085) = 63.55 \text{ u}$$

Notice that the average atomic mass is closer to 63 u than it is to 65 u. This makes sense because there's more of the mass ~63 u isotope than there is of the mass ~65 u isotope.

2.2 THE NATURE OF DECAY

Since humans can't sense radioactive decay, we need some other way to measure it. There are a variety of radiation detectors, as discussed in Chapter 7. For now, we just need to know that it can be done, albeit not always with high efficiency. Radioactive decay is **isotropic**, which means that particles and/or photons are emitted equally in all directions by a sample of a radioactive nuclide. Detectors can only "see" radiation that interacts with it, typically a small fraction of all radiation being emitted at the time, that is, those that happen to enter the detector instead of flying off in some other direction. Data recorded by detectors are given units of count rate, such as counts per minute (cpm). If it were possible to detect *all* the decays from a source over time, then the decay rate, or **activity**, would be measured. Decay rate is often given units of decays per minute (dpm) or decays per second (dps). Those who work with radioactive materials often use count rate and decay rate interchangeably. This is understandable, since they are proportional to each other. Readers of this book would do well to know the difference.

Over time, as a sample of a radioactive nuclide decays, it becomes less radioactive. We could make measurements of this change with a detector and a clock. Once we've collected these data, it could be plotted as shown in Figure 2.1a. Mathematically, radioactive decay is an exponential process, meaning that for equal intervals of time, the count rate will decrease by an equal percentage. This also means that if instead of plotting count rate versus time, we plot the logarithm of the count rate versus time, we'll get a straight line, as shown in Figure 2.1b.

Two different forms of the same mathematical equation can describe the change in radioactivity of a source:

$$\ln \frac{A_1}{A_2} = \lambda t \tag{2.1}$$

$$A_2 = A_1 e^{-\lambda t} \tag{2.2}$$

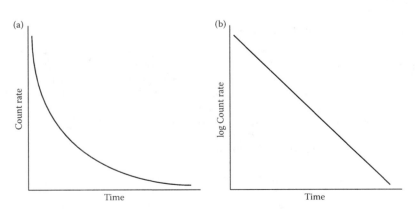

FIGURE 2.1   (a) The change in count rate over time for radioactive decay. (b) The change in the logarithm of the count rate with time.

where A represents activity or count rate so long as it is applied consistently in one equation. A_1 is the activity at the beginning of the time elapsed (t), and A_2 is the activity at the end of that time. ln is the natural logarithm function, and e is natural antilogarithm. λ is the **decay constant** and has units of reciprocal time (time^{-1}). The decay constant is an indication of how quickly a source will decay and is sometimes represented by the letter k. The decay constant (λ) is also equal to the natural logarithm of two divided by half-life ($t_{1/2}$):

$$\lambda = \frac{\ln 2}{t_{1/2}} \tag{2.3}$$

Half-life is defined as the amount of time it takes for the decay rate (or count rate) to reach half of its original value. If a nuclide starts out with an activity of 1000 dpm, it will be 500 dpm after one half-life, 250 dpm after two, and so on. Half-life is commonly used in nuclear science because it is a bit more conceptually tractable than the decay constant. A nuclide with a very long half-life will decay slowly and be radioactive for a very long time. In contrast, a nuclide with a short half-life will decay away quickly.

Example 2.2

A radionuclide has a half-life of 6.01 hours. If its activity is currently 5430 dpm, what will its activity be 4.00 hours from now?

$$A_2 = A_1 e^{-\lambda t} = A_1 e^{-\frac{\ln 2}{t_{1/2}}t} = 5430\ \text{dpm} \times e^{-\frac{\ln 2}{6.01\ \text{h}} \times 4.00\ \text{h}}$$
$$= 5430\ \text{dpm} \times e^{-0.461} = 3423\ \text{dpm}$$

Less than one half-life has gone by, so we expect the answer to be greater than half of the original. Notice also that the time units need to match within the exponent (both are hours in this example), and the activity units for A_1 and A_2 must also match. If your instructor is a stickler for significant figures, then the answer should probably be rounded to 3420 dpm. Generally speaking, three significant figures are the best that can be expected from most detector measurements.

Example 2.3

A radioactive nuclide had an activity of 1.38×10^5 dpm exactly 60 days ago but now has an activity of 6.05×10^4 dpm. What is its half-life?

$$\ln \frac{A_1}{A_2} = \lambda t$$

$$\ln \left(\frac{1.38 \times 10^5\ \text{dpm}}{6.05 \times 10^4\ \text{dpm}} \right) = \frac{\ln 2}{t_{1/2}}(60\ \text{d})$$

$$0.825 = \frac{41.6\,\mathrm{d}}{t_{1/2}}$$

$$t_{1/2} = 50.4\,\mathrm{d}$$

In this example, approximately one half-life has passed, and the activities differ by (roughly) a factor of two. Notice that units for half-life ($t_{1/2}$) come from the elapsed time (t).

Activity is also commonly expressed in units of Becquerels (Bq) or Curies (Ci). The Becquerel (Bq) is named for Henri Becquerel, the French physicist who first observed radioactivity. It is the *Systéme International d'Units* (SI) unit and is equal to 1 dps. This is a tiny value, so megabecquerel (MBq) or kilobecquerel (kBq) are often used. Remember that the kilo prefix means 10^3, and the mega prefix multiplies by 10^6. The Curie is named for Marie Curie, the Polish scientist who performed much of the early groundbreaking work with radioactive materials. One Curie is defined as exactly 3.7×10^{10} dps, which is the rate of decay of exactly 1 g of ^{226}Ra. This is a rather large number, so millicuries (mCi = 10^{-3} Ci) or microcuries (μCi, 1 μCi = 10^{-6} Ci = 2.22×10^6 dpm) are often used for smaller sources.

Example 2.4

A ^{133}Ba source was calibrated to emit 36.9 kBq of radiation. If its activity today is measured at 5.55×10^5 dpm, how long ago was it calibrated?

First, convert the initial activity units from kBq to dpm:

$$36.9\,\mathrm{kBq} \times \frac{1000\,\mathrm{Bq}}{\mathrm{kBq}} \times \left(\frac{1\,\mathrm{dps}}{1\,\mathrm{Bq}}\right) = 3.69 \times 10^4\,\mathrm{dps}$$

$$\frac{3.69 \times 10^4\,\mathrm{decays}}{\mathrm{s}} \times \frac{60\,\mathrm{s}}{\mathrm{min}} = 2.21 \times 10^6\,\mathrm{dpm}$$

Now that the activity units are all on the same page, we can use Equation 2.1 to solve for time:

$$\ln\frac{2.21 \times 10^6\,\mathrm{dpm}}{5.55 \times 10^5\,\mathrm{dpm}} = \frac{\ln 2}{10.54\,\mathrm{a}}t$$

$$1.384 = 0.06576\,\mathrm{a}^{-1} \times t$$

$$t = 21.04\,\mathrm{a}$$

In this example, it's necessary to convert one of the two given activities to match the units of the other. The choice to convert A_1 to dpm was arbitrary. A_2 could've been converted to kBq instead. It's also necessary to look up the half-life for ^{133}Ba (10.54 years, Appendix A). Finally, notice that roughly two half-lives have passed, and the activity is roughly one-fourth of its original value.

As mentioned at the beginning of this chapter, detectors typically only "see" a fraction of all decays. The percentage of all decays observed is called the **percent efficiency**:

$$\% \text{ efficiency} = \frac{\text{cpm}}{\text{dpm}} \times 100\% \qquad (2.4)$$

Example 2.5

A ^{210}Pb source was calibrated at 0.0202 µCi 31.0 years ago. Under your detector, it records 2.16×10^3 cpm. What is your detector's percent efficiency?

First calculate the calibrated activity in units of decays per minute:

$$0.0202 \, \mu\text{Ci} \times \frac{2.22 \times 10^6 \, \text{dpm}}{\mu\text{Ci}} = 4.48 \times 10^4 \, \text{dpm}$$

After looking up the half-life of ^{210}Pb in Appendix A, calculate what the *activity* is today. In other words, how much has the source decayed since it was calibrated?

$$A_2 = 4.48 \times 10^4 e^{-\left(\frac{\ln 2}{22.3a}\right)31.0\,a} = 1.71 \times 10^4 \, \text{dpm}$$

Finally, use the count data to determine the percent efficiency:

$$\% \text{ efficiency} = \frac{\text{cpm}}{\text{dpm}} \times 100\% = \frac{2.16 \times 10^3 \, \text{cpm}}{1.71 \times 10^4 \, \text{dpm}} \times 100\% = 12.6\%$$

In addition to the limited size of the detector, there are a number of experimental conditions that can affect the percent efficiency and, therefore, the observed count rate. These are discussed in Chapter 7.

So long as the conditions of the experiment remain constant, the count rate will be proportional to activity and can be used in the equations presented at the beginning of this chapter. There's another important proportionality that needs to be considered. If there's more radioactive material, there should be more activity. The mathematical relationship is remarkably simple:

$$A = \lambda N \qquad (2.5)$$

where A is activity (it cannot be count rate in this equation), λ is the decay constant, and N is the number of atoms of the radioactive nuclide. The more you've got, the toastier it is. This is a simple, yet powerful little equation that relates the amount of material with its activity. If the number of atoms is known, it can be converted to moles of atoms using Avogadro's number (6.022×10^{23} atoms/mole) and then into mass using the atomic mass (grams/mole). Likewise, if the mass is known, the number of atoms can be calculated, as well as activity.

Example 2.6

What is the mass of 45.5 kBq of ^{137}Cs?

First convert activity to dpm:

$$45.5 \text{ kBq} \times \frac{1000 \text{ Bq}}{\text{kBq}} \times \left(\frac{1 \text{ dps}}{1 \text{ Bq}}\right) \times \frac{60 \text{ s}}{\text{min}} = 2.73 \times 10^6 \text{ dpm}$$

Next, calculate the number of ^{137}Cs atoms (after looking up the half-life for ^{137}Cs), but first we'll have to rearrange Equation 2.5 and then substitute ln $2/t_{1/2}$ for λ:

$$A = \lambda N \quad N = \frac{A}{\lambda} = \frac{A \cdot t_{1/2}}{\ln 2} = \frac{(2.73 \times 10^6 \text{ dpm})(30.07 \text{ a})}{\ln 2}$$

But wait! Notice the problem with the time units? Decays/minute doesn't cancel out when multiplied by years. To fix this, it's handy to know that there are approximately 5.259×10^5 minutes in a year.

$$N = \frac{(2.73 \times 10^6 \text{ dpm}) \times (30.07 \text{ a}) \times \left(\dfrac{5.259 \times 10^5 \text{ min}}{\text{a}}\right)}{\ln 2} = 6.23 \times 10^{13} \text{ atoms}$$

Convert to mass:

$$(6.23 \times 10^{13} \text{ atoms}) \times \frac{\text{mol}}{6.022 \times 10^{23} \text{ atoms}} \times \frac{137 \text{ g}}{\text{mol}} = 1.42 \times 10^{-8} \text{ g}$$

Wow! That's a very small mass for something spitting out radiation over 2 million times each minute! A little radioactive material can often go a long way. Notice also that 137 g/mole was used as the atomic mass. In a pinch, the mass number (A) of a nuclide can be used as an approximate atomic mass. We could've easily used a more precise value from Appendix A, but an approximate value is adequate for this problem.

Because of the direct proportionality of activity (A) and number of atoms (N), the decay equations given at the beginning of this chapter can be rewritten as:

$$\ln \frac{N_1}{N_2} = \lambda t \tag{2.6}$$

$$N_2 = N_1 e^{-\lambda t} \tag{2.7}$$

This is particularly handy if we know the mass or number of atoms of a radioactive nuclide.

2.3   SPECIFIC ACTIVITY

Activity can easily be determined using a detector with a known efficiency. First measure the count rate, then use Equation 2.4 to solve for activity. Mass is also easily measured using an appropriate balance. Why is the equation $A = \lambda N$ such a big deal? Even if atoms of a particular radioactive nuclide are painstakingly purified, it is not possible to exactly determine the mass of the nuclides that have not yet decayed. In other words, there will always be some contribution to the mass by the product nuclide(s). Additionally, radioactive nuclides are often incorporated into molecules or other nonradioactive materials. If mass is determined, it is the mass of all atoms present, not just the radioactive ones. Therefore, determining mass (and thereby N) of a radioactive nuclide is not as simple as throwing it onto a balance.

Occasionally, it is important to know how radioactive something is per gram of sample, regardless of what other stuff might be present. **Specific activity** is the activity per unit mass of a substance, whether it is atoms of a single nuclide, a compound, or some mixture of compounds. The SI units for specific activity are Bq/kg, but in the United States, variations on Ci/g are often used. Different activity and mass units can be combined. For our first example, let's keep it simple and calculate the specific activity of a pure nuclide.

Example 2.7

Calculate the specific activity of ^{133}Xe in units of dpm/g.

Notice that the problem doesn't give us a mass. The specific activity will be the same no matter how much ^{133}Xe there is. When we get to the last step, it'd be convenient if we had assumed there's exactly 1 g of ^{133}Xe. Since the mass doesn't matter, why not go with something that will make the math easier later? Assuming exactly 1 g, let's calculate N.

$$1\,g \times \frac{mol}{133\,g} \times \frac{6.022 \times 10^{23}\ atoms}{mol} = 4.53 \times 10^{21}\ atoms$$

Now calculate activity:

$$A = \lambda N = \frac{\ln 2}{5.243\,d} \times (4.53 \times 10^{21}\ atoms) = 5.99 \times 10^{20}\ decays\,per\,day$$

Note that, because the half-life has units of days, the units for activity are decays per *day*. Calculating specific activity from here involves conversion of activity to units of decays per minute, then dividing by the total mass of the sample, which we cleverly chose as exactly 1 g:

$$\frac{\dfrac{5.99 \times 10^{20}\ decays}{day} \times \dfrac{d}{24\,h} \times \dfrac{h}{60\,min}}{1\,g} = 4.16 \times 10^{17} \frac{dpm}{g}$$

Example 2.8

Calculate the specific activity (dpm/g) of a gas mixture containing 78 g N_2, 21 g O_2, and 2.01 grams of ^{133}Xe.

There are no naturally occurring radioactive isotopes of nitrogen or of oxygen; therefore, ^{133}Xe is the only hot component of this mixture. The activity of any mass of ^{133}Xe can be calculated using its specific activity (conveniently determined in the previous example):

$$\frac{4.16 \times 10^{17} \text{ dpm}}{g} \times 2.01\,g = 8.36 \times 10^{17} \text{ dpm}$$

Okay, maybe that was trivial, but this *is* only Chapter 2 … the final step is to divide by the total mass of the sample:

$$\frac{8.36 \times 10^{17} \text{ dpm}}{78\,g + 21g + 2.01g} = 8.27 \times 10^{15} \frac{\text{dpm}}{g}$$

Notice that specific activity decreases as a radioactive nuclide becomes more "dilute" (Examples 2.7 and 2.8) within a sample.

2.4   DATING

The fact that radioactive decay can easily be predicted through mathematics opens up some interesting practical applications. One application is **radioactive dating**—the determination of the age of an object by measuring how much radioactive material is left in it. The most common form of dating is carbon dating. Small amounts of radioactive ^{14}C are continuously produced in the upper atmosphere through the bombardment of cosmic rays (Section 9.5). ^{14}C has a half-life of 5715 years, so it is also continually decaying. The rates of production have long since balanced out with the rate of decay, meaning that there has been a small but constant amount of ^{14}C on the planet for almost all of its history. Carbon is an essential element for all living things; therefore, a small fraction of their carbon is ^{14}C. Once an organism dies, it no longer takes in ^{14}C from the biosphere, and the ^{14}C in the dead organism slowly begins to decay. Over time, the ratio of ^{14}C to nonradioactive carbon (mostly ^{12}C) gradually decreases.

The specific activity of carbon is a convenient way to express the relative amount of ^{14}C present in all carbon. Prior to 1950, the specific activity of carbon on Earth was roughly 14 dpm/g. It varied somewhat before that time because cosmic ray bombardment is not constant but fluctuates through fairly regular cycles. The specific activity of carbon has varied in recent decades primarily because of the tremendous amount of fossil fuels being burned. Fossil fuels are made up of carbon from living things that died quite a long time ago. Since the time they died, all the ^{14}C that was once part of them has decayed away. Burning fossil fuels, therefore, pumps the atmosphere full of nonradioactive isotopes of carbon, lowering the specific activity of all carbon. It causes other problems too, but those are topics for other books.

Carbon dating works well for dating stuff that has been dead less than ∼60,000 years (∼10 half-lives). If it's been dead longer than that, it may be difficult to accurately measure the amount of ^{14}C present because it's almost all decayed away. Carbon dating was famously used to date the Shroud of Turin, which is believed by some to be the burial shroud of Jesus.

Example 2.9

The specific activity of carbon from the Shroud of Turin is 12.8 dpm/g today. When were the plants killed to make this cloth?

The equation we'd like to use here is Equation 2.1, but we'll need to calculate the decay constant from the half-life of ^{14}C first, using Equation 2.3.

$$\lambda = \frac{\ln 2}{t_{1/2}} = \frac{\ln 2}{5715\ a} = 1.213 \times 10^{-4} a^{-1}$$

First, we rearrange Equation 2.1, solving for t and then substituting in the appropriate values:

$$\ln \frac{A_1}{A_2} = \lambda t \quad t = \frac{\ln \dfrac{A_1}{A_2}}{\lambda} = \frac{\ln\left(\dfrac{14\ \text{dpm/g}}{12.8\ \text{dpm/g}}\right)}{1.213 \times 10^{-4} a^{-1}} = \frac{0.090}{1.213 \times 10^{-4} a^{-1}} = 740\ a$$

Therefore, the cloth was made about 740 years ago, or approximately 1300. Note that the answer is only expressed to two significant figures. As mentioned earlier, the specific activity of carbon has fluctuated in the past, so the answer is an approximate value. Generally speaking, ^{14}C-dating calculations done at this level of sophistication are only good to about two significant figures. Notice also that we can use specific activity values for A_1 and A_2 because they are both for pure carbon; therefore, the gram units in their denominators are equivalent (grams of pure carbon for both) and cancel out.

Other nuclides can be used for similar dating methods. For example, the age of rocks containing ^{238}U can be determined by measuring the ^{238}U:^{206}Pb ratio, assuming that all of the ^{206}Pb now present in the rock comes from the ^{238}U decay series (Figure 1.7). The same can be done when ^{87}Rb is found in a rock.

Example 2.10

A meteorite contains 0.530 g of ^{87}Rb and 0.042 g of ^{87}Sr. If 5 mg of the ^{87}Sr was present when the rock was formed, how old is the rock?

Over time, ^{87}Rb undergoes beta decay to ^{87}Sr:

$$^{87}_{37}Rb \rightarrow {}^{87}_{38}Sr + {}^{0}_{-1}e$$

How much ^{87}Rb was present in the rock when it was formed? It should be the sum of what's there now and what has decayed to ^{87}Sr since the rock was formed:

$$0.530 \text{ g } ^{87}\text{Rb} + \left(0.042 \text{ g } ^{87}\text{Sr} - 0.005 \text{ g } ^{87}\text{Sr}\right) \times \frac{\text{mol } ^{87}\text{Sr}}{871 \text{g } ^{87}\text{Sr}}\right.$$

$$\left.\times \frac{1 \text{mol } ^{87}\text{Rb}}{1 \text{mol } ^{87}\text{Sr}} \times \frac{87 \text{ g } ^{87}\text{Rb}}{\text{mol } ^{87}\text{Rb}}\right) = 0.567 \text{ g } ^{87}\text{Rb}$$

Notice that the conversion from grams of ^{87}Sr to grams of ^{87}Rb is really unnecessary since they have the same mass number, and one ^{87}Sr is produced for every ^{87}Rb that decays. This is often the case in dating problems but not always.

Now the math is the same as in the ^{14}C example, except here we use the gram ratio instead of the activity ratio (since $A \propto N$, and N is proportional to mass so long as the nuclide is the same):

$$t = \frac{\ln\dfrac{N_1}{N_2}}{\dfrac{\ln 2}{t_{1/2}}} = \frac{\ln\left(\dfrac{0.567 \text{ g}}{0.530 \text{ g}}\right)}{\dfrac{\ln 2}{4.9 \times 10^{10} \text{ a}}} = 4.8 \times 10^9 \text{ a}$$

2.5 BRANCHED DECAY

Radioactive decay is often a messy process. As we've already seen in the naturally occurring decay series (Figure 1.7), some nuclides make matters worse by decaying by more than one pathway. ^{64}Cu is a classic example of a nuclide that has more than one decay mode. It can decay by β^- to ^{64}Zn, or by β^+ or electron capture to ^{64}Ni. The half-life for ^{64}Cu is given as 12.701 h, but this accounts for all decays. Information on the relative percentages of each form of decay, or **branch ratio**, is given in Table 2.1. These data are from the National Nuclear Data Center at Brookhaven National Laboratory (http://www.nndc.bnl.gov/chart/). Table 2.1 tells us that ^{64}Cu decays via β^- to ^{64}Zn only 38.5% of the time, and the remaining 61.5% of the time, it decays via positron emission or electron capture to ^{64}Ni.

We can experimentally determine half-life a couple different ways. We can observe the change in count rate over time and use:

$$t_{1/2} = \frac{\ln 2}{\ln\dfrac{A_1}{A_2}} \times t$$

TABLE 2.1
Branch Ratios for ^{64}Cu Decay

Decay Mode	Branch Ratio	Nuclide Produced
β^-	38.5%	^{64}Zn
β^+	17.6%	^{64}Ni
EC	43.9%	^{64}Ni

which is a rearrangement of Equation 2.1. Using this method does not require application of the branch ratio, that is, the change in activity (or count rate) with time does not depend on the branch ratio.

The other way to find half-life is to determine the activity and mass of the nuclide (convert to N) and use:

$$t_{1/2} = \frac{\ln 2 \times N}{A}$$

which is a rearrangement of $A = \lambda N$ (Equation 2.5). As you might guess, the first method is more useful when the nuclide has a short half-life (a change in activity can be observed in a reasonable amount of time), and the second method works best for nuclides with long half-lives (measure mass and activity—take your time, the decay rate's not changing any time soon). If this latter method is used with a nuclide that undergoes branched decay, the branch ratio must be taken into account to properly calculate activity. Activity results from all decays. If only one mode is observed during an experiment, branch ratio and efficiency must be used to calculate activity. It's a bit like trying to figure out how quickly a bucket of water will drain through a hole based on the amount of water in the bucket and the flow rate out of the hole. The problem is that the bucket has another hole that you can't see. The bucket will always drain faster than you think it should, based on the flow rate from the hole you can see.

Example 2.11

A sample of copper is known to contain 9.2×10^7 atoms of ^{64}Cu, and registers 2320 cpm on a detector that can only detect beta particles (β^-). If this detector has an efficiency of 7.20%, what is the half-life of ^{64}Cu?

First calculate the observed activity in dpm using Equation 2.4 and the percent efficiency:

$$dpm = \frac{2320 \text{ cpm}}{7.2\%} \times 100\% = 3.2 \times 10^4 \text{ dpm}$$

Next, account for the branch ratio to determine the total activity. We know the decay rate is only for β^-, so the decay rate for all decays must be higher, so we divide by 0.385 (the fractional branch ratio from Table 2.1):

$$\frac{3.2 \times 10^4 \text{ dpm}}{0.385} = 8.4 \times 10^4 \text{ dpm}$$

Now plug into $A = \lambda N$:

$$8.4 \times 10^4 \text{ dpm} = \lambda \times (9.2 \times 10^7 \text{ atoms})$$
$$\lambda = 9.1 \times 10^{-4} \text{ min}^{-1}$$

Because the units for activity were decays per minute, the units on the decay constant are reciprocal minutes. Now convert decay constant to half-life:

$$t_{1/2} = \frac{\ln 2}{\lambda} = \frac{\ln 2}{9.1 \times 10^{-4}\,\text{min}^{-1}} = 7.6 \times 10^2\,\text{min}$$

It looks like this half-life would be more conveniently expressed in hours:

$$t_{1/2} = 7.6 \times 10^2\,\text{min} \times \frac{1\,\text{h}}{60\,\text{min}} = 13\,\text{h}$$

With only two significant figures to work with, this is pretty close to the actual value of 12.701 h.

If we had left out the branch ratio in these calculations, the half-life would be 33 hours—quite a way off! A half-life calculated using data from one branch of the decay is called a **partial half-life**. A partial half-life is always longer than the true or total half-life. If we know the fraction of the decays that go by each branch, we can calculate the partial half-life by simply dividing the total half-life by the fraction. Since the true half-life for ^{64}Cu is 12.701 h, the electron capture (EC, the other decay pathway for ^{64}Cu) partial half-life is:

$$\frac{12.701\,\text{h}}{0.439} = 28.9\,\text{h}$$

This makes sense! Since only about half of the decays are EC, the partial half-life should appear to be about twice as long.

Example 2.12

^{192}Ir decays via β⁻ emission 95.2% of the time. If its β⁻ partial half-life is observed to be 77.5 d, what is its true half-life?

$$77.5\,\text{d} \times 0.952 = 73.8\,\text{d}$$

2.6 EQUILIBRIA

The naturally occurring decay series that begins with ^{238}U starts with an alpha decay forming ^{234}Th (Figure 1.7). ^{234}Th is also unstable toward decay, spitting out a beta particle, forming ^{234}Pa. These two decay processes can be combined into a single equation:

$$^{238}_{92}\text{U} \xrightarrow{\ \alpha\ } {}^{234}_{90}\text{Th} \xrightarrow{\ \beta^-\ } {}^{234}_{91}\text{Pa}$$

In any decay, the original nuclide is called the **parent** nuclide, and the product nuclide is called the **daughter**. In the case of a decay series, all nuclides formed after the parent decays are called daughters. Here, ^{238}U is the parent, and ^{234}Th and ^{234}Pa are daughters. A decay series could also be generically represented by:

$$A \xrightarrow{t_{1/2(A)}} B \xrightarrow{t_{1/2(B)}} C \cdots$$

Since daughter B is constantly being formed by decay of parent A and is constantly decaying to daughter C, how can we figure out how much B exists at any one moment in time? More generally, how can we determine the total activity of the sample or the amount of any one nuclide at a particular point in time? It seems almost too complex to comprehend, but we can look at three simple scenarios where the math becomes more accessible.

2.6.1 SECULAR EQUILIBRIUM

If the half-life of the parent is very long relative to the half-life of the daughter, and the parent is isolated at time $= 0$, then the following equations describe the system *at any time* after the parent is isolated:

$$A_B = A_A(1 - e^{-\lambda_B t}) \qquad (2.8)$$

$$N_B = \frac{\lambda_A}{\lambda_B} N_A(1 - e^{-\lambda_B t}) \qquad (2.9)$$

The subscripts A and B refer to parent and daughter, respectively. Therefore, A_B is the activity of the daughter, and λ_A is the decay constant for the parent. These two equations are mathematically identical. To derive the second from the first, substitute $\lambda_X N_X$ ($X = A, B$) for the corresponding activities.

If the activities of both parent and daughter are plotted after the isolation of pure parent, something like Figure 2.2 will result. Because the parent half-life is so incredibly long, its activity is constant over the period of observation. Daughter activity starts at zero because the parent was chemically isolated. As daughter is produced, its activity gradually increases. However, as the amount of daughter builds up, its rate of decay gradually becomes more significant. When the rate of formation of B is equal to the rate of decay of B, it has reached **secular equilibrium**, much like the ^{14}C in our biosphere discussed in Section 2.4 of this chapter. Secular equilibrium is reached roughly seven daughter half-lives after the parent is isolated. At this point, the activity of the daughter is equal to the activity of the parent, and the activity equation given above (Equation 2.8) can be simplified to $A_B = A_A$. This is true for all radioactive daughters (not just B) of a long-lived parent A. If instead, C is a stable nuclide (i.e., A and B are the only radioactive species present), the total activity after secular equilibrium is reached is twice that of either the parent or the daughter, as illustrated in Figure 2.2.

Because $A_B = A_A$ at any point in time *after* secular equilibrium is reached, we can substitute in λN for the activities and generate more equations such as the following:

$$\lambda_A N_A = \lambda_B N_B \qquad (2.10)$$

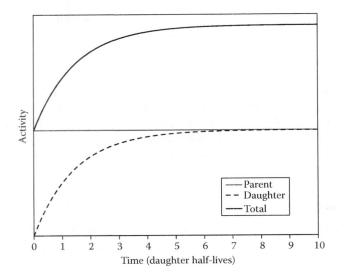

FIGURE 2.2 Secular equilibrium.

$$\frac{N_{\text{B}}}{N_{\text{A}}} = \frac{\lambda_{\text{A}}}{\lambda_{\text{B}}} = \frac{t_{1/2(\text{B})}}{t_{1/2(\text{A})}} \tag{2.11}$$

These equations can only be applied after secular equilibrium is reached. Secular equilibrium is very handy for solving all kinds of problems. It is especially useful with the three naturally occurring radioactive decay series. As you may recall, these decay series are a significant source of many of the radionuclides found in nature. Since they all start with a long-lived nuclide and have been around for a few billion years, secular equilibrium applies.

1. Since the activities of all nuclides in secular equilibrium are equal, the mass ratios ($N_{\text{B}}/N_{\text{A}}$) can easily be determined.
2. We have another method for determining the half-life of a long-lived nuclide. As we saw in Section 2.5, it is difficult to determine a long half-life by sitting around waiting to observe a change in activity over time. If we know the half-life of one of the daughters and its mass ratio to the parent, we can calculate the half-life of the long-lived parent using Equation 2.11.
3. The mass of the parent can be determined from the activity of any daughter.
4. The mass of the parent can be used to calculate the activity of a daughter.

Example 2.13

10.2 kg of uranium is isolated from some ore, which also contains 3.33×10^{-3} g of ^{231}Pa. What is the half-life of ^{235}U?

$$t_{1/2(U\text{-}235)} = \frac{N_{U\text{-}235}}{N_{Pa\text{-}231}} t_{1/2(Pa\text{-}231)}$$

$$= \frac{10{,}200\ g \times \dfrac{mol\ U}{238.0\ g} \times \dfrac{6.022 \times 10^{23}\ atoms}{mol} \times \dfrac{0.7204\ atoms\ {}^{235}U}{100\ atoms\ U}}{(3.33 \times 10^{-3}\ g) \times \dfrac{mol\ {}^{231}Pa}{231g} \times \dfrac{6.022 \times 10^{23}\ atoms}{mol}}$$

$$\times (3.28 \times 10^{4}\ a)$$

$$= 7.02 \times 10^{8}\ a$$

It is important to realize that uranium isolated from ore will contain all naturally occurring isotopes of uranium in their normal proportions. To do this problem, the number of ${}^{235}U$ atoms within 10.2 kg (10,200 g) of uranium must be calculated. ${}^{235}U$ makes up only 0.7204% of all uranium. Notice that the conversion from moles to atoms (Avogadro's number) is unnecessary, in both the numerator and the denominator, as it cancels out.

Example 2.14

What is the activity of ${}^{222}Rn$ in an ore sample containing 3.00 g of uranium?

$$A_{Rn\text{-}222} = A_{U\text{-}238} = \lambda_{U\text{-}238}N_{U\text{-}238}$$

$$= \left(\frac{\ln 2}{4.468 \times 10^{9}\ a}\right) \left(\begin{array}{c} 3.00\ g\ U \times \dfrac{mol\ U}{238.0\ g\ U} \times \dfrac{6.022 \times 10^{23}\ atoms\ U}{mol\ U} \\[2mm] \times \dfrac{99.2742\ atoms\ {}^{238}U}{100\ atoms\ U} \end{array} \right)$$

$$= 1.17 \times 10^{12}\ \text{decays per year}$$

$$= 2.22 \times 10^{6}\ \text{dpm}$$

In this example, the correct decay series must first be determined. ${}^{222}Rn$ is only observed in the ${}^{238}U$ decay series; therefore, it is only produced by ${}^{238}U$, which is 99.2742% abundant.

The previous discussion of secular equilibria assumes that the parent does not undergo branched decay and that no branched decay happens between parent and a daughter that doesn't also go through that daughter. Close examination of the decay series for the nuclides involved in Examples 2.13 and 2.14 should bring a sigh of relief. No branches involving or between these nuclides. If the parent (A) undergoes branched decay, then the activity of the daughter (A_B) is equal to the activity of the parent (A_A) times the (fractional) branch ratio (BR):

$$A_B = A_A \times BR \tag{2.12}$$

2.6.2 Transient Equilibrium

Transient equilibrium is like secular equilibrium in that the half-life of the parent is longer than the half-life of the daughter. The difference is the time of observation.

A nuclear equilibrium is considered transient if significant parental decay occurs during the time of observation (usually more than 7–10 daughter half-lives). When transient equilibrium occurs, the parent and daughter half-lives tend to be closer together in value.

If the parent is isolated at time = 0, then as the daughter is produced and begins to decay, the parent is also decaying. Parental activity will decrease in an exponential fashion, as shown in Figure 2.3. Keep in mind that the parent's half-life is longer than the daughter's, so a dramatic decrease in parent activity is not observed in Figure 2.3 (i.e., it is only slightly curved). Daughter activity grows at first, peaks, and then decreases, eventually paralleling the decrease in parent activity. Notice that daughter activity eventually exceeds parent activity. It is after ~7 daughter half-lives that the relative activities of parent and daughter stabilize, establishing transient equilibrium. Again, assuming parent (A) and daughter (B) are the only significant radioactive species present, total activity starts at the same point as parent activity, increases as the amount of daughter increases, and then parallels the decrease in parent and daughter activity. *After* equilibrium is achieved, the following equations can be applied:

$$\frac{A_A}{A_B} = 1 - \frac{t_{1/2(B)}}{t_{1/2(A)}} \tag{2.13}$$

$$\frac{N_B}{N_A} = \frac{t_{1/2(B)}}{t_{1/2(A)} - t_{1/2(B)}} \tag{2.14}$$

Again, these two equations are mathematically related through the awesome power of $A = \lambda N$. The choice of which to use is largely determined by the problem. Does it involve activities or numbers of atoms? The former is much more likely.

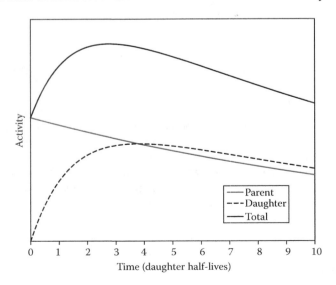

FIGURE 2.3   Transient equilibrium.

Example 2.15

A sample of ^{47}Ca is isolated from other radionuclides. Twenty-eight days later, the activity due to ^{47}Ca is 2.19×10^4 dpm. What activity would you expect for ^{47}Sc, also 28 days after the original sample of ^{47}Ca was isolated?

The half-life for ^{47}Ca is 4.536 d, and for ^{47}Sc, it is 3.349 d. This sample has attained transient equilibrium. The equilibrium is transient because a lot of the isolated ^{47}Ca will have decayed after 28 days, that is, it's been sitting around for 28 d/4.536 d = 6.2 parent half-lives; 28 days is also long enough for equilibrium to be established because 28 d/3.349 d = 8.4 daughter half-lives. Rearrange Equation 2.13 to solve for A_B, and plug in the numbers.

$$A_{\text{Sc-47}} = \frac{A_{\text{Ca-47}}}{1 - \dfrac{t_{1/2(\text{Sc-47})}}{t_{1/2(\text{Ca-47})}}} = \frac{2.19 \times 10^4 \text{ dpm}}{1 - \dfrac{3.349 \text{ d}}{4.536 \text{ d}}} = 8.37 \times 10^4 \text{ dpm}$$

For all transient equilibria, daughter activity is maximized at time $t_{\max}$, which can be calculated from the parent and daughter half-lives using Equation 2.15.

$$t_{\max} = \left[\frac{1.44 t_{1/2(A)} t_{1/2(B)}}{(t_{1/2(A)} - t_{1/2(B)})}\right] \times \ln \frac{t_{1/2(A)}}{t_{1/2(B)}} \tag{2.15}$$

Example 2.16

In the previous example, how long did it take the daughter to reach maximum activity?

$$
\begin{aligned}
t_{\max} &= \left[\frac{1.44 t_{1/2(\text{Ca-47})} t_{1/2(\text{Sc-47})}}{(t_{1/2(\text{Ca-47})} - t_{1/2(\text{Sc-47})})}\right] \times \ln\left(\frac{t_{1/2(\text{Ca-47})}}{t_{1/2(\text{Sc-47})}}\right) \\
&= \left[\frac{1.44 \times 4.536 \text{ d} \times 3.349 \text{ d}}{(4.536 \text{ d} - 3.349 \text{ d})}\right] \times \ln\left(\frac{4.536 \text{ d}}{3.349 \text{ d}}\right) \\
&= 5.59 \text{ d}
\end{aligned}
$$

2.6.3 No Equilibrium

When the half-life of the parent is shorter than the daughter, no equilibrium is reached. If pure parent is again separated at time = 0, the parent activity will decrease rapidly, as shown in Figure 2.4. Daughter activity will increase until the parent is nearing depletion, then it will gradually decrease. When the parent is gone (>7 parent half-lives), the daughter's decrease in activity will be based entirely on its own half-life ($t_{1/2(B)}$). Since no equilibrium is reached, no special equations are necessary to describe the change in daughter activity after the parent is gone. The activity equations given at the beginning of this chapter (Equations 2.1 and 2.2) are

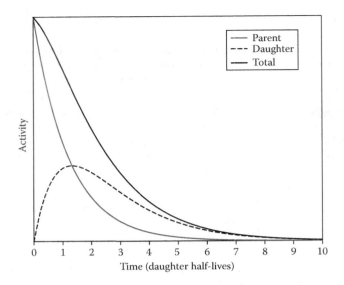

FIGURE 2.4 Equilibrium is not established when $t_{1/2A} < t_{1/2B}$.

sufficient. However, Equation 2.15, used to calculate maximum daughter activity for transient equilibria, also applies here:

$$t_{max} = \left| \frac{1.44 t_{1/2(A)} t_{1/2(B)}}{(t_{1/2(A)} - t_{1/2(B)})} \right| \times \ln \frac{t_{1/2(A)}}{t_{1/2(B)}}$$

Nuclear equilibria are sometimes confusing to students. Perhaps it's because the term "equilibrium" is not the best one to describe what is actually happening. We say the nuclear equilibria are reached when the relative rate of formation of daughter equals its relative rate of decay (A → B → C...). For most students, the more familiar use of the word "equilibrium" likely comes from chemistry. In a chemical reaction, equilibrium is reached when the rate of formation of products equals the rate of formation of reactants. In other words, reactants are combining to form products at the same rate that products are combining to form reactants (A ⇌ B). To extend this analogy to a radioactive decay would mean that the daughter would have to "decay" back into the parent, which is not possible. "Steady-state" would be a better term to describe nuclear equilibria. Steady-state is like pouring water into a bucket that's already full—the rate water is being added is equal to the rate it is spilling out, but we always have the same amount of water in the bucket. Nuclear equilibria are all about the defined mathematical relationships between the amount of parent and daughter. Parent is constantly converting to daughter, and daughter is constantly decaying, but the ratio between parent and daughter is constant at equilibrium.

Two other points of confusion are trying to decide if a system under investigation is at equilibrium and what kind of equilibrium it is. In this text, the passage of seven daughter half-lives is consistently used as the point where equilibrium is first reached.

Other texts state that equilibrium is only attained after 10 daughter half-lives have elapsed. If the point in time is within the controversial $7\text{--}10 \times t_{1/2(B)}$ period, then it is largely a matter of the level of precision required. For most work, seven half-lives will be sufficient, and that's why it is used as a point of demarcation in this book. However, if very precise measurements are being made, then waiting until 10 half-lives have passed before declaring the system to be at equilibrium may be necessary.

The confusion in determining whether secular or transient equilibrium is appropriate to a particular problem is understandable. The difference can be ambiguous. The main distinction is that the half-life of the parent needs to be long compared to the time of observation for the equilibrium to be secular. In other words, the parent cannot decay significantly ($<5\%\text{--}10\%$) during the time of observation. Again, the exact percentage of decay that is tolerable will depend on the level of precision required by the problem.

2.7 STATISTICS

Radioactive decay is a random event. We can't say when an unstable nucleus will decay, only that it will probably decay within seven or so half-lives. If we've got a lot of an unstable nuclide, the number of them that decay in a certain amount of time will vary, even if the number of these nuclides does not change significantly. Table 2.2 gives count data for a ^{204}Tl source counted 20 times for 1 minute each. With a half-life of 3.8 years, it did not decay appreciably during the 20 minutes these data were collected. The lowest count rate is 2071 cpm, and the highest 2217 cpm—a difference of 146 cpm. What is the true count rate for this sample?

As you might guess, the best we can do is average our data and hope it is close to the true value. How much error is in this estimate? Statistics can give us the answers!

The probability that we will get a particular count rate can be approximated by a Gaussian (or normal) distribution. A Gaussian is a curve, symmetric about the mean ($\bar{x}$) value as shown in Figure 2.5. The vertical axis of the plot corresponds to

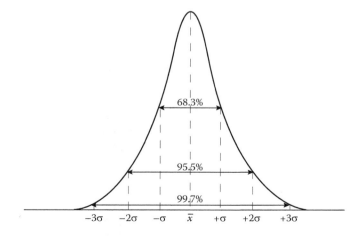

FIGURE 2.5 A Gaussian distribution.

the probability that a value along the horizontal axis will be observed. The curve is highest at the mean and tapers away in both directions, so we are more likely to get a count rate close to the mean rather than far from it.

Because radioactive decay is a perfectly random event, the uncertainty associated with our measurement (the standard deviation or σ) can be approximated as the square root of the mean value:

$$\sigma \approx \sqrt{\bar{x}} \qquad\qquad (2.16)$$

This equation applies no matter how many measurements are made, even if there is only one. This calculation of standard deviation is likely a little different than the one on your calculator or in Microsoft® Excel, as it applies only to statistically random processes (such as radioactive decay). Most nuclear scientists use uncertainties expressed to 2σ, which means that 95.5% of the time an additional count will fall within this range. For our sample data (Table 2.2), we could state that the count mean is 2142 ± 92 cpm to a confidence level of 95.5% ($\bar{x} \pm 2\sigma$). The 2σ range of these counts is therefore 2050 cpm to 2235 cpm. In other words, if we collected data for another 1-minute count, 95.5% of the time, it will fall between 2050 cpm and 2235 cpm. Note that none of the 20 counts in Table 2.2 exceed these limits.

TABLE 2.2
Count Data

Run	cpm
1	2123
2	2189
3	2120
4	2167
5	2125
6	2217
7	2192
8	2098
9	2154
10	2130
11	2112
12	2151
13	2191
14	2124
15	2071
16	2085
17	2179
18	2134
19	2096
20	2187

The mean and standard deviation allow us to look at a single set of data and characterize its precision. A relatively large standard deviation means poor precision, and a relatively small σ means good precision. Notice that the magnitude of the standard deviation depends on the magnitude of the mean value (Equation 2.16). For example, let's say we've collected another set of data from a toastier source. Its mean count rate is 4284 cpm, twice what we observed from the data in Table 2.2. This second set of data then has a value for 2σ of 131 cpm $\left(2\sqrt{4284} = 131\right)$. This is larger than the 2σ value we observed (92 cpm) when the average count rate was 2142 cpm. Which is more precise?

A good way to compare the precision for two or more sets of data is to calculate the relative standard deviation (RSD). RSD is determined by dividing the standard deviation by the mean value:

$$\text{RSD} \approx \frac{\sigma}{\overline{x}} \qquad (2.17)$$

The RSD for the data in Table 2.2 is 0.022, while the RSD for the data at the higher count rate is 0.015. Relatively speaking, the higher count rate data have less uncertainty and are, therefore, more precise. This is generally true; higher count rates have lower relative uncertainty.

Sometimes we would like to compare the average count of one set of data points with the average count from another set of data points. If we repeat the above experiment (Table 2.2) exactly as it was originally performed and obtain a new set of 20 1-minute counts, the mean value of the new set could be compared with the mean value of the first set using the standard deviation of the mean (σ_{mean}). It is obtained by dividing the standard deviation by the square root of N, where N is the number of data (runs) in the set:

$$\sigma_{\text{mean}} = \frac{\sigma}{\sqrt{N}} \qquad (2.18)$$

For the data given in Table 2.2:

$$\sigma_{\text{mean}} = \frac{46}{\sqrt{20}} = 10\,\text{cpm}$$

This would, therefore, be 20 cpm at the 95.5% ($2\sigma_{\text{mean}}$) confidence level. We could then quote the mean and *its* uncertainty as 2142 ± 20 cpm. This suggests that the new mean value, calculated from the second set of data, should fall in the range of 2122 to 2162 cpm. Note that the uncertainty will become smaller as the number of runs increases.

Experimentally, the mean value obtained in a series of 20 1-minute counts should have the same uncertainty as that obtained from a single 20-minute count. To calculate σ_{mean} for a single count, we can simply replace the number of runs (N), in Equation 2.18 with the time, in minutes, for the single count.

As an example, let's say the total counts for the lone 20-minute count was 43,000 counts, using the same experimental setup as was used to generate the data in Table 2.2. The average or mean count rate is obtained (as before) by dividing the total counts by the total time of the counts. The average count rate ($\bar{x}$) is calculated as 43,000 counts/20 minutes = 2150 cpm. The standard deviation is calculated as before:

$$\sigma \approx \sqrt{\bar{x}} = \sqrt{2150} = 46$$

For this count, the mean value with uncertainty is 2150 ± 92 cpm to a confidence level of 95.5%. In other words, if we collected an additional 1-minute count using the same setup, we would expect it to be somewhere between 2058 and 2242 cpm. Pretty much the same result as with the 20 1-minute count set of data in Table 2.2 (2050–2235 cpm).

What we would like to do now is compare ranges established by the σ_{mean} values for the two sets of data we have collected. The standard deviation of a mean obtained for a count longer than 1 minute is determined by dividing the standard deviation by the square root of the time it took to collect the data (in minutes):

$$\sigma_{mean} = \frac{\sigma}{\sqrt{t}} \tag{2.19}$$

For our single, 20-minute count, the standard deviation of the mean is:

$$\sigma_{mean} = \frac{46}{\sqrt{20}} = 10$$

Therefore, the mean value for the single 20-minute count is 2150 ± 20 cpm to a certainty of 95.5%. This is a range of 2130–2170 cpm. If we now look at the mean ranges for the two data sets, we have 2122–2162 cpm and 2130–2170 cpm. Since these ranges overlap, the two mean values are not significantly different from each other. Numerically, the two mean values are different, but this difference is not large enough to be statistically significant. Many more runs or a much longer counting time (lowering σ_{mean}) would be needed to determine if the mean values are, in fact, different.

If we had a very large sample of data, we could calculate the true mean (usually represented as μ). The standard deviation (σ) would then be exactly equal to the square root of the true mean ($\sqrt{\mu}$). In a limited number of count runs, like the 20 in our first data set, we can calculate a good value of σ by using the average of our counts ($\bar{x}$) instead of the true mean μ, but we should keep in mind that this is an approximation because $\bar{x}$ is only an approximation of the true mean μ.

Readers should take care not to confuse the meanings of σ and σ_{mean}. Both are expressions of uncertainty and reflect on the precision of the data. Standard deviation (σ) establishes the range within which we can expect another *single* value to appear. For the data in Table 2.2, it tells us that there's a 95.5% chance that another 1-minute count run will fall between 2050 and 2235 cpm. The standard deviation of the mean (σ_{mean}) establishes the range within which we can expect another *set* of runs to produce a mean value. For the data in Table 2.2, it tells us that there's a 95.5% chance that the mean value of another set of 1-minute counts will fall between 2122 and 2162 cpm.

We can still calculate an uncertainty for a single count x by assuming it is a reasonable approximation of the average of many counts $\bar{x}$. The uncertainty of a single count is found by $\sigma \approx \sqrt{x}$. For a confidence level of 95.5% or 2σ, this is $2\sqrt{x}$, that is, two times the square root of a count is the uncertainty of each count. Please note that σ must be calculated using the raw count data. For example, if a source exhibits 2345 counts over a 20-second period, then $2\sigma = 2 \times \sqrt{2345} = 97$. When the count rate is converted to a rate in units of counts per minute, the count data (here, counts per 20 s) is multiplied by 3. If we want an idea of the precision of the count rate in terms of counts per minute, the value for 2σ must also be multiplied by 3. Other (such as dead time and background) corrections are applied to the count rate but not to the uncertainty (2σ). The final count rate (cpm) and random error for this example are 7068 ± 291 cpm (try the math yourself), or about a 4% uncertainty. Remember when this text stated that count data are usually only good to three significant figures? These numbers suggest that three is a little generous, at least at this count rate and data collection time.

As we discovered earlier, higher count data (longer count times or a higher activity sample) will have a smaller uncertainty and should be favored in drawing the best-fit line to experimental data. When a computer draws such a line, it typically takes each point as equally valid. You may find that you get better results if you eliminate points with the lowest count rates. They can be eliminated because of their higher uncertainty (low precision). Points that are more than $\pm 2\sigma$ from the expected count are sometimes viewed with some skepticism by students, but remember that 4.5% of all points are expected to fall outside of this range ($100\% - 95.5\% = 4.5\%$). Points that are beyond 3σ of the expected value (a 99.7% confidence level) *could* be eliminated from most data as they are probably due to mistakes made in collecting the data. Students are cautioned from simply discarding data because they are outside of some level of precision. Unless you have some other reason to suspect there is a problem with a particular data point, it should be retained.

As stated previously, radioactive decay data closely approximate a Gaussian distribution. In reality, radioactive data are better fit to the similar, but mathematically more complex, Poisson distribution. To determine whether a set of data follows a Poisson distribution, we need to calculate χ^2 (chi-squared):

$$\chi^2 = \frac{1}{\bar{x}} \sum (\bar{x} - x)^2 \tag{2.20}$$

To find χ^2, we'll need to find the square of the difference of every individual value in the data set with the mean value (($\bar{x} - x)^2$), add them up ($\Sigma(\bar{x} - x)^2$), and then multiply by the reciprocal of the mean ($1/\bar{x}$). This sounds crazy, but it works. Try calculating this value using the data in Table 2.2. χ^2 for those data should equal 14.9.

Is this good? To interpret a chi-squared value, we'll need to use a table of chi-squared like the one found in Appendix B. To use this table, you need to know that there are 20 data points; therefore, the data have 19 degrees of freedom (N–1). Reading down the first column to 19, there are several values of chi-squared listed on this row of the table to be compared with the calculated value. For 19 degrees of freedom, there is 1 chance in 100 that chi-squared will be smaller than 7.633 and also 1 chance in 100 that it will be larger than 36.19. Our value of 14.9 certainly falls within that range.

A more reasonable set of limits is the 80% chance. There is a 10% chance that chi-squared will be smaller than 11.65 for 19 degrees of freedom. This is obtained by reading down the 0.90 column to 19 degrees of freedom. There is also a 10% chance that chi-squared will be larger than 27.20. This is obtained by reading down the 0.90 and 0.10 columns to 19 degrees of freedom. Therefore, there is an 80% chance that chi-squared will fall between 11.65 and 27.20 for a Poisson distribution. An 80% chance means that 8 out of 10 times this will happen. Once again, our value of 14.9 fits within the range, giving us confidence that our data fit a Poisson distribution, as well they should.

There have probably been many times in your life when you did not have a chi-squared table at hand. If you look down the 0.50 column, you will note that chi-squared increases with the number of trials N. The value of the chi-squared is roughly N–2. If you ever need a ballpark estimate of chi-squared for, say, 31 determinations, it would be about 29. The entry in the table for 30 degrees of freedom, that is 31 data at the 50% probability, is 29.34. Darn close.

So far, we've only discussed the inherently random nature of radioactive decay, which leads to a certain amount of uncertainty in data collected from radioactive sources. This is truly random error, but the wonderful thing about it is that we can always calculate it. What if we carefully calculate the random error associated with some decay measurements, but it is clear some additional error is also present? That error would be systematic error, and it is likely due to some external change during the experiment. Perhaps the source or the detector moved, or some shielding between the source and detector was added or removed. It could be a number of factors. A good scientist will take great care in making measurements and consider many possibilities if systematic error is apparent.

QUESTIONS

2.1 Without using the fact that one unified atomic mass unit is equal to 1.66×10^{-24} g, calculate the mass of exactly 1 g of ^{12}C in unified atomic mass units. Does this value look familiar? Briefly explain.

2.2 Calculate the average atomic mass of naturally occurring Mg.

2.3 Naturally occurring rubidium is a mixture of only two isotopes: ^{85}Rb (84.9118 u) and ^{87}Rb (86.9092 u). If the average atomic mass for Rb is 85.4678 u, calculate the percent abundance of its two isotopes.

2.4 Calculate the number of ^{81}Br atoms that are present in a sample of 1.56 g of pure bromine.

2.5 What is the difference between count rate and decay rate? How are they related, mathematically?

2.6   Derive Equation 2.3 from Equation 2.1.

2.7   What fraction of the original activity is left after three half-lives have passed? After five?

2.8   What is the activity of a ^{60}Co source calibrated at 95 kBq 2016 days ago?

2.9   Counts from a radionuclide decrease from 5718 dpm to 515 dpm in 24.0 hours. What is the half-life of this nuclide?

2.10 A sample of ^{99}Mo has an activity of 15,000 dpm at noon today. Calculate its activity 30.0 days from now.

2.11 A sample of ^{204}Tl was calibrated at 0.501 µCi. If its current activity is 1.48×10^5 dpm, when was it calibrated?

2.12 In 1971, Darleane Hoffman reported the detection of 20 million atoms of ^{244}Pu, isolated in an ore sample from a California mine. Assuming two significant figures, what mass of ^{244}Pu would've been necessary 4.6 billion years ago to produce 20 million atoms in 1971?

2.13 If you observe 2554 cpm for a ^{137}Cs source on the first shelf of your detector, and the actual activity is 10,756 dpm, what is the percent efficiency for the first shelf of the detector?

2.14 Using a ^{204}Tl source that was calibrated at 361 kBq on May 15, 2013, you observe 347,816 cpm, 236,560 cpm, and 135,284 cpm on the first, second, and third shelves below your detector on May 26, 2017. Calculate the percent efficiency on all three shelves.

2.15 A sample of ^{90}Y records 1500 cpm on the first shelf of the detector. If the percent efficiency of the first shelf is 3.45%, what will the activity of the source be 5.00 days from now?

2.16 Calculate the activity due to the decay of ^{40}K in your body. It will help to know that there are 140 g of potassium in a person weighing 70 kg.

2.17 Calculate your specific activity based on hydrogen, carbon, and potassium. The natural percent abundances of these nuclides are listed in Table 2.3. It might also be useful to know that humans are 18.6% carbon, 9.5% hydrogen, and 0.35% potassium by weight.

2.18 Calculate the specific activity (MBq/g) of 0.353 g of NaTcO$_4$. Assume all the technetium is ^{99}Tc, and it is the only radioactive nuclide present.

2.19 Calculate the specific activity (µCi/g) of naturally occurring platinum.

2.20 An Internet newspaper reported that the Japanese government was allowing fertilizer with a specific activity of less than 200 Bq/kg to be sold on the public market shortly after the Fukushima meltdowns. The implication was that

---

TABLE 2.3

Percent Abundances for Problem 17

Nuclide	% Abundance
^{3}H	5.0×10^{-15}
^{14}C	1.2×10^{-10}
^{40}K	0.0117

the government was nefariously allowing its people to consume contaminated agricultural products. A common fertilizer used by home gardeners is 20% (by weight) K_2O. Assuming two significant figures, calculate the specific activity of the fertilizer in Bq/kg. Why is this a conservative value?

2.21 A piece of wood taken from an Egyptian tomb shows 55.9% of the ^{14}C activity than a living piece of wood did during the approximate time the tomb was constructed. When did the tree containing this piece of wood die?

2.22 Samples of bone and charcoal found near some pottery shards in China are 19,000 years old using carbon dating. What would you expect the specific activity of carbon in these samples to be?

2.23 A rock is found to contain 0.092 g of ^{40}K and 0.530 g of ^{40}Ar. How old is the rock? What assumption(s) are you making to solve this problem? Do you believe they are valid?

2.24 What are the partial half-lives for the two major modes of decay for ^{36}Cl?

2.25 A sample of copper known to contain 9.76×10^{-15} g of ^{64}Cu gives a count rate of 2315 cpm on a detector that is 7.20% efficient toward this energy of beta particle. Assuming the detector can only detect beta decay, calculate the half-life of this nuclide.

2.26 Determine if each of the following are at secular equilibrium, transient equilibrium, or neither.
   a. A rock containing ^{235}U and ^{231}Pa.
   b. A sample of ^{47}Ca is isolated and observed for 35 days.
   c. A sample of ^{72}Zn is isolated and observed for 2 days.
   d. A sample of ^{19}F is isolated and observed for 40 days.

2.27 Some cave paintings in northern Spain have been dated by determining the $^{230}Th/^{238}U$ activity ratio in the calcite that later covered them. The idea is that uranium is naturally separated from the rest of its decay series in concentrated calcium carbonate solutions, which deposited on the paintings shortly after they were made. If the activity ratio is 0.290, how old are the calcite deposits on top of this painting? Assume that the ^{230}Th activity has been corrected to account only for the ^{230}Th formed from ^{238}U decay. Why is this correction important?

2.28 A rock contains 10.1 g of ^{232}Th. What is the activity of ^{212}Po in the same rock?

2.29 A rock contains 1.77 g of ^{232}Th. What is the total activity, from all radioactive nuclides, of this rock?

2.30 What is the mass ratio of ^{230}Th to ^{234}Th in a sample of uranium ore?

2.31 Which of the following are true of secular equilibria?
   a. $t_{1/2(A)} \gg t_{1/2(B)}$.
   b. $k_{(A)} \gg k_{(B)}$.
   c. The activity of the parent will remain constant over the period of observation.
   d. It is reached at $5 \times t_{1/2(B)}$ after isolation of the parent.

2.32 A sample of ^{99}Mo with an activity of 7.67×10^{13} dpm is isolated. After 56 hours, what activity of ^{99m}Tc would be observed? Assume that only 86% of all ^{99}Mo decays to ^{99m}Tc, and the remainder decays directly to ^{99}Tc.

TABLE 2.4
Count Data for Problem 37

2018	2098	2217	2075	2087
2112	2207	2117	2045	2062
2056	2039	2139	2153	2222
2122	2048	2143	2225	2110

TABLE 2.5
Count Rates for Problem 38

Run 1 (cpm)	85	83	97	82	84	80	94	92
	85	97	104	88	93	83	94	
Run 2 (cpm)	1967	1908	1954	1916	1928	1887	1904	1964
	1896	1897	1922	1849	1880	1855	1837	

2.33 A 7.0 μg sample of ^{32}Si is isolated. What will be the activity of ^{32}P in this sample after four weeks? Which type of nuclear equilibrium does this represent?

2.34 How many grams of radium are in a sample of ore containing 1.00 g of Th? What is the activity of ^{219}Rn in this sample?

2.35 A 7.0 mg sample of ^{48}Cr is isolated. When will its daughter reach maximum activity? Which type of nuclear equilibrium does this represent?

2.36 Is it surprising that all 20 count rates listed in Table 2.2 fall within two standard deviations of the mean ($\overline{x} \pm 2\sigma$)? Briefly explain.

2.37 A sample was counted 20 times for 1 minute each in the same geometry. The results (cpm) are shown in Table 2.4. Calculate the mean, standard deviation, RSD, and σ_{mean}. How many (%) data are within 1σ and outside of 2σ? Do these data appear to fit a Gaussian distribution? Do they fit a Poisson distribution? Briefly explain your answers.

2.38 Calculate the percent error (to 2σ) and RSD for the 15 data given in Table 2.5 for "Run 1." Repeat with the data from "Run 2." Do you observe a difference? Briefly explain.

3 Energy and the Nucleus

3.1 BINDING ENERGY

In Chapter 1, we learned that the strong force holds the nucleus together. The strong force is energy, and energy is required to pull a nucleus apart (thank goodness!). The energy required to separate a nucleus into individual nucleons is the **nuclear binding energy**. Nuclear binding energy could also be defined as the amount of energy released by the reverse process, that is, when a nucleus is assembled from its individual nucleons. Energy doesn't just suddenly appear. It must come from somewhere. In our theoretical construction of a nucleus from protons and neutrons, it comes from mass. You may have learned that mass is always conserved or that energy is always conserved. However, mass and energy can be converted one into the other; therefore, they are *collectively* conserved. If matter is destroyed, energy is created. Likewise, it is possible to create matter from energy. Sounds like *Star Trek*, but it's true! The mathematical equation that relates matter and energy is very simple, and very well known:

$$E = mc^2 \tag{3.1}$$

Energy is equal to mass times the square of the speed of light (2.998×10^8 m/s). When using Equation 3.1, the appropriate unit for mass is the kilogram (kg), which makes the energy unit the joule ($J = kg \cdot m^2/s^2$). Nuclear scientists like to think on the atomic scale and prefer to use the unified atomic mass unit (u) as a unit of mass. How does that convert to energy?

$$1\,u = 931.5\,MeV$$

One unified atomic mass unit (u) of mass is equal to 931.5 million electron volts (MeV) of energy. The electron volt (eV) is also a unit of energy, just like the joule. The joule is rather large when thinking on the atomic scale ($1\,J = 6.242 \times 10^{18}$ eV), so the electron volt is more convenient. An **electron volt** is defined as the amount of energy required to move one electron across a 1-volt potential—in other words, not much.

What about nuclear binding energy? We can calculate it by looking at how much energy is released when a nucleus is assembled from protons and neutrons. Let's start small, like an alpha particle ($^4_2$He). The mass of a proton is 1.007276 u, and the mass of a neutron is 1.008665 u, so the mass of an alpha particle should be:

$$(2 \times 1.007276\,u) + (2 \times 1.008665\,u) = 4.031882\,u$$

At this point, it'd be nice to compare this mass to the mass of an alpha particle. If we look up ^{4}He in Appendix A or in a chart of the nuclides, an *atomic* mass is given, which means electrons are included in the mass. To make comparisons easier, let's add the two electrons to our alpha particle, making it a ^{4}He atom:

$$4.031882\,u + (2 \times 0.000549\,u) = 4.032980\,u$$

The electrons are lightweights, but they do make a difference. This is the expected atomic mass, and it is clearly larger than the actual atomic mass of 4.002603 u, meaning that some mass is converted to energy during the assembly of a ^{4}He atom. This difference in mass is called the **mass defect**.

$$4.032980\,u - 4.0026032\,u = 0.030377\,u$$

It should now be apparent why a crazy number of significant figures were necessary at the beginning of these calculations. Once the masses are subtracted, the number of significant figures can go way down (from 7 to 5 in our case). A good rule of thumb is to express masses six places past the decimal in problems like this. Converting the mass defect into energy gives the nuclear binding energy for ^{4}He:

$$0.030377\,u \times \frac{931.5\,\text{MeV}}{u} = 28.30\,\text{MeV}$$

That wasn't so hard; let's try another one!

Example 3.1

Calculate the mass defect and nuclear binding energy for ^{56}Fe.
We can save time by combining the proton and electron masses, that is, using the atomic mass of ^{1}H (1.007825 u):

$^{56}_{26}\text{Fo}_{30}$	$26(^1_1\text{H})$	$26(1.007825\,u) = 26.203450\,u$
	$30(^1n)$	$30(1.008665\,u) = 30.259950\,u$
	expected atomic mass	$56.463400\,u$
	actual atomic mass	$55.934937\,u$
	mass defect	$0.528463\,u$
	convert to energy	$\times 931.5\,\text{MeV/u}$
	binding energy	$492.3\,\text{MeV}$

Wow! The mass defect and binding energy are enormous for ^{56}Fe when compared those to ^{4}He, but we really should've expected it. Every nucleon will lose a little mass in forming a nucleus; the more nucleons that are glued together, the more total mass will be lost—to make more glue. The more mass lost, the greater the binding energy. This also makes sense from the perspective of taking the nucleus apart. The bigger it is, the more energy should be required to pull it completely apart.

Is there any way to compare the nuclear binding energy values for ^{4}He and ^{56}Fe? Yes! Dividing the nuclear binding energy (or the mass defect) by the number of nucleons gives the average amount of energy it takes to pluck a single nucleon from that nucleus. In other words, how tightly held is each nucleon? This quantity is known as the **binding energy per nucleon** and can allow comparisons between nuclei.

$$^4_2\text{He} \quad \frac{28.30 \text{ MeV}}{4 \text{ nucleons}} = 7.074 \frac{\text{MeV}}{\text{nucleon}} \qquad ^{56}_{26}\text{Fe} \quad \frac{492.3 \text{ MeV}}{56 \text{ nucleons}} = 8.790 \frac{\text{MeV}}{\text{nucleon}}$$

It's a heck of a lot harder to pry a nucleon off a ^{56}Fe nucleus than off a ^{4}He nucleus. If these calculations are repeated for all naturally occurring nuclides, a graph such as Figure 3.1 could be produced. Those nuclides with relatively high values for their binding energy per nucleon could be considered more stable than those with lower values. In fact, if two light nuclides with low values were to combine (fusion), or one heavy nuclide were to split (fission) into two, a great deal of energy could be released as nuclides with greater stability are formed. These two processes will be discussed in more detail in Chapter 10. For now, it is simply important to understand that changes in binding energy per nucleon are the energetic driving force behind fission and fusion.

Roughly speaking, binding energy per nucleon is strongest for nuclides with mass numbers between 50 and 100, and it is weaker for nuclides above and below that range. Remember that the strong force is about 100 times stronger than the Coulomb repulsion (electrostatic) of protons. When we add another nucleon to a very small nuclide, we generally get a lot of strong force with it and very little Coulomb repulsion (because there are few protons in the nucleus), the net result being higher binding energy per nucleon. This works until the nuclide starts getting too big. Then the range of the

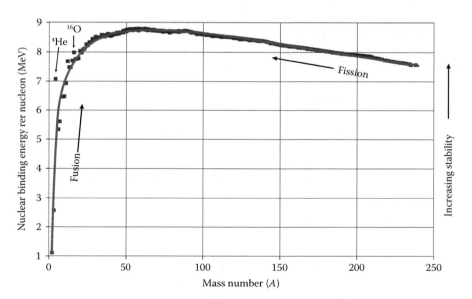

FIGURE 3.1 Relative stabilities of the naturally occurring nuclides.

strong force for each nucleon becomes smaller than the nucleus. Gradually, Coulomb repulsion becomes more and more significant, and the nucleons are held less tightly.

Another way to look at Figure 3.1 is by focusing on the fact that, unlike the Force (in *Star Wars*), the strong force has a limited, rather fixed range. Imagine you're a proton (because you like to stay positive!) in a relatively small ($A < 50$) nucleus. You feel the attraction to all the other nucleons, but there aren't that many, so you are not held all that tightly. Now imagine you're a proton in a nucleus with 50–100 other nucleons—ahhh, the sweet spot! You feel attraction with all the other nucleons because the nucleus is just the right size for every proton and neutron to feel the attraction of all the others. Since you've got a lot more nucleons in the nucleus than before, the attraction is stronger—you're held more tightly. Subatomic life is good. Now, imagine you're a proton in a larger nucleus ($A > 100$). You still feel a strong attraction to the nucleons near you, but you can no longer feel much, if any attraction of the protons and neutrons on the far side. The nucleus is now too big. The other disadvantage for large nuclei is that they have a lot of protons packed into a tiny space. As the nucleus gets bigger and bigger, attraction due to the strong force remains more or less constant, but more protons are added, increasing the pressure to push the nucleus apart, lowering binding energy per nucleon.

There are some interesting anomalies in the low mass end of Figure 3.1. Notice that ^{4}He sits well above the curve, indicating that its nucleons are bound especially tightly and, therefore, has some unusual stability. This is because ^{4}He has two protons and two neutrons—both are "magic numbers" (filled nuclear shells) that impart extra stability. Other nuclides with filled shells, such as ^{16}O, also exhibit unusually high nuclear stability.

Another way to understand the instability of large nuclides toward fission and the instability of small nuclides toward fusion is to look at the changes in mass for the nucleons. Similar to the calculations done above for binding energy per nucleon, we can also calculate the average mass of a nucleon for any given nuclide. Simply divide the mass defect by the mass number (number of nucleons) and subtract from the mass of a neutron or proton. These calculations are performed here for our two example nuclides of ^{4}He and ^{56}Fe.

Average mass of a proton:

$$^4\text{He}: 1.007276\,\text{u} - \frac{0.030377\,\text{u}}{4} = 0.999682\,\text{u}$$

$$^{56}\text{Fe}: 1.007276\,\text{u} - \frac{0.528463\,\text{u}}{56} = 0.997839\,\text{u}$$

Average mass of a neutron:

$$^4\text{He}: 1.008665\,\text{u} - \frac{0.030377\,\text{u}}{4} = 1.001071\,\text{u}$$

$$^{56}\text{Fe}: 1.008665\,\text{u} - \frac{0.528463\,\text{u}}{56} = 0.999228\,\text{u}$$

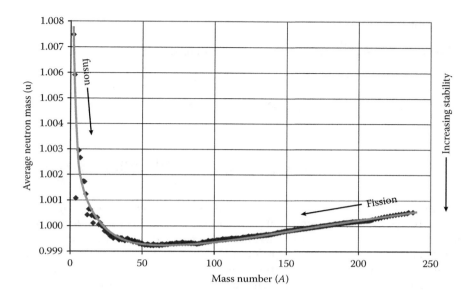

FIGURE 3.2   Average mass of a neutron in the naturally occurring nuclides.

The average mass of the nucleons is lower in ^{56}Fe than in ^{4}He. Where did that extra mass go? It is converted to energy and is used to hold nucleons in ^{56}Fe a little tighter than in ^{4}He. If we again repeat these calculations for all naturally occurring nuclides, a graph similar to Figure 3.1 can be generated (Figure 3.2).

Figure 3.2 is a mirror image of Figure 3.1! This shouldn't be too surprising, as Figure 3.1 looks at average energy, and Figure 3.2 looks at average mass. If energy is higher, there must be less mass around, and if mass is higher, there's less energy around—the two interconvert nicely in nuclei. Figure 3.2 also explains instability toward fission and fusion but from a different perspective. When a large nuclide undergoes fission, the products' nucleons will have less mass. Likewise, when very small nuclides combine to form larger ones, each nucleon loses a little mass (a curious weight-loss plan!). Instability toward fission and fusion means that the individual nucleons in certain nuclei can lose mass when they undergo one of these processes. Once the nuclei hit the mass number range of ~50 to ~100, no more mass can be lost by the nucleons, and they are stable toward fission and fusion. That mass lost during fission and fusion means energy (often a lot) is released.

3.2   TOTAL ENERGY OF DECAY

We'll get to the energetics of fusion and fission in Chapter 10, but the conversion of matter to energy also applies to decay processes. How much energy are we talking about? How can we calculate it? It is simply a matter of figuring out how much mass is lost during the course of the decay, then converting to energy. To calculate the mass lost, add up the masses of the products, and subtract that from the mass of the parent.

Example 3.2

How much energy is released when a ^{222}Rn nuclide decays?

First, write out the balanced decay equation:

$$^{222}_{86}\text{Rn} \rightarrow {}^{218}_{84}\text{Po} + {}^{4}_{2}\text{He}$$

Look up the masses, and do the math:

$$[222.017578\,\text{u} - (218.008966\,\text{u} + 4.002603\,\text{u})] \times \frac{931.5\,\text{MeV}}{\text{u}} = 5.597\,\text{MeV}$$

For decay processes, this energy is the **total energy of decay**. Notice that it is a little different from the experimentally determined 5.590 MeV listed in Appendix A. Remember that masses are also determined by experiment, and all experimentally determined values have some error associated with them. Where does this energy go? Most of it goes into the kinetic energy of the alpha particle. According to Appendix A, the most common alpha particle emitted by ^{222}Rn has an energy of 5.4895 MeV. Sounds about right. The rest is kinetic energy (recoil) of the daughter. We'll look at this in more detail in Chapter 5.

Keep in mind that 5.597 MeV is the amount of energy released for a *single* decay of ^{222}Rn. What if a whole mole of ^{222}Rn nuclides decayed?

$$\frac{5.597\,\text{MeV}}{\text{decay}} \times \frac{1.60 \times 10^{-13}\,\text{J}}{\text{MeV}} \times \frac{6.022 \times 10^{23}\,\text{decay}}{\text{mol}} = 5.39 \times 10^{11}\,\frac{\text{J}}{\text{mol}}$$

For comparison purposes, let's consider a common chemical reaction that cranks out a lot of heat. Burning a mole of methane (CH_4, the main component of natural gas) produces only 9×10^5 J, almost a million-fold less than the decay of ^{222}Rn. Energy is produced by this chemical combustion reaction. Is mass lost? You bet. If you do the math correctly, it should come out to be $\sim 10^{-8}$ g. It is such a small amount that it would be difficult to measure. Therefore, mass is essentially conserved in chemical reactions.

Let's take this one step further. If we could harness all the energy from the decay of 1 mole of ^{222}Rn as heat energy (it is easy!), how much ice could it melt? We'll have to dredge up some thermochemistry to answer this question. Remember that ice at 0°C still needs heat applied to melt to liquid water at 0°C. This is called the heat of fusion and is equal to 6.30 kJ/mole for water:

$$5.39 \times 10^{11}\,\text{J} \times \frac{\text{kJ}}{1000\,\text{J}} \times \frac{\text{mol}}{6.30\,\text{kJ}} \times \frac{18.0\,\text{g}}{\text{mol}} \times \frac{\text{kg}}{1000\,\text{g}} = 1.54 \times 10^6\,\text{kg}$$

One and a half million kilograms! That's a lot of ice. Radioactive decay clearly produces a lot of heat. Believe it or not, this helps explain why the Earth's core is so hot.

Given the age of the Earth ($\sim$4.5 billion years) and the rate at which it is cooling (as a rock in space ... we're not getting into global warming here!), it should've cooled off a lot more than it has. The reason why is that there are, and have been, lots of radioactive materials inside the Earth which have been busy decaying. All the energy from these decays is absorbed by the planet as heat, which helps the planet's core remain toasty warm.

Radioactive decay within the planet also explains where helium comes from. Because of its low atomic mass, helium can escape into space when released into the atmosphere. Where do we get all the helium to fill our balloons? When a radioactive nuclide undergoes alpha decay, the alpha particle quickly picks up a couple of electrons and becomes a helium atom. If this happens below the surface of the Earth, the helium can become trapped underground. The helium we fill balloons with comes from wells, usually drilled looking for natural gas or oil.

3.3 DECAY DIAGRAMS

Nuclear decay is always **exothermic**; in other words, it always gives off energy. Nuclear scientists sometimes use the term **exoergic** to mean the same thing. Nuclides always decay from a less stable nuclide to a more stable one. The atomic number (Z) also changes (except for gamma decay), so we can illustrate decay by plotting Z vs. energy such as in Figure 3.3.

The horizontal lines in Figure 3.3 represent nuclidic states—a nuclide in a particular energy state. The arrow begins at the parent and points toward the daughter. Notice that the daughter is always lower in energy than the parent, indicating that energy is given off during the decay process. The vertical displacement between parent and daughter is the total energy of decay, also known as the energy of transition, and corresponds to the mass lost in the decay. The horizontal position of the daughter relative to the parent varies depending on the type of decay. Two protons are lost during alpha decay, so the daughter is two units of Z lower than the parent. A neutron is converted to a proton during beta decay, so the daughter is one unit higher than the parent. The opposite nucleon conversion happens during positron and electron capture decay, so the daughter ends up one lower in Z. There is no change in atomic number during gamma decay. Drawing these diagrams helps us to see the (sometimes complex) energy changes involved in nuclear decay. These diagrams are also known as energy diagrams or decay schemes.

Let's look at some examples, starting with the three flavors of beta decay.

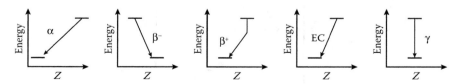

FIGURE 3.3 Generic decay diagrams for alpha, beta, positron, electron capture, and gamma decay.

Example 3.3

^{35}S decays to ^{35}Cl by emitting a beta particle. Draw a decay diagram, and write out the balanced equation. Also, calculate the total energy of decay.

The decay equation is:

$$^{35}_{16}\text{S} \rightarrow {}^{35}_{17}\text{Cl} + _{-1}\text{e}$$

The atomic mass of ^{35}S is 34.969032 u, and the atomic mass of ^{35}Cl is 34.968853 u. The total energy of decay (E) is therefore:

$$(34.969032\,\text{u} - 34.968853\,\text{u}) \times \frac{931.5\,\text{MeV}}{\text{u}} = 0.167\,\text{MeV}$$

According to Appendix A, the total energy of this decay is 0.1672 MeV. Our result is spot on. Appendix A also gives the energy of the beta particle as 0.1674 MeV. In this particular decay, it should be the same as the total energy of decay. Is this a problem? No, not really. Subtle differences like this are usually an indication that the two numbers were measured independently, and, again, since all measured values have some error associated with them, small discrepancies can be ignored. The decay diagram can now be drawn and is pictured in Figure 3.4, using the energy values from Appendix A.

Careful examination of Figure 3.3 reveals something strange about positron decay. It is the only one with a bent arrow. As we'll see later, the mass of two electrons must be created in every positron decay, which requires 1.022 MeV of energy. This energy value is obtained by multiplying the mass of two electrons (5.486×10^{-4} u) by our handy mass/energy conversion constant (931.5 MeV/u):

$$[2 \times (5.486 \times 10^{-4}\,\text{u})] \times \frac{931.5\,\text{MeV}}{\text{u}} = 1.022\,\text{MeV}$$

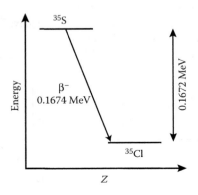

FIGURE 3.4   Decay diagram for ^{35}S.

Since this is required for every positron decay, we account for it by drawing a line straight down from the parent nuclide. The slanted arrow then represents the energy carried off by the positron. Why is mass created in positron decay but not in other forms of decay? The answer will have to wait until Chapter 5. Stay tuned!

Example 3.4

^{17}F undergoes positron decay to ^{17}O. Write out the balanced equation, and draw a decay diagram. If the total energy of decay is 2.7605 MeV, and the atomic mass of the daughter is 16.999132 u, calculate the atomic mass of ^{17}F.

The decay equation is:

$$^{17}_{9}F \rightarrow \, ^{17}_{8}O + \, _{+1}e$$

The decay diagram is shown in Figure 3.5. The positron energy value is from Appendix A. Notice how this value differs from the total energy of decay by 1.02 MeV.

The mass of an atom of ^{17}F can be calculated by taking advantage of the fact that the total energy of decay is related to the difference in mass between parent and daughter, just as it was for regular beta decay, except we're doing this problem backwards, starting at total energy of decay and working back to the masses. The total mass lost in this decay is:

$$2.7605 \, \text{MeV} \times \frac{u}{931.5 \, \text{MeV}} = 0.002963 \, u$$

which allows us to calculate the mass of the parent:

$$16.999132 \, u + 0.002963 \, u = 17.002095 \, u$$

The actual mass is reported as 17.0020952 u in Appendix A. Another excellent result! Next, let's look at electron capture.

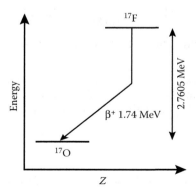

FIGURE 3.5   Decay diagram for ^{17}F.

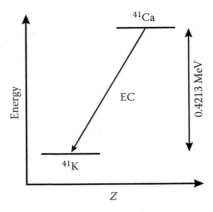

FIGURE 3.6 Decay diagram for ^{41}Ca.

Example 3.5

^{41}Ca undergoes electron capture decay to ^{41}K. Write out the balanced equation, and draw a decay diagram. If the total energy of decay is 0.4213 MeV, and the atomic mass of ^{41}Ca is 40.962278 u, calculate the atomic mass of ^{41}K.

The balanced equation is given below—remember that the electron in electron capture decay equations is an orbital electron:

$$^{41}_{20}\text{Ca} + {}_{-1}\text{e} \rightarrow {}^{41}_{19}\text{K}$$

The decay diagram, shown in Figure 3.6, is similar to that for positron decay, except that a straight arrow is drawn, and no energy value is placed next to the decay mode. With electron capture, there is no emitted particle; therefore, no energy value can be listed next to "EC."

Calculating the atomic mass of the daughter is very similar to the previous example. Keep in mind that the mass of the daughter is always less than the parent.

$$0.4213\,\text{MeV} \times \frac{u}{931.5\,\text{MeV}} = 0.0004523\,u$$

$$40.962278\,u - 0.0004523\,u = 40.961826\,u$$

Once again, we have excellent agreement with the value given in Appendix A (40.9618258 u). Let's crank it up a notch. What if we wanted to include more than one decay in our decay diagram? No problem, just draw it in.

Example 3.6

Draw the decay diagram for the decay of ^{32}Si to ^{32}S. If the total energy of decay for the beta decay of ^{32}Si to ^{32}P is 0.2243 MeV, that for ^{32}P to ^{32}S is 1.7105 MeV, and the mass of ^{32}S is 31.972071 u, calculate the atomic mass of ^{32}Si.

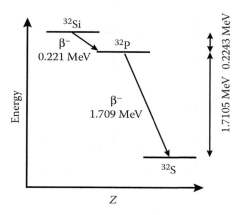

FIGURE 3.7   Decay diagram for ^{32}Si.

The problem doesn't ask, but the two decays can be represented as:

$$^{32}_{14}\text{Si} \rightarrow \,^{32}_{15}\text{P} + \,_{-1}\text{e}$$

$$^{32}_{15}\text{P} \rightarrow \,^{32}_{16}\text{S} + \,_{-1}\text{e}$$

The decay diagram is shown in Figure 3.7. The beta particle energies are from Appendix A and match the total energies of decay reasonably well.

Combining the two decay energies and converting to mass allows us to calculate the atomic mass of ^{32}Si:

$$31.972071\,\text{u} + \left[(0.2243\,\text{MeV} + 1.7105\,\text{MeV}) \times \frac{\text{u}}{931.5\,\text{MeV}}\right]$$

$$= 31.974148\,\text{u}$$

For our final example, let's look at gamma decay. Even though no nucleons change identity, and no particle is spit out of the nucleus during gamma decay, the nucleus still manages to lose mass. Energy in the form of a gamma photon is produced, and, therefore, mass must be destroyed in the process.

Example 3.7

Write out the decay equation, and draw a decay diagram for the decay of $^{77\text{m}}$Br to ^{77}Br. If the mass of $^{77\text{m}}$Br is 76.921494 u, and ^{77}Br is 76.921380 u, calculate the total energy of decay.

Hopefully, this is starting to get a bit easier—the decay equation is:

$$^{77\text{m}}_{35}\text{Br} \rightarrow \,^{77}_{35}\text{Br} + \gamma$$

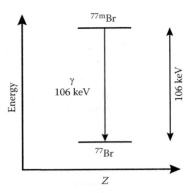

FIGURE 3.8 Decay diagram for ^{77m}Br.

Total energy of decay is calculated the same way it is for all forms of beta decay—from the mass difference between parent and daughter:

$$(76.921494 \text{ u} - 76.921380 \text{ u}) \times \frac{931.5 \text{ MeV}}{\text{u}} = 0.106 \text{ MeV}$$

Energy values for gamma rays are usually expressed in units of kiloelectron volts (keV). Conversion is straightforward, knowing that there are 1000 keV in 1 MeV.

$$0.106 \text{ MeV} \times \frac{1000 \text{ keV}}{\text{MeV}} = 106 \text{ keV}$$

A little research on ^{77m}Br reveals that the only gamma ray observed in its decay has an energy of approximately 106 keV. We can now draw its decay diagram (Figure 3.8) with confidence.

In most of the examples we've seen so far, the total energy of decay is taken away from the nucleus by particles or photons. As noted earlier, some of the energy of alpha decay is left behind with the nucleus as recoil. This introduces a minor complexity to the energetics of alpha decay. We won't be looking at an example of alpha decay here—it will be discussed in Chapter 5.

It should also be noted that the decay examples used in this chapter were carefully selected for their simplicity. Most alpha, beta, positron, and electron capture decay processes initially lead to one or more excited state(s) of the daughter, instead of directly to the ground state as pictured in the examples shown here. This additional level of complexity is also covered in Chapter 5. Finally, it is suggested that the beta and positron particles have the energies shown in Figures 3.4, 3.5, and 3.7. This is not entirely correct. As you will learn in Chapter 5, this is the *maximum* energy these particles can have.

QUESTIONS

3.1 Using Equation 3.1, show that 1 u = 931.5 MeV.
3.2 Calculate the mass defect for ^{72}Ge. Define mass defect, and explain what it represents.

3.3 Calculate the binding energy per nucleon for ^{239}Pu. Would you expect this nuclide to undergo fission, fusion, or neither? Briefly explain.

3.4 Calculate the average mass of a proton in ^{6}Li, ^{66}Zn, and ^{266}Bh.

3.5 Calculate the average binding energy per nucleon and the average mass of a neutron in ^{99}Tc. Would you expect this nuclide to undergo fission, fusion, or neither? Briefly explain.

3.6 Calculate the total energy of decay for ^{226}Ra. How many kilograms of ice could a mole of ^{226}Ra decays melt? How many moles of methane would need to be burned to melt the same mass of ice?

3.7 Using information from Question 2.17, calculate how much mass you lose each month due to radioactive decay of ^{3}H, ^{14}C, and ^{40}K.

3.8 Your body is 18.6% (by weight) carbon. Using 1.2×10^{-10} as the percent natural abundance of ^{14}C, calculate the activity (dpm) of ^{14}C in your body. How much energy (J) is being released in your body every minute because of ^{14}C decay?

3.9 Radioactive materials have sometimes been used directly as a power source. Thermal energy, produced during the decay process, is converted to electrical energy using a thermocouple. How much energy (J) could be generated from the decay of a mole of ^{238}Pu? How much energy (J) is generated per minute by 2.0 kg of ^{238}Pu?

3.10 Define nuclear binding energy, mass defect, binding energy per nucleon, and total energy of decay.

3.11 Using information provided in Appendix A, draw the decay diagrams for the following nuclides:

a. ^{33}P

b. ^{55}Ni

c. ^{49}V

d. ^{172m}Lu

e. ^{190}Pt

3.12 $^{165}_{68}$Er undergoes isobaric decay to $^{165}_{67}$Ho. The atomic mass of the parent is 164.930723 u, and the daughter is 164.930322 u. Using only the information given here, write out the decay equation, draw a decay diagram, and calculate the total energy of decay.

3.13 ^{89m}Y decays to its ground state. The total energy for the decay is 0.9091 MeV, and the atomic mass of the daughter is 88.905848 u. Using only the information given here, write out the decay equation, draw a decay diagram, and calculate the atomic mass of ^{89m}Y.

3.14 ^{90}Sr decays to ^{90}Y with a total energy of decay of 0.546 MeV. This is also the maximum energy of the beta particle emitted. If the mass of the parent is 89.907738 u, calculate the mass of the daughter, and draw a decay diagram. Use only the information given here.

3.15 Using information given in Appendix A, draw a decay diagram for ^{13}N. Show how its atomic mass can be calculated from the mass of the daughter.

3.16 Draw a decay diagram for the main decay pathways for ^{68}Ge to a stable nuclide.

4 Applications of Nuclear Science I
Power and Weapons

Now that we've covered some of the basics of nuclear science, we can begin to discuss a few of its applications. This chapter, and Chapter 8, are specifically designed for those who wish to limit their study of the science and would like to focus on how modern society uses its knowledge of ionizing radiation and nuclear processes. Some of the topics covered in this chapter are covered in more detail in Chapters 10 and 11, so, those of you continuing your study of nuclear science may wish to skip this chapter for now and come back to it when you are ready to begin looking at these applications. This "double coverage" of some of the topics below results in some redundancies between this chapter and later ones—which hopefully serves only to reinforce important concepts rather than prove a distraction.

The first three chapters have taught us a bit about ionizing radiation, nuclear decay, and the tremendous energies resulting from nuclear transformations. The penetrating power of ionizing radiation is used in hospitals, clinics, and private medical practices throughout the world to diagnose and treat disease and to preserve food (Chapter 8). The awesome amount of energy released by nuclear fission is used to provide electricity from hundreds of nuclear power plants around the world, and has been developed into horrific weapons of mass destruction. This chapter provides an introduction to understanding nuclear reactors and weapons, in the context of what we've already learned about that weird thing we call a nucleus.

4.1 NUCLEAR POWER

Approximately 12% of the world's electricity is produced from roughly 450 nuclear power reactors. About 200 more are under construction or are in the planning stages (at the time of this writing), with more than 300 additional plants being proposed for future construction (by 2035). Much of the new construction is occurring in China, Russia, and India. The recent increase in nuclear power plant construction and the proposed significant expansion of electrical production from nuclear has been called the "nuclear renaissance." This term comes from the fact that there was a great deal of nuclear power plant construction in the 1960s and 1970s but very little in the following three decades. During the first couple of decades of the twenty-first century, there's been a renewed interest in new construction.

The recent renewed interest in nuclear power stems from increasing demand for electricity tempered by concern over global climate change. Unlike the other major forms of electricity production, nuclear generates almost no carbon dioxide and, until the Fukushima meltdowns of 2011, was increasingly perceived as safe, reliable, and cost-effective. While perceptions are not always fully reflective of reality, they often have an influence on decision making and must be acknowledged. After Fukushima, some perceptions shifted, dampening some of the enthusiasm for the nuclear renaissance. During the same time, there was dramatic drop in natural gas prices, encouraging construction of natural gas power plants. At the time of this writing (2017), new plant construction was barely outpacing the number of old plants shutting down.

Approximately 20% of the U.S. and 16% of Canadian electrical production is nuclear, while in many European countries and in South Korea, it accounts for more than 30% of all power generation. France leads the world with a little more than 70% of all electrical generation coming from nuclear, although the French government plans to decrease to 50% nuclear by 2025. In the United States, nuclear power plant construction came to a virtual standstill in the 1980s and has remained that way for almost 30 years. The long hiatus was a result of the relatively high cost of building new nuclear power plants, the accidents at Three Mile Island and Chernobyl, and the federal government's lack of resolve in coming up with a plan to deal with spent fuel. As a result, the existing U.S. plants tend to be rather old. Because they have a limited lifetime, some are decommissioned, while others have sought or are seeking license renewals to continue beyond their originally intended period of operation.

The impact of the 2011 meltdowns at the Fukushima plants is being felt most strongly in Japan. Prior to the massive earthquake and tsunami in March 2011, Japan had roughly 50 nuclear power plants in operation, producing about 30% of its electrical needs. By mid-2017, only five plants had returned to operation, accounting for only 2% of the nation's electricity. Fukushima's effects were also felt in Europe, where Germany decided to return to its 1998 plan to phase out nuclear power.[1] In the short term, Germany's nuclear phase out has largely been replaced with increased fossil fuel generation. Switzerland is planning a more gradual phase out. They plan to run their existing reactors so long as their regulatory agency considers them safe. Belgium and South Korea also appear to be on a course toward a non-nuclear future, while Sweden is reducing its fleet of reactors from nine to five by 2020. As already mentioned, France is reducing its nuclear share of electrical generation from about 70% to 50% by 2025. Other nations with nuclear power plants appear to be moving forward with plans for construction of new nuclear plants and license renewals for older plants. In China, Russia, and India, a total of 33 new plants are under construction in mid-2017. Additionally, about a dozen countries without any nuclear generation (as of 2017) are building or planning construction of nuclear power plants.

The U.S. government began to quietly encourage the licensing and construction of new plants in the late 1990s and early 2000s. This change in attitude resulted from increasing costs of oil and natural gas during this period, which currently account for about 35% of all electricity production in the United States, and increasing concerns that burning fossil fuels is negatively affecting the global environment. Ironically, nuclear power, which has long been opposed for environmental concerns (potential

---

[1] Germany had canceled the phase-out in 2009.

for a meltdown, dangers from waste), is now being promoted as a relatively green power source because it does not produce greenhouse gases. Regardless of how it is viewed, new nuclear power plant construction in the United States became a reality in the early 2010s. Four plants were under construction in early 2017 in South Carolina and Georgia, and were expected to be operating by 2020. Construction of the two South Carolina plants was abruptly halted in August 2017 following the bankruptcy of the main contractor, Westinghouse.

The focus of the remainder of this section will be on how nuclear reactors work, as well as the science underlying such contentious issues such as reactor safety and cost, proliferation of nuclear weapons, and nuclear waste.

4.1.1 NUCLEAR FISSION

Before we can understand how a nuclear power reactor works, we need to learn a little bit about nuclear fission. **Nuclear fission** is the splitting of a nucleus into two (or more—rare!) fragments and usually occurs when a neutron hits the nucleus of certain large atoms. Nuclear scientists often imagine the nucleus as a more fluid object rather than a collection of hard spheres (the nucleons). In fact, it is more like a drop of water suspended in air or some Silly Putty® in the hands of a small child.[2] The nucleus can elongate and deform, taking on shapes (such as an egg, Frisbee®,[3] football,[4] or even a bow tie) other than a perfect sphere. When struck by a neutron, enough energy can be added to certain nuclei that they stretch and separate into two, roughly equal halves. The trick is to provide enough energy. With relatively slow neutrons (i.e., "thermal" neutrons—see Chapter 9) present in almost all nuclear reactors, only three nuclides commonly undergo nuclear fission: ^{233}U, ^{235}U, and ^{239}Pu. Nuclides that undergo fission with thermal neutrons are called **fissile**.

An example of a fission reaction involving ^{235}U is:

$$^{235}_{92}U + ^{1}_{0}n \rightarrow ^{132}_{50}Sn + ^{102}_{42}Mo + 2^{1}_{0}n$$

This is an example of a nuclear reaction because a particle is colliding with a nucleus, making new stuff. Note that it is balanced because the atomic and mass numbers add up to the same values on both sides of the arrow. Notice also that two (new) neutrons are released in this particular reaction. This is common in fission reactions, with two to three neutrons typically being released by each fission reaction that takes place. The two product nuclides (^{132}Sn and ^{102}Mo in the example above) are called **fission products** (also known as fission fragments). Fission is a notoriously messy reaction in that lots of different fission products can be formed. In fact, over 400 different nuclides are observed in the fission of ^{235}U. In other words, if you have a lot of ^{235}U undergoing fission reactions, such as in a nuclear reactor, there are going to be lots of different fission products being made.

How much energy is released by a fission reaction? On average, about 200 MeV are released in a fission reaction. Holy cow! That's 50–100 times more energy than is

---

[2] Silly Putty is a registered trademark of Crayola LLC of Easton, Pennsylvania.
[3] Frisbee is a registered trademark of Wham-O Inc. of Carson, California.
[4] Most of you outside the United States could substitute "rugby ball" here.

released by a typical nuclear decay, which we already know is fairly energetic. This quantity can easily be calculated by finding the difference in mass on either side of the arrow, that is, determine how much mass is lost in this reaction.

Mass of reactants	^{235}U	235.043930 u
	$1_0^1 n = 1(1.008665\,u) =$	+1.008665 u
		236.052595 u
Mass of products	^{132}Sn	131.91774 u
	^{102}Mo	101.91030 u
	$2_0^1 n = 2(1.008665\,u) =$	+2.01733 u
		235.84537 u

Mass lost = 236.052595 u − 235.84537 u = 0.20722 u

The amount of energy released per reaction can then be calculated in the usual manner by converting matter to energy.

$$\text{Total energy released} = \frac{931.5\,\text{MeV}}{u} \times 0.20722\,u = 193.0\,\text{MeV}$$

The fact that, on average, two to three neutrons are produced in each fission reaction is quite useful, since these neutrons can cause further fission reactions to take place provided there are more fissile nuclides around, and they get hit by these neutrons. To encourage this kind of behavior, you'd want to arrange lots of fissile nuclides very close to each other. If every fission reaction produces two to three neutrons, and most of these cause additional fission reactions, a **fission chain reaction** will result. This is illustrated in Figure 4.1. One neutron initiates fission in a single ^{235}U nuclide at the top of the figure, which also produces two neutrons. These two neutrons then initiate fission in two more ^{235}U nuclides, producing two neutrons each. Now there are four neutrons that initiate four more fission reactions, which, again, produce two neutrons each. Now there are eight neutrons that could initiate eight more fission reactions. The number of fission reactions that take place will rapidly increase with each subsequent generation of fission reactions. This could quickly get out of control!

If there were just enough fissile nuclides around for a self-sustained fission chain reaction, then we'd have a **critical mass** of that nuclide. If more than the critical mass is present, then it is called a **supercritical mass**. A supercritical mass can be arranged in different ways to get different results. A very compact arrangement would increase the odds that every neutron produced will cause another fission reaction in a relatively short time. This will lead to near exponential growth in the number of reactions and a rather sudden release of a great deal of energy, that is, a really big explosion. If the supercritical mass is spread out in such a way that, on average, only one of the neutrons released per fission causes another fission reaction, then the reaction rate can be controlled to release a constant amount of energy over a long time. This is the basis of nuclear power reactors.

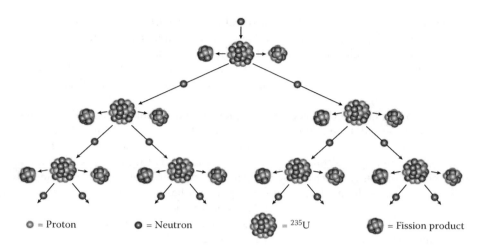

= Proton   = Neutron   = ^{235}U = Fission product

FIGURE 4.1   Nuclear chain reaction.

4.1.2   NUCLEAR REACTORS

The ^{235}U in a nuclear reactor is contained in small cylindrical pellets about the size of the end of your little finger, which are placed end to end in a metal pipe that is then sealed at both ends (**fuel rod**), and these rods are then distributed in the reactor parallel to each other and separated so there's space between them.

A schematic view of a typical nuclear reactor is shown in Figure 4.2. It works the same way that a fossil fuel power plant does. Fossil fuel plants burn materials that contain carbon—such as coal, natural gas, or oil. The energy released from the chemical reaction (combustion) that takes place heats water to steam, which drives a steam turbine that produces electricity. Nuclear reactors use nuclear reactions (fission) to heat the water to steam, which drives a steam turbine that produces electricity. The fundamental difference between fossil fuel and nuclear plants is simply how the water is heated.

The goal of a nuclear power plant is to maintain the rate of fission reactions at a constant level. If there are too many neutrons flying around causing too many fission reactions, too much heat is generated. If the reactor gets too hot, the fuel rods (and the metal assemblies that hold them) will melt (reactor **meltdown**). On the other hand, if there are too few neutrons, the chain reaction will shut down. To maintain the perfect number of neutrons, some will have to be absorbed. This is done with **control rods**, which can move in and out between the fuel rods.

An odd thing about neutrons is that the probability that they will cause a nuclear reaction to take place depends on how fast they are moving, that is, how much energy they have. When they are produced by a fission reaction, they are moving too fast to efficiently cause fission reactions in ^{235}U. This is another reason to put the uranium fuel into thin rods and separate them. This forces the too-fast neutrons to travel through matter before it runs into another ^{235}U atom. We want to be sure that the matter they must travel through will allow them to pass but will also slow them down. This is called moderating the neutrons, and the matter used is called a **moderator**. The most common moderators are water (sometimes called "light-water" by nuclear geeks), heavy water (water with ^{2}H, also known as deuterium, instead of the usual ^{1}H in regular water), and carbon (typically in the form of graphite).

The moderator slows the neutrons, so they are more likely to cause fission, and the control rods soak up extra neutrons to keep the fission reactions occurring at a reasonable rate. Now all we need is some way to transfer the heat from the fuel rods. The **coolant** does this and can be any material that transfers heat well, but the most common (by far) in nuclear reactors is water. Water is nice because it also functions as the moderator, it's cheap, and it has a high heat capacity. Heat capacity is the amount of energy it takes to warm up a certain amount (such as 1 g) of a substance by 1 degree. It takes more energy to warm up water than equivalent amounts of most anything else, so water can hold more heat per gram than other chemicals.

Careful examination of Figure 4.2 shows that the water heated by the nuclear reactions is not the same water (steam) that drives the turbines. The water heated by the nuclear reactions is kept under enough pressure that it remains a liquid. It is pumped through the reactor **core** (where the fuel is), then though a heat exchanger to transfer some of its thermal energy to another closed water loop. This second loop is operated at lower pressure so the water is heated to steam, which then drives the turbines to produce electricity. This secondary loop of water then needs to be cooled (condensed back to liquid), so it runs through another heat exchanger, transferring its excess thermal energy to a third loop of water (labeled "cooling water" in Figure 4.2). Unlike the first two loops, the third loop is not closed, as it transfers excess thermal energy to the environment—a river, a lake, the ocean, or the air. The immense, hourglass-shaped cooling towers that are so often associated with nuclear power plants simply transfer this waste heat to the air. The white clouds that are often seen emerging from them is not smoke but water vapor.

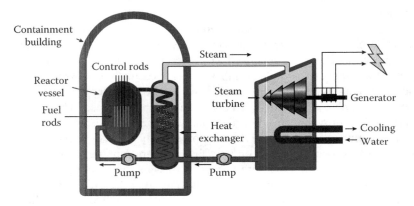

FIGURE 4.2 Schematic drawing of a nuclear power reactor (pressurized water reactor, PWR).

The reactor depicted in Figure 4.2 is called a pressurized water reactor (PWR) because water is used as a coolant, and it is kept under pressure in the reactor core. Roughly 65% of the world's nuclear reactors are a PWR design. Another popular design (~15%) is the boiling water reactor (BWR) (see Figure 4.3). The boiling water reactor doesn't excessively pressurize the water that is pumped into the core, so it boils. The steam that is generated drives the turbines directly instead of transferring its thermal energy to a second water loop. The steam still needs to be condensed and the waste heat transferred to the environment. Notice that another difference is that BWRs have a smaller containment structure which is then enclosed within the reactor building. The reactor building doesn't necessarily have the distinctive silo shape like most other nuclear reactors, and it also acts as secondary containment—another barrier between the reactor core and the environment.

Dealing with waste heat is also an issue for fossil fuel power plants. Coal or natural gas plants also need to be located near large bodies of water to soak up the waste heat or have cooling towers to dump the waste heat into the air. The large cooling towers that so often symbolize nuke plants are neither unique nor necessary to them. If waste heat from any type of power plant is not being dumped into the air, then it is likely going into water nearby. Dumping waste heat into a body of water can have deleterious environmental consequences. To avoid these problems, the cooling water can be pumped into an artificial holding pond before releasing it to natural waters.

As the previous paragraphs imply, not all the energy released by the fission reactions can be converted to electricity, that is, the process is not 100% efficient. In fact, only ~35% of the energy released by the nuclear reactions is successfully converted to electrical energy. This is comparable to efficiencies obtained in most fossil fuel power plants.

Because the BWR has less plumbing than a PWR (no heat exchanger), it might seem that boiling water reactors would be cheaper to build. While this is true, boiling water reactors are a bit more expensive to operate. As the water flows through

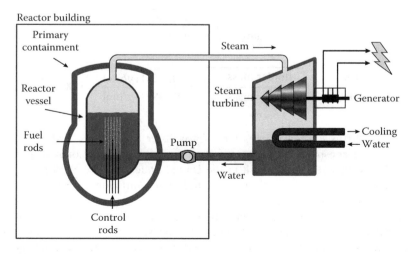

FIGURE 4.3   Schematic for a boiling water reactor (BWR).

the reactor, it becomes slightly contaminated with radioactive materials. Because this water leaves the reactor building, extra shielding is required on any pipes that contain it, and special procedures are necessary for people who work around the turbines to minimize their dose. In the end, BWRs and PWRs cost about the same to build and run.

4.1.3 Nuclear Fuel

Of the three nuclides that commonly undergo fission with thermal neutrons, only ^{235}U can be dug up out of the ground in large enough quantities to power a reactor. Naturally occurring uranium ores are common—the largest proven reserves are currently in Australia, Kazakhstan, Canada, and Russia. The uranium is isolated through physical and chemical processes from some of the other elements present in its ores. It is isolated as U_3O_8, which is also known as "yellow cake," probably because it is bright yellow powder. Naturally occurring uranium is only 0.72% ^{235}U; almost all the rest (99.27%) is ^{238}U, which does not undergo fission with thermal neutrons.

Most nuclear reactors need the uranium to be $\sim$3%–4% ^{235}U. Increasing the percentage of one isotope relative to others is called **enrichment** and typically requires a physical (rather than chemical) process. For many years, the process of choice for uranium enrichment was **gaseous diffusion**. The U_3O_8 is chemically transformed to uranium hexafluoride (UF_6). UF_6 is a solid under normal conditions (1 atm, 20°C), but it sublimes to a gas if the temperature is raised just a bit or the pressure is lowered. Because the $^{235}UF_6$ molecules are a little lighter than the $^{238}UF_6$ molecules, they move (diffuse) a little faster. If we pass UF_6 gas through a very long pipe, the gas that emerges at the far end will have a higher percentage of $^{235}UF_6$ relative to $^{238}UF_6$. The pipe must be pretty darn long, as the difference in mass is quite subtle. The relative diffusion rates can be calculated using Graham's Law (Equation 4.1):

$$\frac{\text{rate}_1}{\text{rate}_2} = \sqrt{\frac{\text{molar mass}_2}{\text{molar mass}_1}} \qquad (4.1)$$

$$\frac{\text{rate}_{235}\text{UF}_6}{\text{rate}_{238}\text{UF}_6} = \sqrt{\frac{\text{molar mass}_{238}\text{UF}_6}{\text{molar mass}_{235}\text{UF}_6}} = \sqrt{\frac{352.04\,\text{g/mol}}{349.03\,\text{g/mol}}} = 1.0043$$

$^{235}UF_6$ only diffuses 1.0043 times (or 0.43%) faster than $^{238}UF_6$. Extremely large facilities were constructed to accommodate the tremendous amount of plumbing required. The first such plant built in the United States was the K-25 facility near Oak Ridge, Tennessee. It was four stories tall, half a mile long (0.8 km), and 1000 feet wide (300 m). A huge building!

Today, enrichment using **gas centrifuges** is more common. UF_6 is placed in a cylinder that is spun around at high speed. The heavier $^{238}UF_6$ tends to collect near the cylinder wall, while the lighter $^{235}UF_6$ tends to collect near the central axis of the cylinder (Figure 4.4). Gas centrifuges are preferred for uranium enrichment because

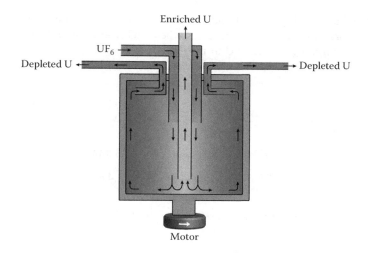

FIGURE 4.4   Flow schematic for a gas centrifuge.

they require only about 3% of the power when compared to an equivalent gaseous diffusion plant. Gas centrifuge plants are also much smaller than gaseous diffusion plants and are, therefore, attractive to those who wish to be somewhat secretive about their enrichment activities.

Once enriched, the uranium is converted back to an oxide (UO_2) and compressed into small pellets for nuclear fuel. If the uranium is to be used in a weapon, it needs to be greater than 80% ^{235}U, which means more enrichment, followed by a chemical transformation to uranium metal. Note that uranium that exceeds 20% ^{235}U is called **highly enriched uranium**, or **HEU**. The 20% threshold may seem oddly arbitrary, but once exceeded, it is relatively easy to get up to weapons-grade uranium (>80% ^{235}U).

The uranium left over from the enrichment process is ~99.7% ^{238}U, and is referred to as **depleted uranium** (or **DU**). Because of its exceptionally high density (1.7 times greater than lead), low cost, and because it is a relatively low radiological danger, depleted uranium is sometimes used as ballast in boats, counterweights, or, ironically, radiation shielding.

Depleted uranium has also been used extensively in armor piercing shells, especially by the United States in its Gulf Wars. There has been some concern that use of DU weapons may result in a significant health hazard to those exposed to the uranium containing vapors or shrapnel when they were used to destroy Iraqi tanks. However, because of the relatively long half-life of ^{238}U, a relatively large amount must be concentrated in one human for it to have a significant radiobiological effect. In fact, uranium will cause kidney failure (due to its chemical toxicity) at levels much lower than when it becomes a concern due to its radioactivity. There is also a concern that shrapnel left in the desert may be an environmental hazard. Again, its long half-life and its natural ubiquity in our environment should tend to ameliorate this concern.

Can other fuels (besides ~3%–4% ^{235}U in ^{238}U) be used in nuclear reactors? Certainly! Any fissile material will work, with ^{235}U, ^{233}U, and ^{239}Pu being the most popular. ^{239}Pu is both produced and "burned" (undergoes fission) in most nuclear power reactors. It is produced in reactors when ^{238}U reacts with a thermal neutron. Instead of undergoing fission, ^{238}U forms ^{239}U:

$$^{238}_{92}\text{U} + {}^{1}\text{n} \rightarrow {}^{239}_{92}\text{U} + \gamma$$

^{239}U then decays to ^{239}Pu through two beta emissions:

$$^{239}_{92}\text{U} \rightarrow {}^{239}_{93}\text{Np} + {}_{-1}\text{e} \quad t_{1/2} = 23 \text{ min}$$

$$^{239}_{93}\text{Np} \rightarrow {}^{239}_{94}\text{Pu} + {}_{-1}\text{e} \quad t_{1/2} = 2.4 \text{ d}$$

Likewise, ^{232}Th (naturally occurring!) can absorb a neutron, then decay to form ^{233}U:

$$^{232}_{90}\text{Th} + {}^{1}\text{n} \rightarrow {}^{233}_{90}\text{Th} + \gamma$$

$$^{233}_{90}\text{Th} \rightarrow {}^{233}_{91}\text{Pa} + {}_{-1}\text{e}$$

$$^{233}_{91}\text{Pa} \rightarrow {}^{233}_{92}\text{U} + {}_{-1}\text{e}$$

Nuclides such as ^{238}U and ^{232}Th, which can produce a fissile nuclide in a nuclear reactor, are called **fertile.**

Reactors can be designed for ^{239}Pu production. Very simply, a slab of ^{238}U (depleted uranium) is placed in a reactor to soak up neutrons to produce enough ^{239}Pu to make its separation from the remaining ^{238}U worthwhile without allowing too much of the ^{239}Pu to undergo fission. In theory, ^{239}Pu could also be separated from the reactor fuel pellets. However, this is a rather awkward way to way to obtain weapons-grade material because it would involve shutting down and refueling the reactor.

Highly enriched uranium or ^{239}Pu are generally used for nuclear weapons (see Section 4.2); however, separating ^{239}Pu from a slab of uranium is much easier than enriching ^{235}U to greater than 80% (which is necessary for weapons). Instead of a tedious physical process, plutonium can easily be separated from uranium using simple chemical processes. Its ease of isolation has made ^{239}Pu the nuclear weapons material of choice and the proliferation of nuclear weapons all too easy.

The dramatic reduction of the U.S. and former Soviet nuclear weapons arsenals following the end of the Cold War has generated a great deal of highly enriched uranium and ^{239}Pu. Much of the nuclear materials from the decommissioned U.S. weapons are being securely stored at the Pantex plant near Amarillo, Texas, and at the Y-12 complex near Oak Ridge, Tennessee. These materials can be used for reactor fuel. For security purposes, the uranium is blended with depleted uranium to bring the ^{235}U down to ~3%–4%, and the ^{239}Pu can be mixed with reactor-grade uranium to produce a mixed oxide (MOX) fuel. While MOX fuel is new to the United States, Europeans and the Japanese have been burning MOX fuel in their reactors for many years.

A high-profile program called "Megatons to Megawatts" was initiated between Russia and the United States in 1993. Under this program, the United States bought highly enriched uranium from Russian warheads and military stockpiles and paid Russia to convert it to reactor fuel for use in U.S. reactors. Approximately 500 tons of Russian highly enriched uranium was converted over the history of the program, which concluded in 2013. At one time, roughly half of the fuel in U.S. nuclear reactors came from the Russian military. This meant that roughly 10% of all electricity in the U.S. at that time originated from material that could've been targeting a U.S. city for destruction, a wonderful irony.

4.1.4   REACTOR SAFETY

While their safety record is excellent when compared to fossil fuel and hydroelectric plants, people generally view nuclear power plants as dangerous.[5] A small portion of this concern is justified, but almost all of it is based on fear and ignorance. Since concerns related to waste and proliferation are discussed later, safety issues related to the mining of uranium and the potential of a catastrophic accident are emphasized here.

Most of the world's uranium is currently mined in Kazakhstan (39% of all uranium mined in 2016), Canada (22%), and Australia (10%). Like other forms of mining, uranium mining is hazardous. In addition to the physical risks shared by all miners, uranium mining also has potential radiological hazards. If not properly ventilated, the air in uranium mines contains relatively high concentrations of radon. Breathing air with too much radon over a long period of time can increase the long-term risk of lung cancer. Uranium miners take preventive measures, and their radiation exposure is carefully minimized and monitored; therefore, they face little additional risk when compared to other miners. Unfortunately, this was not true in the early days of uranium mining. In the 1950s, miners were exposed to and ingested doses that would be unthinkable today. Decades later, these uranium miners now have cancer rates that are significantly higher than the national average.

Perhaps a bigger risk from uranium mining has to do with the mine tailings. Tailings are the solid material left over at the mine after the uranium has largely been removed. In some cases, these tailings have been left on the surface and can leach radioactive nuclides (and heavy metals) into local water supplies. While many of these sites have been or are now being cleaned up, hundreds of people have been exposed, and for a small fraction, this exposure could result in a premature death.

Chernobyl is the only major nuclear power plant accident known to have caused significant human fatalities. Despite the magnitude of this accident and the fear it has engendered, only about 30 deaths immediately followed the accident. Those who died were emergency response personnel who received very large doses while trying to contain the disaster. A couple hundred, to a few thousand, premature deaths are also expected—primarily due to thyroid conditions in those who were children living near the reactor at the time of the accident. As of 2012, approximately 1800 thyroid problems have been attributed to the Chernobyl accident, and only about a

---

[5] In some countries, there is also an acceptance of the necessity of nuclear power (they have few other resources to generate electricity), as well as greater local appreciation of the economic benefits.

dozen of these have resulted in death. The reader should note there is considerable uncertainty in the predicted numbers (a couple hundred to a few thousand), and that the actual numbers are relatively low. Despite the enormous scale of the Chernobyl accident, relatively few people have died from exposure to the radioactive materials that were released.

No deaths are attributable to radiation doses from the accidents at Three Mile Island or Fukushima (at the time of this writing). Based on the doses, very few, if any are expected. However, the population displacements caused by radioactive contamination of large tracts of land around Fukushima and Chernobyl have caused significant emotional stress. There's little doubt that these dislocations have caused real health effects, including early deaths. Some scientists have suggested that, given this disparity of negative health effects, future evacuations following large-scale radioactive contamination be more carefully considered.

So, yes, nuclear power has proven hazardous and may continue to do so for future millennia if our solution to the waste disposal issue proves inadequate (*vide infra*). How does nuclear power stack up against other forms of electrical generation? Tens of thousands of people die each year due to coal's use as the most significant form of power generation in the world today. Coal mining remains one of the most dangerous occupations, and the transportation and burning of the large amounts of coal necessary to keep the plants running remains dangerous for the general public. The burning of coal and other fossil fuels adds to the dramatically increasing concentration of CO_2 in the Earth's atmosphere, which is clearly contributing to global climate change. If just some of the recent meteorological extremes were aggravated by global warming, then the safety record of fossil fuels is much worse than stated at the beginning of this paragraph. A NASA study in 2013 estimates that nuclear power should be credited with saving 1.84 million lives between 1971 and 2009 because the presence of the nuclear plants prevented more fossil fuel plants from being built, resulting in less air pollution. Coal also naturally contains trace amounts of uranium, thorium, and mercury. If not trapped, they can come out of the stack to be distributed downwind. It is an irony that people living downwind of a coal burning plant can receive a greater radiation dose than those living near a nuclear power plant, although neither is significant. However, the tons of mercury that have been pumped into the environment because of burning coal are now rendering many fish species hazardous for human consumption. If fossil fuels were held to the same standards of safety as nuclear power (e.g., careful isolation of all hazardous wastes from the environment), their costs would be driven much higher, making nuclear more economically attractive.

Hundreds of nuclear plants around the world have been operating for decades with excellent safety records. Even so, it might be helpful to more closely examine the three most significant nuclear power plant accidents in recent history—Three Mile Island (TMI) in 1979, Chernobyl in 1986, and Fukushima in 2011. All three accidents were the result of a loss of water coolant, which caused melting of the fuel rods (only partial for TMI). Once the water coolant is lost, the fission chain reaction shuts down. This is because the water also moderates the neutrons, and the fission chain reaction shuts down on its own because of the lower probability of fast neutrons inducing a fission reaction. This happened at TMI and Chernobyl when the reactors lost coolant.[6] However, even after the fission reactions have stopped, there is still enough heat being generated by the decay of the fission products that melting of the fuel rods can occur. Exacerbating this scenario is the exothermic reaction of the zinc cladding on the outside of the fuel rods with steam to form hydrogen gas. It is unfortunate that some of the early nuclear proponents once stated that meltdowns were not possible, as it has reduced the credibility of the industry.

The meltdown at TMI was largely the result of a pressure relief valve that was stuck open and drained coolant from the reactor vessel. Subsequent confusion from conflicting indicators and a bevy of alarms resulted in about half of the fuel melting. The TMI reactor core material did not breach the reactor vessel; however, $\sim10^{12}$ Bq of radioactive gases (mostly Xe and Kr) were vented during the crisis, then again over a year later. Because it was all in the gas phase, and it was composed of elements that are chemically inert, it dispersed quickly resulting in minimal doses to people living nearby. Radiation levels outside of the plant were carefully monitored, and there was never a significant danger to the public. As a result, mandatory evacuation orders were never issued, which was fortunate, as the resulting panic would've undoubtedly resulted in some injury and loss of life.

In an odd twist of fate, the TMI accident occurred shortly after the release of the movie *The China Syndrome*, which depicted an eerily similar loss of coolant accident in a nuclear power plant in Southern California. The movie was fictional and suggested that the ramifications of a nuclear reactor core meltdown would be dire: "...render an area the size of Pennsylvania permanently uninhabitable,[7] not to mention the cancer that would show up later." Many confused the movie with the actual events at TMI (which coincidentally occurred *in* Pennsylvania). This serendipitous coalescence of fiction and reality helped to harden public attitudes in the United States and elsewhere against nuclear power.

The Chernobyl accident was the worst of the three primarily because the reactor building was significantly damaged, directly exposing the core to the atmosphere. This type of reactor suffered from a known flaw in its design that allowed for rapid increases in power output under certain conditions. Unfortunately, these conditions occurred at a time when reactor operators had bypassed safety systems to conduct an experiment. The rapid rise in core temperature resulted in two steam (or a steam and a hydrogen) explosions that blew the roof off the reactor building and started several fires. It was the large explosions, direct exposure of the glowing red-hot reactor materials to the atmosphere, and the fires that led to the extensive distribution of radioactive materials. Approximately 1.4×10^{19} Bq of radioactivity was released. Much of the radioactive material was dropped on northern, eastern, and central Europe, although measurable quantities were spread throughout much of the Northern Hemisphere.

Extinguishing the fires and covering and cooling the exposed core gave some of those involved high, acute radiation doses. Twenty-eight died because of these

---

[6] The Fukushima reactors all automatically shut down the fission chain reaction (scrammed) when the earthquake hit. It wasn't until ~45 minutes later that the tsunami hit and problems began.

[7] The Chernobyl accident caused the greatest evacuation and covered an area of 4300 km², which is only 3.6% the size of Pennsylvania. Many of the areas within the Chernobyl exclusion zone have opened for resettlement.

exposures in the months that followed the accident. Doses to the general public living near the accident were exacerbated by the government's lethargy in ordering an evacuation and its failure to adequately monitor foodstuffs (particularly milk) that were potentially quite contaminated. This was the cause of many of the roughly 1800 cases of thyroid problems mentioned earlier.

Three reactors in Fukushima, Japan suffered fuel meltdowns following a massive earthquake and tsunami. The reactors all weathered the magnitude 9.0 earthquake remarkably well. There is no indication that they were damaged by the quake. However, they were not adequately protected against the ∼14-meter tsunami that flooded the station 45 minutes later. The tsunami knocked out almost all backup power generators, batteries, and electrical switching stations on the site. Without power, the pumps could not circulate enough coolant through the reactor core, and they overheated from the decay of fission products in the fuel rods. Over the following three days, these reactors suffered fuel meltdowns as the water in the core boiled off, exposing the fuel rods. When this happened, some of the hot zinc fuel cladding reacted with the steam to form hydrogen gas. Enough hydrogen gas built up in the buildings surrounding the reactor cores to cause three large explosions (Figure 4.5). Some of the volatile fission products (Noble gases, ^{131}I, ^{137}Cs, and ^{134}Cs) were also dispersed because containment had been breached from the excessively high temperatures and pressures in the reactors.

Estimates have varied greatly since the accident, but somewhere around 9.4×10^{17} Bq of radioactive materials were released from the reactors. The majority was released to the air, and some contaminated land nearby. Despite the severity and scale of the Fukushima meltdowns, the Japanese government did a good job evacuating nearby areas and monitoring food and water. As a result, the doses to the general public are much lower than they were for Chernobyl. It appears that the doses are so low that it is fairly certain that no adverse health effects will be seen. Likewise,

FIGURE 4.5   Damage to the structure surrounding Fukushima Daiichi reactor #4. (Courtesy of Tokyo Electric Power Co.)

the doses to the workers at Fukushima appear[8] to have been much more limited than at Chernobyl, so much so that it is likely that just a handful of workers ($\sim$6 out of 20,000) face only a slightly increased risk of future cancer.

The primary lasting effect of Fukushima appears to be psychological and economic. The Japanese public has shown a strong reluctance toward nuclear power in the years since the accident. Restarting the $\sim$50 reactors in Japan has been a slow and difficult process, primarily due to strong public opposition. As mentioned previously, only five Japanese reactors were producing electricity by mid-2017. Normally responsible for generating roughly 30% of Japan's electricity, their capacity is being made up from conservation and increased burning of fossil fuels (more expensive for Japan). The stabilization of the damaged reactors and the clean-up of contamination, as well as beefing up safety features of other reactors, are also proving costly.

Operation and design of current and future nuclear plants are imminently conscious of mistakes made by operators as well as the design flaws of the TMI and Chernobyl reactors and the Fukushima site. Perhaps the most significant is incorporation of passive shutdown and cooling systems, which could safely stabilize a reactor without power or human intervention. Some new designs incorporate a "core-catcher" to safely contain a core meltdown, preventing contamination outside of the reactor. While these systems can be integrated into new reactor designs, operators are also working to better anticipate potential emergencies at existing reactors so that future significant releases of radioactive materials are prevented.

4.1.5 NUCLEAR WASTE

When fuel rods are first removed from the reactor core, they are still pretty toasty—very radioactive and still producing a fair amount of thermal energy (hot!), both of which are due primarily to presence of short-lived fission products. The fuel rods are stored underwater at the reactor site while the relatively short half-life nuclides decay, and the rods cool off. The water acts as shielding for the radiation as well as a coolant. After they've cooled off a bit, the rods can be placed in dry cask storage. These casks are designed to allow air-cooling (convection) of the rods, provide shielding, and are generally indestructible.

What is spent fuel composed of? Remember that the fuel starts out as $\sim$3%–4% ^{235}U in ^{238}U. When the ^{235}U burns in a nuclear reactor, the percentage of ^{235}U gradually decreases. When fuel rods can no longer be used (**spent**), they still contain roughly 1% ^{235}U and now have $\sim$1% ^{239}Pu (from neutron reaction with some of the ^{238}U, followed by decay), which could still be burned in a reactor. Only $\sim$3% of the spent fuel are fission products, and the remaining 95% is still ^{238}U (Figure 4.6). Since the fission products generally have shorter half-lives than the uranium and plutonium, and the ^{235}U and ^{239}Pu can still be used as fuel, some have suggested that separations be performed on the spent fuel. This is known as **reprocessing** and is the approach some countries[9] have taken toward spent fuel. It should be noted that

---

[8] It has been reported that some of the data for a few workers were attenuated by covering dosimeters with a thin sheet of lead.

[9] Most notably China, France, India, Japan, Russia, and the United Kingdom.

FIGURE 4.6   Composition of spent nuclear fuel.

reprocessing is much more expensive than mining uranium and enriching it to new fuel, and, as discussed later, presents a proliferation problem. The major motivations for reprocessing are environmental. Reprocessing will significantly reduce the volume of waste needing to be stored, allow more of the waste to become cold (no longer radioactive) sooner, and make more extensive use of the fuel.

Since the spent fuel can be recycled, its life and death are often referred to as the **nuclear fuel cycle** (Figure 4.7). As mentioned in Section 4.1.3, the uranium is first mined, undergoes physical and chemical separations, is usually enriched, then is fashioned into fuel pellets (fuel fabrication). These activities are often referred to as the "front-end" of the fuel cycle. The uranium is then used to produce electricity in a reactor (Section 4.1.2). When the spent fuel is removed, it can be reprocessed for additional atom splitting in a reactor or buried. Activities with the fuel after it's removed from the reactor are often referred to as the "back-end" of the cycle. When reprocessing is used, the fuel cycle is often referred to as "closed" because the circle is complete.

Some countries[10] plan to simply dispose of their spent fuel as radioactive waste. This is called the "once-through" or "open" fuel cycle because the fuel is only in the reactor once, and the circle is not complete. Under these circumstances, it's probably better to refer to the spent fuel as "nuclear waste," since it is destined for disposal. The U.S. federal government promised companies running nuclear reactors that it would take all nuclear waste and bury it in a geologic repository. For many reasons

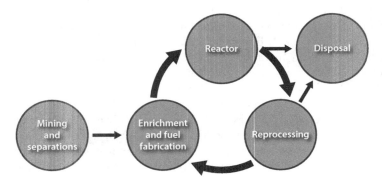

FIGURE 4.7   The nuclear fuel cycle.

[10] Such as Canada, Finland, Spain, Sweden, and the United States.

(some discussed later), the U.S. government was unable to meet its own deadlines to open the repository, and all spent fuel currently remains on site at U.S. reactors. Deciding on the type(s) and location(s) of repositories seems to be a difficult task for the U.S. government. As of 2017, interim storage facilities are being explored to allow the government to fulfill its obligation to the reactor companies without committing to a more permanent solution.

While the United States vacillates, most other countries[11] are moving forward with their nuclear waste disposal plans. Regardless of whether reprocessing is planned, all nuclear waste seems destined to be buried deep underground. Why? Many options have been studied, from dumping in the ocean near a subduction zone, to launching it into the sun. Geologic repositories have been determined to be the safest way to dispose of nuclear waste.

The best support for a geologic repository originally came from a uranium mine. The French used to get some of their uranium from the African nation of Gabon. Back in the early 1970s, a shipment from the Oklo mine in southeastern Gabon arrived at the French enrichment plant with lower than normal percentages of ^{235}U. French scientists started looking into this and found that other isotopic ratios at the Oklo site were not what they should be. Through a great deal of analysis, they reasoned that about 1.7 billion years ago a fission chain reaction took place. In other words, Oklo was the site of a natural nuclear reactor. All the conditions were right. At that time, the ^{235}U natural abundance was $\sim$4%, and there was plenty of water around to moderate the neutrons. It looks like the Oklo reactor ran for several hundred thousand years (off and on, depending on the presence of water), burning about 6 tons of ^{235}U and generating about 1 ton of ^{239}Pu (which underwent fission, or decayed to ^{235}U) and 15,000 MW of power—roughly equivalent to 4 years output of a modern reactor. It also appears that almost all radioactive nuclides produced by the Oklo reactor stayed right where they were made or traveled only a short distance, suggesting that geologic repositories may be an effective long-term storage solution for radioactive wastes. This is a pretty strong suggestion, as the Oklo reactor was located near the surface where there's a lot of water movement. If we can just marginally improve on Nature, we should be safe.

The Swedes opened such a repository in 1988 near Forsmark, Sweden. The repository is located under the Baltic Sea in 1.9 billion-year-old granite. Another opened in 1999 near Carlsbad, New Mexico, in the United States. It is called the "Waste Isolation Pilot Plant" (WIPP), and is constructed in a 200-million-year-old salt formation $\sim$2000 feet underground. Neither of these repositories have accepted spent nuclear fuel, but they do accept less radioactive (low-level) wastes from research and other activities involving radioactive materials. As suggested previously, a good repository is geologically stable and dry. If WIPP had any water flowing through it, the salt would've dissolved, and it is known that the salt at WIPP is over 200 million years old.

Another consideration is security. Having all this waste in one, isolated, place is safer than several, more accessible places. However, it should be emphasized that the waste needs to cool off for 30–50 years before it is placed in a geologic repository.

---

[11] Sweden, Finland, and France are the furthest along in this process.

Once it's stuck in the ground, it's a little more difficult to ensure it can be kept cool enough. Also, radiation levels drop very quickly in the first couple of decades after removing the fuel from the reactor, making it safer to transport. While the United States has decided to leave all the waste at the reactors where it was used (much to the consternation of the utilities that own the reactors), other countries have opted to consolidate their casks in temporary, aboveground waste storage while waiting for the waste to cool enough to be buried.

A common argument against nuclear power is that its waste remains highly radioactive for a very long time. The amount of time given varies, but is typically something rather unimaginable such as tens or hundreds of thousands of years. This simply isn't true. After ~300 years, 98% of the radioactivity is gone, that is, it is only 2% as toasty as when it was pulled out of the reactor.[12] After ~1000 years it is no more radioactive than if we had left the uranium in the ground. As can be seen in Figure 4.6, the waste is mostly (~96%) uranium—pretty much the same as the uranium ore that was mined in the first place. Putting this stuff back in the ground (where it came from) should then make some sense. There are still billions of tons of uranium ore in the soil, some of it not buried particularly well, yet we do not fear it. Why then should we fear carefully placing the waste into carefully selected, geologically stable, repositories? What about the plutonium? Well, plutonium is chemically very similar to uranium. If we put it into the ground in a place where the uranium has been sitting still for the last couple billion years, the plutonium should stick. Extensive research into the geochemistry of plutonium confirms that it will.

Finally, politics are also a consideration. In most countries, volunteer communities are sought among those with suitable geology. Communities often volunteer because of the economic boost a repository will bring to a local area. Repositories can also be forced in areas with less political clout—such was the case with the U.S. repository at Yucca Mountain.

Unfortunately for the United States, WIPP is unsuitable for nuclear waste because the heat from waste will draw water to it, bathing the containers in salt water. Under these conditions, they would quickly corrode and potentially release their contents to the environment. Yucca Mountain, near Las Vegas in Nevada, was selected in 1987 as the sole site for a geologic repository for spent fuel and other more radioactive (high-level) wastes. Yucca Mountain shares many of WIPP's positive attributes except that it is not a salt formation; rather, it is volcanic tuff, formed from eruptions millions of years ago. Additionally, Yucca Mountain is part of the Nevada Test Site, where nearly 1000 nuclear weapons were tested both above and below ground for over 40 years—until the test ban in 1992. This site is large, isolated, secure, and, at the time of its selection, Nevada was not politically prominent. Opening of Yucca Mountain has been delayed, and it is now uncertain whether it will ever receive waste. One of the major causes of the delay has been concern that the geology might not be as stable or dry as originally thought, especially over the long-term. The other cause for delay has been Nevada's increasing political power, partly due to rapid growth in nearby Las Vegas. This has often inhibited key legislation that

---

[12] It is still hot enough that you wouldn't want to snuggle up next to a big pile of the stuff but not anywhere near has "highly radioactive" as some believe.

FIGURE 4.8 A 1.3-million-gallon waste tank under construction. (Courtesy of U.S. Department of Energy.)

would facilitate opening of the site and removed most funding for anything related to the site for several years. Statewide opposition is fairly strong and is at least partially fueled by the fact that the putative repository was forced on Nevada in a rather clumsy political maneuver, rather than agreed on by all parties.

Even though there is a lot of it (thousands of tons), and it is somewhat toasty, spent nuclear fuel is well characterized, rather compact, and quite uniform (it's pretty much the same stuff no matter what reactor it comes from). The other major source of high-level nuclear waste in the United States has none of these positive attributes. It is a legacy of the Cold War with the Soviet Union—making thousands of nuclear weapons generated a lot of radioactive trash, as well as contaminated buildings and grounds. By far, the biggest challenge is cleaning up the 177 underground waste tanks (Figure 4.8) at the Hanford site in south central Washington state. Hanford was started as part of the Manhattan Project and provided the plutonium for the first test weapon as well as the bomb dropped on Nagasaki. During the Cold War, the Hanford site was responsible for producing the plutonium and tritium (^3H) for the U.S. nuclear arsenal. Eight nuclear reactors were built along the banks of the Columbia River, and plutonium separation facilities were placed further inland.

Separation of plutonium is accomplished through a chemical process. At the end of this process, nitric acid solutions containing fission products, as well as traces of uranium, plutonium, other actinides, and lanthanides, needed safe disposal. The tanks were built to accommodate these wastes. The tanks were made of steel and would not hold acidic solutions (low pH) for long, so enough sodium hydroxide (NaOH, also known as lye) was added to make the waste solution basic (high pH) and, therefore, not corrosive to the tanks. Tank volume occasionally became scarce, so the water (not radioactive) was evaporated out of many, allowing additional waste to be added. So much water was evaporated that the tanks now all contain a lot of solids suspended in aqueous slurries (sludge) or hardened into salt formations (salt cake). Radioactive materials are part of or entrapped in both.

Different waste streams were added to different tanks, and sometimes the contents of one tank were pumped into another. Records of these activities are incomplete, leaving each tank with a unique and partially unknown chemical and radionuclide composition. One tank contains a fair amount of organic compounds (containing mostly carbon and hydrogen). The ionizing radiation causes the formation of hydrogen gas (H_2) from these organic compounds and would build up in the sludge and then occasionally "burp" out of the tank. Concerns were raised that the hydrogen could cause the tank to explode during a burp, so a method of stirring the tank contents was developed to allow the slow continuous release of the hydrogen gas.

Most of the materials in these tanks are not radioactive. The three most common ions present are sodium (Na^+), hydroxide (OH^-), and nitrate (NO_3^-). Various methods of separating the hot (radioactive) from the cold (not radioactive) have been researched and some implemented. The idea is to place the cold stuff into a relatively inexpensive waste form, such as concrete (or grout), on site and **vitrify** the relatively small volume of radioactive material. Vitrification means isolating the radioactive stuff in glass and is a relatively expensive process. The radioactive material is distributed throughout a large glass cylinder (called **logs**) on site then shipped to a high-level repository such as Yucca Mountain. Glass is the preferred waste form for long-term storage of radioactive nuclides, as even if water flows over it, it is unlikely any of the radioactive material will leach out.

The engineers who designed the tanks thought they would only be used temporarily, not the 70+ years some have been in use. Some of the tanks have cracked and leaked some of their contents into the ground nearby, creating a more difficult cleanup scenario.

Similar waste tanks also existed at Department of Energy (DOE) facilities near Idaho Falls, Idaho; Oak Ridge, Tennessee; Aiken, South Carolina; and West Valley, New York. These sites only contained a few tanks each; most have already been remediated or will soon be cleaned up.

Hospitals, universities, and private companies produce small amounts of very low-level radioactive wastes. These wastes can often be disposed of in local landfills or landfills specially designed for the disposal of hazardous material, that is, extra effort is taken to isolate the contents of the landfill from the surrounding environment.

4.1.6 COST OF NUCLEAR POWER

When nuclear power plants were first being planned, Lewis Strauss declared "Our children will enjoy in their homes electrical energy too cheap to meter." At the time, Strauss was chair of the U.S. Atomic Energy Commission, so most people understood his comment to be about power from nuclear *fission*. He *may* have been referring instead to power from nuclear *fusion*, but it is not clear. What is clear is that he was speaking for himself and not the nuclear power industry. When nuclear power subsequently became one of the more expensive ways to generate electricity, the quote damaged the credibility of the industry. The high capital costs associated with nuclear power are probably the single biggest factor inhibiting the construction of new plants, especially in the United States. As previously stated, increases in costs of oil and natural gas (prior to 2008), coupled with increasing concerns that the extensive burning of fossil fuels is adversely affecting the global climate, rekindled interest in nuclear power worldwide through the turn of the century. However, the drop in natural gas prices following the widespread application of hydraulic fracturing, along with the relatively low cost in constructing natural gas power plants, has made generation of electricity from natural gas a popular option.

The average U.S. wholesale cost of generating electricity from all sources in 2005 was $50 per MWh (megawatt hour—the production of 1 million watts of power for 1 hour). The same costs for U.S. nuclear power plants varied from $30/MWh to $140/MWh in 2005. Why such a huge spread? The single most important factor in the cost of nuclear power is the time it takes to build the plant. Nuclear is very capital intensive—a lot of money is required up front. Once it is up and running, nuclear plants are relatively cheap to operate because their fuel costs are low compared to fossil fuels—a tremendous amount of energy is generated from very little fuel. Construction time is an issue because most of the capital costs are covered through loans. If the plant takes a long time to build, the interest on the loans can become as large as the loan itself. These costs are recovered only after the plant is built and are passed on to the ratepayers. This was especially bad through the periods of high inflation of the 1970s and 1980s.

Because of its high upfront costs, nuclear tends to fare better in more highly regulated markets with significant baseload electricity demand. A more regulated market insures the return on the high investment costs in building a nuke plant over the long-term. Deregulated markets encourage more short-term (fossil fuel) thinking. Fossil fuel plants, while cheaper to build, experience higher operating costs and are subject to greater volatility in fuel prices. Nuclear plants do better with higher baseload because they are most cost effective running at full capacity 24/7. Fossil fuel plants are a little easier to switch on and off to meet peak or intermittent demand.

The accidents at TMI (1979) and Chernobyl (1986) also contributed to increasing costs of nuclear power generation in the 1980s. Designs to plants under construction at that time were changed, resulting in significant delays and cost overruns. Many existing plants were shut down so new safety equipment could be installed and/or measures instituted. Cost escalations through the 1980s were the most dramatic in the United States but were also observed elsewhere. Reactors starting construction in the 1990s and 2000s in Japan, India, and South Korea generally show decreasing construction costs.

The U.S. government estimates that new nuclear plants constructed in the United States will end up producing electricity at about $95/MWh in 2022, but there remains significant uncertainty in the construction costs. This is nearly two times the price of electricity from a natural gas plant (roughly $55/MWh). It is not surprising that electrical generation from natural gas has nearly quadrupled in the United States from 1990 to 2016. This trend is also observed worldwide. The cost of nuclear energy is also higher than other non-fossil fuel sources such as solar and wind (even without government subsidies). Some European countries are working toward replacing their nuclear capacity with solar, wind, and other renewable sources.

4.1.7 PROLIFERATION OF NUCLEAR WEAPONS

Since the beginning of the nuclear age, scientists and politicians have tried to separate weapons from nuclear power. Unfortunately, the two are closely related and have resulted in nine countries[13] now possessing nuclear weapons. Highly enriched uranium (>80% ^{235}U) or relatively pure ^{239}Pu is required to build nuclear weapons—fortunately they are difficult to come by. The focus for nonproliferation efforts has been to control access to these materials. The "Megatons to Megawatts" program mentioned previously is a good example of a nonproliferation success. Weapons-grade uranium from the former Soviet Union was converted to nuclear fuel and burned in U.S. reactors.

In addition to taking weapons-grade nuclear material out of circulation, nonproliferation efforts have focused on two key components of the nuclear fuel cycle: uranium enrichment and spent fuel reprocessing (Figure 4.7), as both can allow production of weapons material. Uranium enrichment facilities can be used to bring the ^{235}U percentage up to ~3%–4% for use in a reactor, but the process can easily be continued until the uranium is weapons-grade. Therefore, anyone with a uranium enrichment facility has the potential to produce highly enriched uranium. Since every nuclear reactor contains ^{238}U, they also produce ^{239}Pu. Any chemical reprocessing of the spent fuel can lead to the isolation of ^{239}Pu—although this is a rather inconvenient and rarely used method to get ^{239}Pu. More likely, anyone with a nuclear reactor can simply place a slab of ^{238}U into the reactor to let it soak up neutrons and produce ^{239}Pu. That way there's no need to sacrifice the fuel rods to get the plutonium. The U.S. government's opposition to the reprocessing of spent fuel was established primarily on nonproliferation grounds.

One key to limiting proliferation of nuclear weapons is, therefore, some sort of control of enrichment and reprocessing facilities. There aren't many of these facilities, but their numbers grew as the nuclear industry rapidly expanded worldwide through the 1970s and early 1980s. As nuclear power expands into new countries, new enrichment and reprocessing facilities are considered. Consolidation of these facilities where redundant and placement of them under international control has been suggested to ensure they are used only for energy purposes. Unfortunately, national sovereignty and security concerns tend to inhibit this type of cooperation.

Another key is the strengthening or replacement of the United Nation's treaty on the nonproliferation of nuclear weapons (NPT). The NPT went into effect in 1970 and required the nations with nuclear weapons to disarm and prohibited other nations from developing nuclear weapons. In 1970, only five nations (the United States, the United Kingdom, France, the Soviet Union, and China) possessed nuclear weapons. Unfortunately, four new nations (India, Pakistan, Israel, and North Korea) have since created nuclear weapons.[14] These four nations are the only nations not currently signatories to the NPT. Equally unfortunate, the five original nuclear powers still have significant stockpiles of nuclear weapons. Some have suggested that the United States begin a program of unilateral disarmament, allowing the United

---

[13] The United States, Russia, France, the United Kingdom, China, India, Pakistan, Israel, and North Korea all have nuclear weapons as of this writing.
[14] South Africa also created a small number (~6) of nuclear weapons but dismantled them before signing the NPT in 1991.

States to gain some moral leadership while encouraging other nuclear powers to also disarm. They argue that since the end of the Cold War the nuclear arsenal no longer has anything to deter, so why should it be preserved?

The debate over nuclear power is a complex and sometimes contentious one. The interested reader is encouraged to also look at some of the resources listed in the bibliography. Many contain more detailed discussions of this topic.

4.2 NUCLEAR WEAPONS

4.2.1 FISSION BOMBS

In theory, an atomic (fission) bomb is a relatively simple device to build; fortunately, getting the fissile material to put into it is rather difficult. The material should have a high probability of undergoing neutron-induced fission, produce a fair number of neutrons with each fission reaction (on average), and have a reasonably long half-life—after we go to all the trouble of building a bomb, it'd be nice to be able to store it for a while. Finally, the fissile material should be accessible—either something we can dig up or easily make in a nuclear reactor. Three nuclides that fit all the criteria well are ^{233}U, ^{235}U, and ^{239}Pu. ^{235}U occurs naturally but requires high enrichment (>80%) for use in weapons. ^{233}U and ^{239}Pu can be made in a nuclear reactor and then chemically separated (Section 4.1.3).

Despite intense security surrounding the Manhattan Project, the Soviet Union obtained a great deal of information by recruiting several spies from those working on the project. It was only four years after the United States first tested an atomic bomb that the Soviets also detonated one. No spying is necessary today, as good descriptions of nuclear weapons are posted on the Internet.

The basic concept behind a **fission bomb** is also simple—it is based on the accelerating fission chain reaction that occurs when a supercritical mass of fissile material exists. The idea is to quickly create a supercritical mass and then try to keep it contained for as long as possible. Making the supercritical mass quickly means that very little of the fissile material undergoes fission (and releases energy) before the supercritical mass is formed. Keeping it contained means maximizing the number of fission reactions (energy released) before the supercritical mass blows itself apart, ending the chain reaction. The result should be a really big explosion.

The simplest fission bomb design is the "**gun-type**," shown in Figure 4.9. In this weapon, two subcritical masses of ^{235}U are separated in the bomb. One is "shot" (like a gun shooting a bullet) into the other by a chemical (conventional) explosive. When the two masses come together, they form a supercritical mass. To help start the fission chain reaction, a neutron **initiator** is placed where the center of the supercritical mass ends up. The initiator's purpose is to inject a large number of neutrons into the supercritical mass at the very moment it is formed. This can be accomplished with a small amount of polonium-beryllium mixture.

Polonium is used because a couple of its isotopes (^{208}Po and ^{209}Po) decay fairly "cleanly" via alpha; that is, no (or few) other emissions are observed. ^{235}U also decays via alpha, so why bother with polonium? These isotopes of polonium have shorter half-lives than ^{235}U and, therefore, have higher specific activities than ^{235}U.

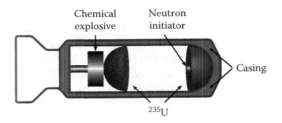

FIGURE 4.9   Schematic of a gun-type fission weapon.

In other words, they crank out more alpha particles per unit mass than the ^{235}U. So why are alpha particles important? If some could be converted to neutrons at the moment the supercritical mass is formed, a lot more fission reactions can take place, and a much larger explosion will result. At the moment the supercritical mass is formed, the polonium gets mashed together with some ^{9}Be. The alpha particles given off at that moment by the polonium can then react with the beryllium to form ^{12}C and a neutron:

$$^9_4Be + ^4_2He \rightarrow ^{12}_6C + ^1n$$

The first atomic bomb used in warfare was a gun-type weapon, called Little Boy, and was dropped by the United States on Hiroshima on August 6, 1945. The Manhattan Project scientists were so confident it would work; they didn't even bother to test a gun-type weapon. Facilitating the decision was the lack of highly enriched uranium—they only had enough to build one bomb. The yield of the bomb was equivalent to 15,000 tons of TNT. Nuclear weapon yield is typically measured in terms of equivalent tonnage of the conventional explosive trinitrotoluene (TNT). For reference, the explosion of 1 ton of TNT produces 4.2×10^9 J of energy. Little Boy was, therefore, referred to as a 15-kiloton device.

The casing around a gun-type bomb can be made from any sturdy material. Remember, its purpose is to try to keep the supercritical mass together as long as possible (even if it is just a small fraction of a second) to maximize the number of ^{235}U fission reactions, thereby increasing the weapon's explosive power (yield). Depleted uranium ($\sim$99.7% ^{238}U) is an excellent choice. Not only is it a strong metal, it will also interact with the fast neutrons produced by the chain reaction in favorable ways. It can reflect some neutrons back into the ^{235}U, and it can undergo fission with others. Both types of interactions will increase the yield of the weapon. The proper term for the ^{238}U used in this way is a **tamper**.

The other kind of fission bomb is the implosion-type (Figure 4.10). It is constructed with a hollow, or "low-density"[15] sphere of ^{239}Pu, which is surrounded by chemical explosives. The idea is to detonate the conventional explosives at exactly the same time, causing the hollow plutonium sphere to collapse to a small solid

---

[15] If a subcritical mass is compressed, its density will increase, and it can become supercritical. Imagine a marshmallow made out of plutonium.

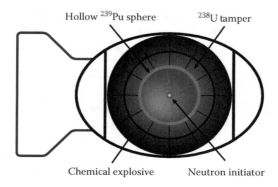

Hollow ^{239}Pu sphere ^{238}U tamper

Chemical explosive   Neutron initiator

FIGURE 4.10   Schematic of an implosion-type fission weapon.

sphere that is now a supercritical mass. Again, a ^{238}U tamper surrounds the pluto-nium sphere to increase the yield.

The Manhattan Project scientists knew that a ^{239}Pu bomb required a different design than ^{235}U. The issue is one of timing. A gun-type weapon does not form the supercritical mass fast enough for ^{239}Pu. Because it is more likely to undergo fission reactions, and because it produces more neutrons per fission (on average), ^{239}Pu will begin to blow itself apart before the supercritical mass is completely formed in a gun-type device. Potentially, this could result in the bomb destroying just itself rather than an entire city. The collapse of a sphere is a faster way to create a supercritical mass. For it to work, however, the timing of the detonations of all the conventional explosives must be perfect. It's a bit like trying to squeeze a water balloon with your hands—tough to do without some of it coming out between your fingers.

The Manhattan Project scientists were so worried about the timing issue that they decided to test an implosion-type device. The codename for the test was "Trinity" and was set up in the Jornada del Muerto (Journey of the Dead Man) desert in southern New Mexico. On July 16, 1945, the first nuclear weapon was detonated, ushering in the atomic age. The power of the explosion exceeded all expectations. The awesome power of the blast moved project director J. Robert Oppenheimer to quote the following passage from the sacred Hindu text, the Bhagavad-Gita:

> If the radiance of a thousand suns
> Were to burst at once into the sky,
> That would be like the splendor of the Mighty One…
> I am become Death,
> The shatterer of Worlds.

A second implosion-type weapon was dropped on Nagasaki on August 9, 1945, bringing World War II to a horrific conclusion. The Nagasaki weapon was given the name Fat Man and had a yield of 22 kilotons.

As described in Section 4.1.3, ^{239}Pu is formed in the fuel rods in nuclear reactors. It can also be made by placing a slab of ^{238}U in the reactor. To optimize ^{239}Pu produc-tion, it is best to run the reactor at low power (low-neutron flux) with fuel that has more ^{238}U and less ^{235}U (lower enrichment than normal). These conditions optimize the opportunity for the ^{238}U to be converted to ^{239}Pu and minimize the deleterious reactions of ^{239}Pu with neutrons, either undergoing fission or forming ^{240}Pu:

$$^{239}_{94}\text{Pu} + {}^1\text{n} \rightarrow {}^{240}_{94}\text{Pu} + \gamma$$

When the plutonium is chemically separated from the uranium, all isotopes are separated together; therefore, there is always some ^{240}Pu mixed in with the ^{239}Pu. This is bad for a weapon because ^{240}Pu is not fissile and has an annoying tendency to decay via spontaneous fission—no neutron required! If, while the supercritical mass is being assembled just prior to detonation, some of the ^{240}Pu undergoes spontaneous fission, a few neutrons will be produced, which can cause a short chain reaction. This will cause the weapon to blow itself apart before many fission reactions have occurred, significantly lowering the yield. In the business, this is known as a **fizzle** and is believed to have happened when North Korea tested its first nuclear weapon in 2006. To avoid this sort of embarrassment, a plutonium device should be $<7\%$ ^{240}Pu.

Fission weapons have been designed with yields ranging from 0.01 to 500 kilotons. The low end of the spectrum represents an extremely inefficient device, that is, only a fraction of the fissile material undergoes fission, and the rest just gets spread over the blast area or is carried away in the fallout. Little Boy was only about 1.3% efficient, meaning that only 1.3% of the ^{235}U underwent fission. Fat Man was about 17% efficient.

Yield is generally related to the amount of fissile material in the weapon. At a certain point, it is no longer possible to pack more fissile material into a bomb without simply spreading the excess around. In other words, only a certain amount of material can undergo fission before the bomb starts blowing itself apart and the chain reaction shuts down. A maximum of roughly 500 kilotons on fission weapon yield is a practical limit. To achieve more destructive power, the very nature of the bomb needs to change.

4.2.2 Fusion Bombs

Fusion weapons (also known as hydrogen bombs, H-bombs, or thermonuclear devices) are *much* more powerful. Yields as high as 60,000 kilotons (that's 60 megatons!) have been made. They are over 4000 times more powerful than the bomb dropped on Hiroshima. As their various names imply, these weapons take advantage of the relatively high amount of energy released per unit mass for fusion reactions. To get fusion to take place, very high temperatures are needed.

A fission device is used as an **initiator** for fusion weapons. It produces the necessary heat, as well as gamma rays and neutrons, which are all helpful in making fusion possible. The fission device is also referred to as the **primary** (Figure 4.11) as it is set off first, and its detonation causes the fusion part of the weapon (the **secondary**) to explode.

A **neutron reflector** is placed behind the primary. Its purpose is to reflect neutrons back into the fissile material of the fission device and toward the secondary. It is made from beryllium because this element has a low probability of absorbing neutrons but a relatively high probability of scattering them, making it a pretty good neutron "mirror." Note that not all neutrons are reflected by the beryllium; it simply has a higher probability of reflecting neutrons than most other materials.

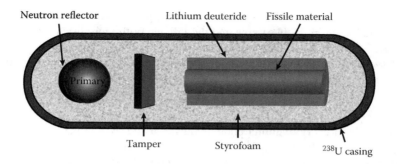

FIGURE 4.11   Schematic representation of a fusion weapon.

A ^{238}U tamper is placed between the primary and the secondary. The fast neutrons produced by the primary can (a) pass right through the tamper, (b) be reflected by the tamper back into the primary, or (c) cause the ^{238}U in the tamper to fission. All are helpful in making a more powerful explosion. The ^{238}U casing also acts as a tamper.

How does the secondary work? As shown in Figure 4.11, it is simply a cylinder of fissile material (such as ^{239}Pu) wrapped with lithium deuteride (LiD) packed in the device with Styrofoam™[16] (and you thought packing peanuts were invented for Grandma to ship you fragile stuff!). The tremendous heat and gamma radiation created by the primary turns the Styrofoam™ into a super-hot plasma, squeezing and heating the secondary. The neutrons that make it through the tamper cause the fissile material inside the secondary to fission, adding more heat ($\sim 10^8$ K!) and pressure to the LiD. Neutrons being released by the fissile material in the primary and the secondary react with the lithium to form tritium (^{3}H):

$$^6_3\text{Li} + {}^1\text{n} \rightarrow {}^4_2\text{He} + {}^3_1\text{H}$$

The tritium can now fuse with the deuterium:

$$^2_1\text{H} + {}^3_1\text{H} \rightarrow {}^4_2\text{He} + {}^1\text{n}$$

The neutrons produced by this fusion reaction are very high energy (14 MeV), and will get some of the ^{238}U casing to fission, as well as cause more fission to occur in the primary, creating an even bigger explosion. The total yield will depend on how much LiD and fissile material are present and whether the casing is made from ^{238}U. The high-yield bombs can get rather large and heavy and would be impractical as a weapon, as there would be no easy way to deliver it to a target.

Even though fusion weapons use fission triggers, these weapons are often considered relatively "**clean**,"[17] whereas fission nukes, especially those with low yields, are called "**dirty**." Adding certain materials can also make a fusion weapon dirty.

---

[16] Styrofoam is a registered trademark of the Dow Chemical Company.

[17] It should be a little odd to consider nuclear weapons as "clean." They are very deadly and cause unbelievable environmental destruction and human suffering. Additionally, they spread their highly radioactive fission products over a broad area. The term is only meant to be relative to other nuclear weapons.

Most commonly, use of ^{238}U as a tamper or casing material will increase levels of ionizing radiation in the blast area and the fallout. With more fission reactions taking place, there are more fission products to spread around. The addition of natural cobalt (which is 100% ^{59}Co) to a fusion weapon results in the formation of ^{60}Co from the reaction with a neutron:

$$^{59}_{27}Co + {}^{1}n \rightarrow {}^{60}_{27}Co + \gamma$$

^{60}Co emits high-energy gamma photons during its decay and has a half-life short enough to give it a high specific activity but long enough to make it frightening to move back to contaminated areas for quite a while after the attack.

$$^{60}_{27}Co \rightarrow {}^{60}_{28}Ni + {}_{-1}e + 2\gamma$$

Such a weapon was called a "doomsday device" because it could effectively evacuate a large area for many years, depending on how much the populace fears ionizing radiation. Many forms of popular fiction picked up on this moniker, although the fictional devices could usually destroy the entire planet, something that is not possible with a single real weapon.

4.2.3 OTHER BOMBS

The **neutron bomb** (or enhanced radiation device) is a small H-bomb designed for maximum neutron production while decreasing blast and heat and is, therefore, considered "clean." This is done by minimizing the amount of fissile material in the primary and the secondary (cuts down on blast and fission products) and replacing any ^{238}U in the bomb with non-fissile materials. The purpose behind a neutron bomb was to act as a deterrent to an enemy with lots of armored vehicles (such as the former Soviet Union) from attacking friendly countries (Western Europe). The intense neutron radiation would easily penetrate the armor and be absorbed by the humans inside them. Supposedly this is a great way to disable a large armored invasion near a friendly city. As clean and tidy as they sound, they would still cause a fair amount of local damage and leave a good deal of radioactive contamination. They were criticized because some have suggested that such a weapon might be too tempting to use in an otherwise conventional war, which could then escalate the conflict to a full-fledged nuclear war.

More recently "**bunker buster**" nukes have been proposed. A nuclear weapon would be placed inside a special casing—designed to hit the ground at high speeds and penetrate the surface to the depth of an underground bunker before detonating. Such weapons exist with conventional explosives. They are designed like a spear (long and narrow) to maximize ground penetration before detonation. However, if a bunker is built deep enough and with sufficiently thick walls and ceiling (hardened), it could easily survive a conventional bunker buster bomb. To destroy such a bunker, some have proposed replacing the conventional explosives with a nuclear warhead. Unfortunately, nuclear warheads have more girth, making the whole weapon too wide to penetrate very deep underground. This means the weapon will detonate near the surface and will cause radioactive fallout above the surface, possibly spreading to nearby areas (civilian populations, neighboring countries, etc.). It also means that a nuclear bunker buster may not be any more effective than a conventional one at destroying its target. Again, some have suggested that developing such a weapon could make it easier for a conventional war to escalate to massive nuclear destruction.

Discussion of another (non-nuclear) kind of **dirty bomb** has sprung up since the late 1990s. Back then, Chechen rebels unsuccessfully tried three times to combine radioactive materials and chemical explosives in their struggle against Russia. The technical name for this type of weapon is *radiological dispersal device* or RDD. The chemical explosive could be as simple as the powder gleaned from fireworks or as powerful as the fertilizer/nitromethane mixture used in the 1995 Oklahoma City bombing. Generally speaking, the purpose behind setting off a dirty bomb is not to cause widespread death and destruction but rather to cause panic and emotional distress, as well as economic dislocation. It is very unlikely that anyone will die from radiation exposure resulting from a dirty bomb, but it will certainly play into the general public's fear of ionizing radiation. It will also contaminate an area that may have to remain evacuated until it can be cleaned-up.

4.3   NUCLEAR FORENSICS

In an era when potential development of nuclear weapons may be a pretext to war, it might be a good idea to have some more conclusive evidence of weapons production before invading. **Nuclear forensics** involves the analysis of radioactive materials or environmental samples containing radioactive materials and/or their byproducts. The goals of analysis are to determine what the material might be, where it came from, and how it is being used. The main areas of interest for nuclear forensics are smuggling of nuclear materials, nuclear terrorism (e.g., dirty bombs), and nuclear weapons proliferation. Perhaps the most dramatic illustration of nuclear forensics in popular culture was in the 1991 Tom Clancy novel (and 2002 movie) *The Sum of All Fears*. In this movie, a terrorist group sets off a small nuclear bomb at a football game in the hopes of starting a nuclear war between the United States and Russia. Fortunately, nuclear scientists quickly analyze the fallout and, from certain nuclide ratios, determine the nuclear material in the bomb originated in the United States.

From the moment uranium ores are taken from the ground, they have a specific signature. The percentage uranium in the ore and the other minerals/ores present are often characteristic of a specific site. After mining, the ore is processed in ways that again vary from place to place, creating material that is unique in its precise elemental and nuclidic composition. The material can then go into a reactor, which will again create a characteristic composition because different reactor designs cause the fuel to burn a little differently. The spent fuel may also be reprocessed, which again makes its exact composition unique, depending on the reprocessing techniques used. The material may also go into a weapon but again be subjected to somewhat unique processes depending on which country is preparing the material. Finally, the material may sit for a time at any stage and decay, again creating a material with unique composition. One of the main tasks of a nuclear forensic analyst is, therefore, a detailed analysis of the exact composition of any material under investigation.

For example, weapons-grade uranium is at least 80% ^{235}U, and weapons-grade plutonium must be greater than 93% ^{239}Pu. The main "impurities" are ^{238}U and ^{240}Pu, respectively. Both impurities are difficult to remove because chemical processes cannot separate isotopes; therefore, more cumbersome physical processes (such as gaseous diffusion) must be used. As a result, different countries adopted different standards for the percentage of fissile materials in their weapons, providing the analyst with a signature. The United States, for example, uses 93% ^{235}U in uranium-based weapons and keeps ^{240}Pu to only 5%–6% in its plutonium-containing bombs.

Along with analyzing the remnants of the fissile material in a weapon, there are other ways to learn about a nuclear bomb after it has blown up. The ratio of the fissile material to fission products can indicate the yield of the device. A higher yield would indicate a more sophisticated design, suggesting the weapon came from the arsenal of one of the more established nuclear powers. Since a burst of neutrons is created at the moment of the explosion, some may interact with the immediate surroundings. Thus, it may be possible to determine whether the bomb was in a wooden or metal crate when it blew. Application of nuclear forensic analysis can also determine the origin of the radioactive material in a dirty bomb (RDD). This material could come from the nuclear fuel cycle or (more likely) from a hospital, clinic, or industrial source.

Nuclear forensics is also useful in illuminating details behind the smuggling of nuclear materials. Since one of the hardest parts of building a nuclear weapon is obtaining the fissile material, attempts to smuggle such material have become more common in popular fiction as well as in the real world. Most such cases are frauds— bad guys pretending to have weapons-grade uranium or plutonium, but in reality only having ^{60}Co, ^{137}Cs, depleted uranium, natural uranium, or ^{241}Am. It's still radioactive but not useful for a nuclear weapon. From 1985 to 2001, there were a little less than 500 seizures of radioactive materials. Of these, fewer than 3% have involved weapons-grade stuff. Even then, they were in quantities too small to build a single device. One positive result of these seizures has been increasing accountability and security for all radioactive materials.

Let's say that you live in a country that neighbors one that is highly secretive, and you think they might be building a uranium enrichment facility, nuclear reactors, a spent fuel reprocessing center, and/or nuclear weapons. They won't answer your calls, so you resort to spying. If you have access to soil, plants, animals, air, or water samples that are leaving your neighbor's country, you might be able to figure out what's going on through nuclear forensics. Once operational, these facilities will have emissions that can tell you what they're up to. Certain elements tend to stick to soils or bioaccumulate in flora and/or fauna. For example, iodine (radioactive, if your neighbor is up to no good) will collect in the thyroid gland of mammals.

If a sample contains a nuclide with radioactive daughters that have sufficiently long half-lives, then knowledge of nuclear equilibria (Section 2.6) can be used to figure out how "old" it is, that is, how long ago the parent was isolated. This only works so long as the ratio of parent to daughter is still changing, that is, it has not yet reached equilibrium. Once at equilibrium, the ratios no longer change, and it's no longer possible to determine how old the sample is.

Example 4.1

A sample of ^{235}U provided for analysis has a ^{231}Pa to ^{235}U atomic ratio of 1.89×10^{-8}. How long ago was the ^{235}U isolated?

This is an example of secular equilibrium since the half-life of the parent (^{235}U) is very long (7.04×10^8 a) relative to our time of observation (likely just a few hours). We're hoping this sample is not yet at equilibrium. To find out, we'll use Equation 2.11 to calculate the ^{231}Pa/^{235}U ratio at equilibrium:

$$\frac{N_{Pa\text{-}231}}{N_{U\text{-}235}} = \frac{t_{1/2(Pa\text{-}231)}}{t_{1/2(U\text{-}235)}} = \frac{3.28 \times 10^4 \text{ a}}{7.04 \times 10^8 \text{ a}} = 4.66 \times 10^{-5}$$

It looks like we're not even close to equilibrium. We can now proceed, using Equation 2.9 to determine how much time has elapsed. First, let's rearrange it so we can use the ratio given in the problem:

$$N_{Pa\text{-}231} = \frac{\lambda_{U\text{-}235}}{\lambda_{Pa\text{-}231}} N_{U\text{-}235}(1 - e^{-\lambda_{Pa\text{-}231}t})$$

$$\frac{N_{Pa\text{-}231}}{N_{U\text{-}235}} = \frac{t_{1/2(Pa\text{-}231)}}{t_{1/2(U\text{-}235)}}(1 - e^{-\lambda_{Pa\text{-}231}t})$$

$$1.89 \times 10^{-8} = \frac{3.28 \times 10^4 \text{ a}}{7.04 \times 10^8 \text{ a}}\left(1 - e^{-\left(\frac{\ln 2}{3.28 \times 10^4 \text{ a}}\right)t}\right)$$

$$4.06 \times 10^{-4} = 1 - e^{-\left(\frac{\ln 2}{3.28 \times 10^4 \text{ a}}\right)t}$$

$$-4.06 \times 10^{-4} = -\left(\frac{\ln 2}{3.28 \times 10^4 \text{ a}}\right)t$$

$$t = 19.2 \text{ a}$$

QUESTIONS

4.1 Define the following: nuclear fission, fissile, fission chain reaction, and critical mass.

4.2 What is the fundamental difference between fossil fuel and nuclear power plants?

4.3 Briefly explain what moderators and control rods do in nuclear reactors.

4.4 What is the difference between a PWR and a BWR?

4.5 Explain why enrichment requires a physical, rather than chemical, process.

4.6 Which would diffuse faster: ^{1}H$_2$O or ^{2}H$_2$O? How much faster?

4.7 Briefly describe how a gas centrifuge enriches uranium.

4.8 ^{233}U can be produced in nuclear reactors in a manner similar to the production of ^{239}Pu. What naturally occurring element must be added to a reactor to produce ^{233}U? Write out all nuclear reactions and decays that are part of ^{233}U production.

4.9   Define the following: depleted uranium, spent fuel, HEU, and reprocessing.

4.10 In the event of another nuclear power plant disaster like Chernobyl, state and local politicians have distributed KI tablets to people living near such facilities. The idea is to take the pill to saturate the thyroid with iodide, thereby preventing thyroid uptake of the radioactive iodide released by the accident. Give two reasons why the distribution of these pills could be considered a political placebo rather than populace protection. Give one reason why distribution of the pills may help protect public health.

4.11 Airborne radioactive materials released from the Fukushima nuclear plants in the spring of 2011 reached the West Coast of the United States about two weeks later. This caused a minor panic as people rushed out to buy potassium iodide pills (see Question 4.10 above), even though radiation levels were only a small fraction above normal background. If you had been living in California at that time, would you have taken KI? Briefly explain.

4.12 Assuming all the radioactive material released by the Chernobyl accident to be ^{137}Cs, what mass (kg) was released?

4.13 Calculate the percent abundance of ^{235}U on Earth 1.7 billion years ago. Assume the uranium is only composed of ^{235}U and ^{238}U.

4.14 If you were either for or against nuclear power before reading this chapter, give four to five bullet points outlining the opposite viewpoint. If you had no opinion, flip a coin—heads, write in favor, tails opposed.

4.15 From a public safety perspective, does it make sense to replace nuclear with natural gas power plants?

4.16 What does a tamper do in a nuclear weapon? A neutron initiator?

4.17 Do you think that unilateral nuclear disarmament by the United States would be a positive step for the nonproliferation of nuclear weapons? Explain.

4.18 What mass of ^{239}Pu was present in the bomb dropped on Nagasaki? What happened to the plutonium that did not fission?

4.19 What is an RDD? What are its likely effects?

4.20 What would the ratio of ^{231}Pa to ^{235}U be in a sample of ^{235}U that was last purified 27.4 years ago?

4.21 Homeland security agents intercepted an 8.96 kg chunk of metal from a nefarious foreign organization. The sample is alleged to be ^{235}U. Your analysis shows an alpha count rate about 10 times lower than would normally be expected for bare uranium metal. The density is also a little low for uranium. Finally, the energies of the most prominent alpha particles observed are 4.197 and 4.147 MeV. What can you report to the homeland security agents? What tests would you recommend be performed next?

5 Radioactive Decay
The Gory Details

WARNING! This chapter is not intended for the casual reader. Proceed only if you are truly curious about better understanding the different forms of decay, or your instructor is telling you that you better read and understand this stuff.

5.1 ALPHA DECAY

This form of decay is most commonly observed for nuclides heavier than ^{208}Pb. The following equation can generically represent alpha decay:

$$_Z^A X \rightarrow {}_{Z-2}^{A-4}Y^{2-} + {}_2^4 He^{2+}$$

Basically, the nucleus is blowing chunks; in this case a ^{4}He nucleus, otherwise known as an alpha particle. The parent is just too big, and the best way it can move toward a stable nuclide is to get rid of four of its nucleons. In terms of stuff spit out by an unstable nucleus (such as beta particles, positrons, and photons), alpha particles are quite large. Notice that, in the equation above, charge is also accounted for. Assuming all the electrons stay on the daughter, it should have a 2– charge because two protons just left the nucleus. The 2+ charge on the alpha particle provides charge balance. It is not always necessary to indicate the electrical charges in decay equations, so long as they are understood.

Alpha decay can also be represented using a somewhat simpler notation:

$$_Z^A X \xrightarrow{\ \alpha\ } {}_{Z-2}^{A-4}Y$$

Because of the relatively large size and high charge of the alpha particle, it has a high probability of interacting with matter. In other words, it is very easy to shield alpha radiation. Alpha particles are also *monoenergetic* because the parent and the daughter have very specific (quantized nuclidic) energy states. This means that if the energy states of the parent and the daughter are always the same, the alpha particle will always have the same energy. The transition from parent to daughter always involves a specific amount of energy, as we've already seen in drawing simple decay diagrams.

The fly in this ointment is the fact that the daughter isn't always produced in the same energy state. If the daughter is initially formed in an excited state, then the alpha particle emitted in that decay will have less energy. The excited states are also quantized, so even if several different alpha energies are possible for the decay of a particular nuclide, only those energies will be observed. How can this be represented in a decay diagram? It'll take a while to answer that. First, let's review a bit of what we learned in Chapter 3 by looking at the decay of ^{246}Cm.

Example 5.1

When ^{246}Cm decays, alpha particles with energies of 5.386 MeV and 5.342 MeV are observed. Gamma photons with an energy of 44.5 keV are also observed. Write out the decay equation and calculate the total energy of decay.

The decay equation (without charges) is written as:

$$^{246}_{96}\text{Cm} \rightarrow\ ^{242}_{94}\text{Pu} +\ ^{4}_{2}\text{He}$$

Calculate the total energy of decay, just like in Chapter 3 (mass values are from Appendix A).

$$[246.067224\,\text{u} - (242.058743\,\text{u} + 4.002603\,\text{u})] \times \frac{931.5\,\text{MeV}}{\text{u}} = 5.475\,\text{MeV}$$

Notice that the total energy of decay is greater than the energy values of both alpha particles observed. This should be expected, as part of the total energy of decay goes into the kinetic energy of the daughter. How much? The answer has to do with momentum. Momentum is mass times velocity and is conserved when a projectile is shot out of a larger object. When a bullet is shot from a gun, there is a certain amount of recoil (or kick) in the gun. The gun wants to move in the opposite direction of the bullet. The amount of recoil depends on the masses of the bullet and gun. The momentum for each is the same, but the velocity is higher for the bullet because it has a lower mass. The same is true for alpha decay (Figure 5.1) where the alpha particle is the bullet, and the daughter nucleus is the gun.

The alpha particle moves a lot faster because it has a lower mass. Since kinetic energy is one-half times the mass times the velocity squared, the alpha particle will also have a greater kinetic energy than the daughter. Classical physics tells us that the kinetic energy of the alpha particle (or any smaller object shot out of a larger object) can be calculated using the formula:

$$K_{\alpha} = K_{T}\left(\frac{M}{M+m}\right) \tag{5.1}$$

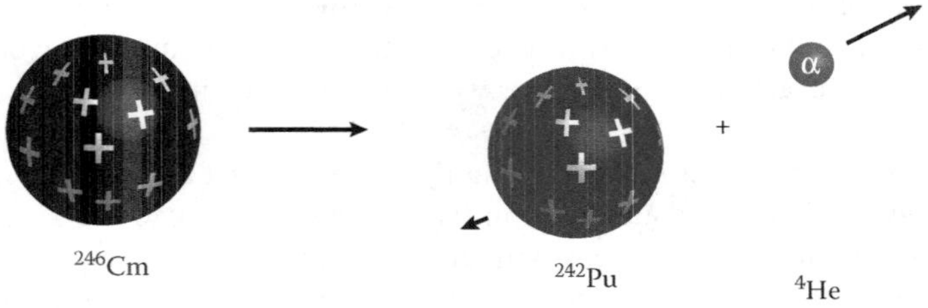

FIGURE 5.1   Recoil in the alpha decay of ^{246}Cm.

where M is the mass of the daughter, m is the mass of the alpha particle, and K_T is the energy of transition. The energy of transition is the energy difference between the parent nuclide, usually in the ground state, to the resulting daughter state (could be excited or ground). If the resulting daughter state is the ground state, then K_T is equal to the total energy of decay, and K_α represents the maximum energy value for an alpha particle for this decay.

Even though this formula comes from classical physics and is meant to be applied to things such as firing balls from a cannon, it works well enough in the subatomic universe. However, you should keep in mind that it is a bit of an approximation, as the subatomic universe follows different rules than our (macro) world.

The kinetic energy of the daughter (or any larger object) can also be determined using the following closely related formula:

$$K_F = K_T \left(\frac{m}{M + m} \right) \tag{5.2}$$

Example 5.2

Using data from Example 5.1, calculate the maximum energy of the alpha particle and the resulting recoil energy of the daughter in the decay of ^{246}Cm.

The maximum energy alpha particle can only result from decay straight to the ground state of the daughter (^{242}Pu). Therefore, K_T is equal to the total energy of decay, which we've already calculated to be 5.475 MeV.

$$K_\alpha = 5.475 \, \text{MeV} \left(\frac{242}{242 + 4} \right) = 5.386 \, \text{MeV}$$

$$K_F = 5.475 \, \text{MeV} \left(\frac{4}{242 + 4} \right) = 0.0890 \, \text{MeV} \approx 89 \, \text{keV}$$

Using mass numbers for the masses is sufficient for the level of accuracy needed. Remember, these formulas are not exact for subatomic cannons. If you have any doubts about using mass numbers, go ahead and replace them with exact masses and see if there is a significant effect on the results.

Notice that the two energy values determined in Example 5.2 add up to the total energy of decay. Energy is conserved as it is partitioned between the daughter and the alpha particle:

$$K_T = K_a + K_F = 5.386 \, \text{MeV} + 0.089 \, \text{MeV} = 5.475 \, \text{MeV}$$

What does "maximum energy of the alpha particle" mean? It means that the total energy of decay is exactly partitioned between the alpha particle and the daughter. In other words, a ^{246}Cm decay that emits a 5.386 MeV alpha particle corresponds to a transition from the ground state of ^{246}Cm to the ground state of ^{242}Pu. If ^{246}Cm emits an alpha particle with anything less than the maximum energy, then the transition must end at an excited state of ^{242}Pu, not the ground state. We're finally ready to draw the decay diagram for the decay of ^{246}Cm. It is shown in Figure 5.2.

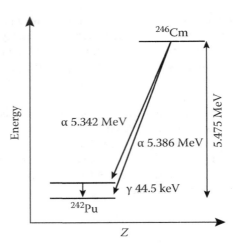

FIGURE 5.2   Decay diagram for ^{246}Cm.

We've already confirmed that the 5.386 MeV alpha emissions correspond to decay from ground state ^{246}Cm to ground state ^{242}Pu. The lower energy alpha particle emitted must, therefore, lead to an excited state of ^{242}Pu. In fact, it should be roughly 44 keV higher in energy than the ground state since that is the difference in energy between the two alpha emissions.

$$5.386 \text{ MeV} - 5.342 \text{ MeV} = 0.044 \text{ MeV} = 44 \text{ keV}$$

Fortuitously (it would seem), this is very close to the energy of the gamma photon that is also observed in this decay (44.5 keV). This strongly suggests that it is the observed gamma photon following the emission of a 5.343 MeV alpha particle from ^{246}Cm.

To properly calculate the difference in energy between the ground and the excited states of ^{242}Pu in Figure 5.2, we must first calculate the energy of transition from ground state ^{246}Cm to excited state ^{242}Pu. First, rearrange Equation 5.1 so that it is solved for K_T, and then plug in the energy of the alpha particle for this transition:

$$K_\alpha = K_T \left(\frac{M}{M+m} \right) \quad K_T = K_\alpha \left(\frac{M+m}{M} \right)$$

$$K_T = 5.343 \text{ MeV} \times \left(\frac{242+4}{242} \right) = 5.431 \text{ MeV}$$

The difference in energy for the nuclidic states of ^{242}Pu is the same as the difference in the two K_T values:

$$5.475 \text{ MeV} - 5.431 \text{ MeV} = 0.044 \text{ MeV} = 44 \text{ keV}$$

Again, we have good agreement with the experimentally observed value of 44.5 keV. Keep in mind that all energy values determined above came from experiments (were empirically determined), and all have some error associated with them. Small differences in calculated and observed values can usually be ignored in drawing decay diagrams.

Finally, notice that the arrows from ^{246}Cm do not quite reach the two nuclidic states of ^{242}Pu. This is to show that the energy of each transition is greater than the energy of the individual alpha particles (the remainder is daughter recoil). Now you know why the alpha arrow didn't touch the daughter nuclidic state in Figure 3.3. Note that other forms of decay in Figure 3.3 all have arrows touching their daughter nuclidic states.

Example 5.3

Show that mass and energy are collectively conserved in the alpha decay of ^{246}Cm.

Before decay takes place, all we have is the mass of the parent. If mass and energy are conserved, then we need to end up with most of that mass plus enough energy to make up the mass that is lost:

mass of parent = mass of daughter + mass of α + energy of α + recoil energy

$$246.067224\,u = 242.058743\,u + 4.002603\,u + (5.386\,\text{MeV} + 0.089\,\text{MeV})$$

$$\frac{u}{931.5\,\text{MeV}}$$

$$= 246.067224\,u$$

Excellent agreement! Note that this only works with the energy for the maximum energy alpha particle. If we wanted to use the 5.342 MeV alpha particle, we'd also need to include the 44.5 keV gamma ray on the right-hand side of the equation. We'd also have to calculate recoil for this transition—there's no guarantee it'll be the same as with the higher energy alpha emission.

Let's look at how the N/Z ratio changes in the decay we've been studying:

$$^{246}_{96}\text{Cm}_{150} \xrightarrow{\ \alpha\ } \ ^{242}_{94}\text{Pu}_{148}$$

$$N/Z = 1.56 \qquad\qquad 1.57$$

It *increases* while forming a lower Z nuclide! As you remember from Figure 1.4, the curvature in the belt of stability means that it is generally better to form a smaller nuclide with a lower N/Z ratio. After a series (or even just one or two) of alpha decays, the ratio will get to the point where beta decay is necessary because the N/Z ratio is now too high. Examination of Figure 1.7 illustrates this point nicely. The increase in N/Z in alpha decay requires each of the three naturally occurring decay series to be

a mix of alpha and beta decay. Notice also that alpha decay does not affect the even
or odd nature of the number of protons or neutrons. Both parent and daughter in our
example are even-even (*ee*). If the parent had been *eo*, the daughter would also be *eo*.

One final point about alpha decay: Just because a nuclide is larger than ^{208}Pb
doesn't mean it'll decay by alpha. Large nuclides with poor *N/Z* ratios *can* also decay
by beta, electron capture, or positron. In some cases, branched decay (alpha plus one
or more flavors of beta) is observed. Likewise, a nuclide doesn't have to be huge to
decay via alpha emission. ^{144}Nd (and a few others below ^{208}Pb) decays via alpha, and
^{8}Be splits into two alpha particles when it decays.

5.2 BETA DECAY

There are some problems with the rather simple way beta decay has been represented
in this book. Don't worry; our representation has just been incomplete, not inaccu-
rate. As represented so far, beta decay violates three conservation laws:

1. *The Law of Conservation of Particles.* The beta particle is created by the
   nucleus but from what? Flipping a quark? A particle cannot be created all
   by itself; an antiparticle must also be created at the same time.
2. *The Law of Conservation of Angular Momentum.* Angular momentum can
   be thought of simply as spin. Every subatomic particle and every nucleus
   has a spin value. It can be a whole number or a whole number divided by 2.
   This law states that the difference in spin between reactants and products in
   a nuclear equation must be a whole number. Let's look at the spin values of
   the known participants in the decay of ^{14}C:

$$^{14}\text{C} \rightarrow {}^{14}\text{N} + {}_{-1}\text{e}$$
Spin values :   0   1   ½

   The sum of the spins on the product side is 1-1/2. Since the spin of the
   parent is 0, the change in spin will also be 1-1/2, not a whole number and,
   therefore, is in violation of the law. Written in this way, every beta decay
   will appear to violate the Law of Conservation of Angular Momentum.
3. *The Law of Conservation of Energy.* Energy can be converted into different
   forms of energy or even into matter, but it can't just go away or be created
   from nothing. The problem with beta decay is that the beta particles are not
   monoenergetic like in alpha decay. A broad energy distribution is observed
   for all beta particles instead of a relatively narrow peak, as illustrated in
   Figure 5.3. It is not possible for the same nuclear transition to take place but
   different energies be produced.

The answer to all our problems would be an antiparticle that is emitted at the same
time as the beta particle. This antiparticle should have a spin value of 1/2, and it should
share decay energy with the beta particle in a relatively random fashion. Finally, this
antiparticle should have a very low probability of interacting with matter, as it has
proven very difficult to detect. An antineutrino ($\overline{\nu}$) has all these attributes. It has no

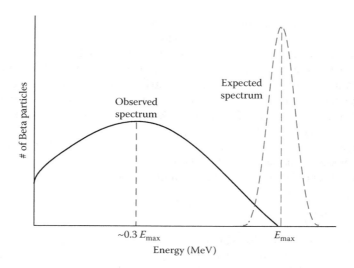

FIGURE 5.3 Expected (dashed curve) and observed (solid curve) energy spectra for beta particles.

charge, very low mass, and an extremely low probability of interacting with matter. We can now write a complete equation for the decay of ^{14}C:

$$^{14}_{6}\text{C} \rightarrow {}^{14}_{7}\text{N} + {}_{-1}\text{e} + \bar{\nu}$$

We can also write a generic beta decay equation (including charges) as:

$$^{A}_{Z}\text{X} \rightarrow {}^{A}_{Z+1}\text{Y}^{+} + {}_{-1}\text{e}^{-} + \bar{\nu}$$

The positive charge on the daughter (Y) results from it having the same number of electrons as the parent (X), yet one more proton. The negative charge on the beta particle balances this out. In general, beta decay equations will not be written in this text with explicit charges or antineutrinos. The reader should understand they are implied. The main reason for leaving the antineutrino off is that it remains a rather elusive and poorly understood antiparticle.

Beta decay equations can also be written using a similar shorthand as we saw with alpha decay:

$$^{A}_{Z}\text{X} \xrightarrow{\beta^{-}} {}^{A}_{Z+1}\text{Y}$$

The ultimate example of beta decay is the decay of free neutrons. On their own, neutrons have a half-life of about 10 minutes. It is a good thing they are stable when incorporated into nuclei!

$$^{1}_{0}\text{n} \rightarrow {}^{1}_{1}\text{p} + {}_{-1}\text{e}^{-} + \bar{\nu}$$

All the stuff in this equation has a spin value of 1/2. Is the Law of Conservation of Angular Momentum violated in this decay?

The maximum beta energy (E_{max}) is given in Appendix A and on charts of nuclides, and it is often quoted as the energy of the beta particle. Theoretically, this could only happen if the energy of the antineutrino emitted in the same decay is zero. Examination of Figure 5.3 shows that beta particles never have this energy, since it always shares some with the antineutrino. According to Figure 5.3, the most probable energy for a beta particle is $\sim 0.3E_{max}$.

We've already seen how to draw an alpha decay diagram that incorporates a gamma emission. The same can be done with beta decay. In fact, it is very unusual to observe alpha or beta decay without gamma emission.

Example 5.4

^{47}Sc decays with the emission of two beta particles with E_{max} values of 0.439 MeV and 0.600 MeV. A gamma ray with an energy of 159.4 keV is also observed. Draw a decay diagram that accounts for all this information.

The complete decay equation is:

$$^{47}_{21}\text{Sc} \rightarrow \ ^{47}_{22}\text{Ti}^+ + \ _{-1}e^- + \bar{\nu}$$

The total energy of decay is calculated in the usual manner, after looking up the atomic masses of parent and daughter (Appendix A):

$$(46.952408\,\text{u} - 46.951763\,\text{u}) \times \frac{931.5\,\text{MeV}}{\text{u}} = 0.601\,\text{MeV}$$

Since we are using the atomic masses of the parent and daughter nuclides, we don't need to add in the mass of the beta particle—it is already included in the atomic mass of ^{47}Ti (remember it is formed as a monocation). The mass of the antineutrino is so small it is negligible, so it does not need to be included in the calculation.

The difference in energy between the two beta particles is close to the energy of the observed gamma ray:

$$0.600\,\text{MeV} - 0.439\,\text{MeV} = 0.161\,\text{MeV} = 161\,\text{keV}$$

We're ready to draw! The 0.600 MeV beta particle is very close in energy to the total energy of decay and, therefore, corresponds to a transition between the ground states of the parent and the daughter. The sum of the lower energy beta particle and the energy of the gamma ray is also pretty close to the total energy of decay, so odds are good that the 0.439 MeV beta emission leads to an excited state of ^{47}Ti, which is 0.1594 MeV higher in energy than its ground state. The complete diagram is shown in Figure 5.4.

The most common decay pathways for radioactive nuclides do not always involve a direct transition between ground states and can involve more than one gamma emission. The most prominent beta particle observed in the decay of ^{43}K has a maximum energy of 0.83 MeV, but its total energy of decay is 1.82 MeV. As part of this pathway,

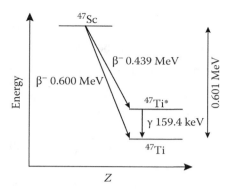

FIGURE 5.4 Decay diagram for ^{47}Sc.

gamma photons with energies of 372.8 keV and 617.5 keV are also observed. These three emitted energies add up to the total energy of decay:

$$0.83\,\text{MeV} + 0.3728\,\text{MeV} + 0.6175\,\text{MeV} = 1.82\,\text{MeV}$$

The primary decay pathway for ^{43}K *could* therefore look like Figure 5.5. Based on the information provided, we can't decide on the order of the two gamma emissions. The only way to know for sure is to consult a higher authority. A convenient source is the interactive chart of the nuclides posted by the International Atomic Energy Agency at www-nds.iaea.org. They provide a decay diagram, which shows that the decay of ^{43}K involves a few other minor pathways as well as affirming Figure 5.5 as correct for the major pathway.

A significant part of the discussion of alpha decay earlier in this chapter was the issue of recoil. Thus far into our look at beta decay, it hasn't been mentioned. Are the calculations above done properly? Let's find out.

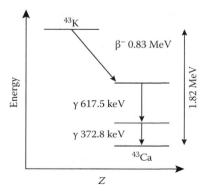

FIGURE 5.5 Partial decay diagram for ^{43}K.

Example 5.5

Calculate the recoil energy resulting from beta decay for a free neutron.

This is a good example because neutrons are relative lightweights, and the total energy of decay is equal to E_{max} (0.782 MeV). If recoil is ever observed in beta decay, we should see it here. We can use the same equations as for alpha decay, except m is now the mass of an electron:

$$K_F = K_T\left(\frac{m}{M+m}\right) = 0.782 \, \text{MeV} \times \frac{0.000549 \, \text{u}}{1.01 \, \text{u}} = 0.000425 \, \text{MeV}$$

The recoil energy in beta decay is very small compared to the total energy of decay and can easily be neglected in beta decay energy calculations. The reason beta decay is so different from alpha decay is because the mass of an electron is so much smaller than the mass of an alpha particle. A smaller bullet produces less recoil.

5.3 POSITRON DECAY

Just like beta (minus) decay, positron decay has been simplified in Chapters 1–3. Without the addition of a neutrino to the product side of the decay equation, positron decay would also violate the same conservation laws as beta decay. The complete, generic representation of positron decay is:

$$^A_Z\text{X} \rightarrow \, ^A_{Z-1}\text{Y}^- + \, _{+1}\text{e}^+ + \nu$$

A neutrino is formed here, rather than the antineutrino formed in beta decay. This is because the positron itself is an antiparticle; therefore, a particle also needs to be created to properly conserve particles. The positron is the corresponding antiparticle to the electron. Every particle (such as protons, neutrons, and neutrinos) has a corresponding antiparticle. These antiparticles are collectively known as antimatter. That's the fuel for starships in the stunningly successful *Star Trek* sagas. It is also used to provide the devastatingly explosive force in their photon torpedoes. As is often the case, there is some basis in fact here. When antimatter comes into contact with (regular) matter, they annihilate each other, forming energy. When an electron and a positron annihilate each other, two *annihilation photons* are formed, each with an energy of 0.511 MeV. Let's do the math.

Example 5.6

Calculate the energy released from a positron/electron annihilation.

The equation is:

$$_{+1}\text{e} + \, _{-1}\text{e} \rightarrow 2\gamma$$

The electron and the positron both have a mass of 5.486×10^{-4} u, and the photons have no mass. Therefore, $2(5.486 \times 10^{-4}$ u$)$ is destroyed, and the energy produced is:

$$2 \times (5.486 \times 10^{-4} \text{ u}) \times \frac{931.5 \text{ MeV}}{\text{u}} = 1.022 \text{ MeV}$$

This energy is divided equally between the two annihilation photons, meaning each will be 0.511 MeV. This is a lot of energy released by a small amount of matter. It wouldn't take much antimatter to make a pretty powerful explosive. Interestingly, the two annihilation photons are emitted approximately 180° apart. This forms the basis of positron emission tomography (PET), which is discussed in more detail in Chapter 8. Simultaneous detection of the two photons from annihilation allows for stunning 3-D images of people's innards for medical diagnostic procedures.

Let's get back to the generic positron decay at the beginning of this section. Paying close attention to charge, the daughter (Y) is formed as a monoanion (−1 charge), assuming it has the same number of electrons as the parent. This negative charge is balanced by the positive charge of the positron, so charge is balanced in this decay.

What about mass balance? Since the daughter is produced as an anion *and* a positron is emitted, the total mass on the product side is equal to the atomic mass of the daughter plus the mass of two electrons. Compare this with beta decay. There, the difference in mass is simply the difference in atomic masses of the parent and daughter nuclides.

This means we have some additional energy accounting to do in positron decay. The total energy of decay is equal to the (kinetic) energy of the positron plus the energy to "create" two electrons (1.022 MeV, the same as you get when they are destroyed). In beta decay, the total energy of decay can equal the maximum possible beta energy for the transition, something that can never be true in positron decay. To calculate maximum possible positron energy for a particular decay, we have to account for the extra mass:

$$[\text{at. mass X} - (\text{at. mass Y} + \text{ mass } _{-1}e + \text{ mass } _{+1}e)] \times \frac{931.5 \text{ MeV}}{\text{u}} \qquad (5.3)$$

Example 5.7

^{18}Ne decays with the emission of two positron particles with E_{max} values of 3.42 MeV and 2.38 MeV. A gamma ray with an energy of 1.04 MeV is also observed. Draw a decay diagram that accounts for all this information. Also calculate the maximum possible positron energy for this decay:

$$^{18}_{10}\text{Ne} \rightarrow \ ^{18}_{9}\text{F} + \ _{+1}e$$

The total energy of decay can still be calculated in the usual manner—from the differences in masses between parent and daughter:

$$(18.005697 \text{ u} - 18.000938 \text{ u}) \times \frac{931.5 \text{ MeV}}{\text{u}} = 4.433 \text{ MeV}$$

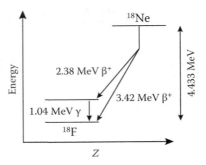

FIGURE 5.6 Decay diagram for ^{18}Ne.

This value differs somewhat from what is recorded in Appendix A. Should we be worried? No way, we did the math correctly—the difference is likely due to experimental errors. The maximum energy for a positron is:

$$[18.005697\,u - (18.000938\,u + 5.486 \times 10^{-4}\,u + 5.486 \times 10^{-4}\,u)] \times \frac{931.5\,\text{MeV}}{u}$$

$$= 3.411\,\text{MeV}$$

Notice it matches one of the E_{max} values given in the problem quite closely. This means the 3.42 MeV positron gives ^{18}F in its ground state. We can also guess that the 1.04 MeV gamma photon is emitted following the 2.38 MeV because the difference in the E_{max} values for the two positrons is 1.04 MeV:

$$3.42\,\text{MeV} - 2.38\,\text{MeV} = 1.04\,\text{MeV}$$

The maximum energy positron plus the energy to "create" two electrons should equal the total energy of decay:

$$3.42\,\text{MeV} + 1.022\,\text{MeV} = 4.44\,\text{MeV}$$

Close enough. The decay diagram is given in Figure 5.6. The short vertical line from ^{18}Ne represents the 1.022 MeV required by all positron decays. If the total energy of decay is less than 1.022 MeV, positron decay is not possible.

What about recoil? The results should be the same as with regular beta, so the answer is no, it is insignificant. Also, like regular beta decay, the shorthand shown can represent positron decay:

$$^{A}_{Z}X \xrightarrow{\ \beta^{+}\ } {}^{A}_{Z-1}Y$$

5.4   ELECTRON CAPTURE

Electron capture probably seems like the strangest form of radioactive decay ever, as it is the only one that is not strictly a nuclear change since it requires an orbital electron. Its generic representation is quite similar to positron decay:

$$^{A}_{Z}X + {}^{0}_{-1}e \rightarrow {}^{A}_{Z-1}Y + \nu$$

Remember that the electron on the left side of the arrow is an orbital electron. The neutrino on the product side balances the loss of a particle on the reactant side. Since the parent uses one of its own electrons to convert a proton to a neutron, the daughter is formed with one less proton and one less electron. Therefore, the daughter is neutral, not a cation or an anion. Because the electron is not originally part of the nucleus, some prefer to write electron capture equations with the following shorthand:

$$_Z^A X \xrightarrow{\text{EC}} {}_{Z-1}^A Y + \nu$$

Example 5.8

Draw a decay diagram for ^{72}Se. It decays via electron capture with a total energy of decay of 0.33 MeV. A gamma ray with an energy of 46 keV is observed in all decays.

Since the gamma photon is observed in all decays, the electron capture must lead to an excited state of the daughter, which then decays to the ground state by spitting out the gamma (Figure 5.7).

It can sometimes be difficult to decide how to draw the electron capture arrow. In Example 5.8 it is unambiguous, but if information from a printed chart were used, you wouldn't know that the gamma photon is observed in each decay.

Positron and electron capture decays both do the same thing. They both produce a daughter with one less proton and one more neutron than the parent. Can we tell if a particular nuclide will prefer to decay by electron capture or positron emission? Sort of. Electron capture is more likely with high Z nuclides, while positron decay is more likely for light nuclides, but that's about the best we can do. For most proton-rich nuclides, either or both forms of decay are observed. The only time we can be definitive is when the total energy of decay is less than 1.022 MeV. In these cases, positron decay cannot occur, leaving electron capture as the only decay option.

To better understand electron capture, we'll need to know a bit about orbital electrons. Electrons occupy a volume of space (orbital) around the nucleus that is (more or less) defined by the solution to the mathematically complex Schrödinger equation. The solution generates a set of four numbers that are unique for each electron in an atom. These numbers are called quantum numbers, and individually they represent main energy level, orbital shape, orbital orientation, and electron spin. We'll only be concerned with the quantum number that describes the main energy level. This

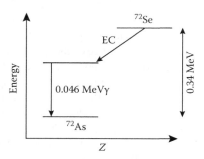

FIGURE 5.7   Decay diagram for ^{72}Se.

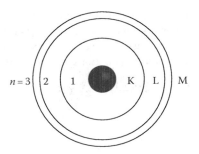

FIGURE 5.8   The Bohr atomic model.

quantum number is known as the principal quantum number and is represented by n. It has positive integral values (1, 2, 3, …). Since we're not concerned with orbital shape or orientation, we can represent the different energy levels (shells) as concentric circles around the nucleus. This type of atomic model is called the Bohr model, named after the Danish physicist Niels Bohr who first proposed it. An example is pictured in Figure 5.8. The energy levels $n = 1$, 2, and 3, are often represented by the letters K, L, and M, respectively. The letter designations are older but are more commonly used in medical applications, so we will continue their use in this text. The L shell is further subdivided into three subshells (L_1, L_2, L_3), and M into five. There is only one subshell for the K shell, so it is simply represented as K.

Which electrons crash into the nucleus? Electron capture is most likely to take place with a K shell (K capture) electron. This makes sense, as the K shell is the lowest energy shell, and they spend the most time closest to the nucleus. Electron capture from the L (L capture) or M (M capture) shells is also possible, but not probable. For example, K capture happens in ∼90%, L capture in ∼9%, and M capture in ∼1% of all electron capture decays of ^{41}Ca. L and M capture increase in probability with Z, which can be rationalized by the fact that all electron shells move closer to the nucleus with increasing Z. The same rationale can be applied to the relative probabilities of positron and electron capture decays. A larger Z means the K shell electrons are closer to the nucleus, so electron capture tends to be more popular with heavier nuclides. In reality, it is a bit more complex. For instance, a large value for the total energy of decay tends to favor positron decay, and high decay energies are more often observed for low Z nuclides.

Superficially, electron capture could be considered a stealth mode of decay. If electron capture occurs without a gamma photon emitted, how can we tell that decay has taken place? According to the generic decay equation at the beginning of this section, only a neutrino is emitted, and neutrinos are extremely difficult to detect. Fortunately, the atom's electrons can tell us. After K capture has occurred, the K shell will be short one electron. This electron vacancy means that the daughter is created with its electrons in an excited state. To get back to the ground state, an electron from a higher level drops down to fill the vacancy. When this happens, energy must be emitted. Because the electron energy levels are quantized, the energy emitted for a particular atom is always exactly the same, so long as the electronic transition is between the same two levels. Because the K shell is so much lower in energy than the others, the energy given off when a K vacancy is filled is often an X-ray photon.

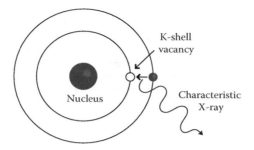

FIGURE 5.9 The aftermath of electron capture—Part I.

Like gamma rays, X-rays are high-energy photons (see Figure 1.1). Gamma and X-ray photons can have the same energy, and to a detector, they are indistinguishable. The difference between the two lies solely in how they are generated. Gamma rays are generated in the nucleus as a result of a transition from an excited nuclear state to one lower in energy. X-rays are generated by electrons in one of two ways. First, as we've seen here, an outer shell electron fills an inner shell vacancy (there are other ways to create these vacancies). This type of X-ray is termed a **characteristic X-ray** because they are produced with specific energies. Second, high-speed (free) electrons can interact with matter by slowing down as it travels near a nucleus. This type of X-ray is called **continuous X-ray** because the amount of energy released will depend on how close the electron gets to the nucleus, producing a broad spectrum of photon energies. It is also known as ***bremsstrahlung***, which is German for "slowing-down radiation."

We can finally draw a picture of what happens after electron capture (Figure 5.9). A K shell vacancy is created by the electron capture decay. An L shell electron then fills this vacancy. Keep in mind that any higher-energy shell could fill the vacancy; however, an L shell electron is the most probable. Because of the electronic transition, a characteristic X-ray is emitted that has an energy corresponding to the difference in electron binding energy between the L and K shells in the daughter.

Characteristic X-rays are designated according to the subshells that mark the beginning and the end of the electronic transition. The most commonly observed transition is labeled $K_{\alpha 1}$ which corresponds to an electronic transition from the L_3 subshell to the K shell. $K_{\alpha 2}$ is the $L_2 \rightarrow K$ electronic transition, and $K_{\alpha 3}$ is the $L_1 \rightarrow K$ transition. The $K'_{\beta 1}$ represents a combination of all the transitions from the M shell to the K shell. There is generally little difference in the energies of the various M subshells, so there is also very little difference in the individual K_β values. The energy values for characteristic X-rays correspond to differences in energy between the (sub)shells. These energy values are called **electron binding energies** because they represent the amount of energy required to remove an electron from an atom. These energy levels can be represented in an energy level diagram such as Figure 5.10. This diagram is for the element holmium. Similar diagrams for other elements will vary—typically the numbers will increase with atomic number. This makes sense; it should be more difficult to remove a K shell electron from an element with 92 protons than one with only 12 protons. An approximate energy value is given for the five M subshells in Figure 5.10. They actually range in value from 1.35 keV to 2.13 keV.

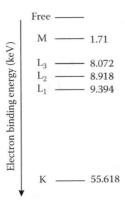

FIGURE 5.10   Energy level diagram for Ho.

Example 5.9

Using Figure 5.10, calculate the energy values of the following characteristic X-rays for holmium: $K_{\alpha 1}$, $K_{\alpha 2}$, and $K'_{\beta 1}$.

$K_{\alpha 1}$	$L_3 \rightarrow K$	$55.618\,\text{keV} - 8.072\,\text{keV} = 47.546\,\text{keV}$
$K_{\alpha 2}$	$L_2 \rightarrow K$	$55.618\,\text{keV} - 8.918\,\text{keV} = 46.700\,\text{keV}$
$K'_{\beta 1}$	$M \rightarrow K$	$55.618\,\text{keV} - 1.71\,\text{keV} = 53.91\,\text{keV}$

Instead of emitting a characteristic X-ray, an atom can get rid of the energy generated by the electronic transition by spitting out one of its electrons, as illustrated in Figure 5.11. This process is known as the **Auger** (oh-ZHAY) **effect**, and the electron that gets booted is known as an **Auger electron**.

Auger electrons are designated by: (1) the shell or subshell where the vacancy was originally created, (2) the subshell of the electron that drops to fill the vacancy, and (3) the subshell of origin for the Auger electron. For example, if the electron dropping into the K shell in the left side of Figure 5.11 is from the L_2 subshell, and the Auger electron is from the L_3 subshell, then the emitted electron would be called a KL_2L_3 Auger electron. The kinetic energy of the KL_2L_3 Auger electron is equal to the energy gained by the atom when the electron drops from the L_2 subshell to the

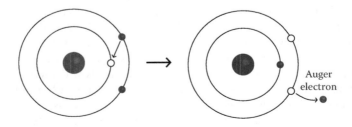

FIGURE 5.11   The aftermath of electron capture—Part II—the Auger effect.

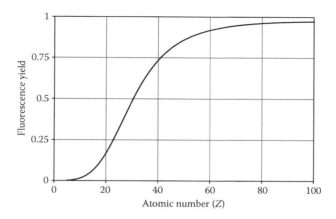

FIGURE 5.12 Fluorescence yield as a function of atomic number. (Data from Hubbell, J. H., *J. Phys. Chem. Ref. Data*, 23, 339, 1994.)

K shell, minus the energy required to remove an L_3 electron. For holmium (Figure 5.10) it is:

$$KL_2L_3 = 55.618 \text{ keV} - 8.918 \text{ keV} - 8.072 \text{ keV} = 38.628 \text{ keV}$$

The characteristic X-ray and Auger electron calculations done here are reversed from how they would normally be performed. Typically, the energies of the characteristic X-rays and Auger electrons are determined experimentally, and the electron binding energies are calculated from these energies.

Multiple Auger effects are possible in a single decay, potentially leaving the daughter as a highly charged cation. Since chemical bonds often involve the sharing of electrons between two atoms, electron capture can be especially disruptive when it occurs within a molecule.

Is it possible to tell whether characteristic X-rays or Auger electrons will be emitted because of the electron capture decay? Not really. Both are always observed when electron capture is a decay mode for a particular nuclide. The probability that X-rays will be produced following electron capture is called **fluorescence yield**. It is symbolized by the Greek letter omega (ω). With a K subscript (ω_K), it refers to fluorescence yield when the original vacancy is in the K shell. Figure 5.12 shows a graph of fluorescence yield (ω) versus atomic number (Z). The probability that characteristic X-rays will be emitted increases with atomic number. In other words, the probability that Auger electrons will be emitted decreases with atomic number. This makes sense; more protons in the nucleus means that electrons in the same shell will be bound more tightly. In this case, the atom is more likely to shed its excess energy by emitting an X-ray rather than try to boot one of the tightly bound electrons.

5.5 MULTIPLE DECAY MODES

We've already seen that proton-rich nuclides can decay by both positron emission and electron capture. Some nuclides, especially those with moderate atomic numbers

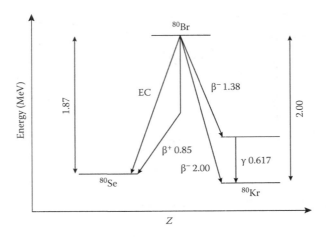

FIGURE 5.13   Decay diagram for ^{80}Br.

in the middle of the belt of stability and odd numbers of both protons and neutrons, can decay by all three beta modes. Can a decay diagram be drawn for these schizophrenic nuclides? You bet.

Example 5.10

^{80}Br decays via all three beta modes. Using the information provided here, draw its decay diagram. The total energy of decay for β^- is 2.00 MeV, and it is 1.87 MeV for β^+ and electron capture. The most common beta particles emitted have E_{max} values of 2.00 MeV and 1.38 MeV, while the most common positron has an E_{max} of 0.85 MeV. A 617 keV gamma photon is also prominently observed during β^- decay.

The decay diagram is shown in Figure 5.13. All energy values are given in MeV. Note that all values add up to appropriate values; for example, the 1.38 MeV beta particle combines with the 0.617 MeV gamma photon to roughly equal the total energy of beta decay. Since no gamma ray is associated with electron capture, its arrow is drawn all the way to the ground state of the ^{80}Se daughter. If you're curious, the branch ratios observed for ^{80}Br are 91.7% for β^- decay, 6.1% electron capture decay, and 2.2% positron decay.

Multiple decay modes are also often observed for heavy ($Z > 82$) nuclides with poor N/Z ratios. These nuclides typically decay via alpha and some form of beta. Note that heavy, neutron-rich nuclides are more likely to decay via β^- than heavy, proton-rich nuclides decaying via electron capture or β^+ emission. This is likely because alpha decay also increases the N/Z ratio (see Section 5.1).

5.6   THE VALLEY OF BETA STABILITY

All three flavors of beta decay involve no change in the mass number—they occur along isobars. Let's take a closer look at a set of isobars, in particular the eight nuclides closest to the belt of stability when $A = 99$ (Table 5.1).

TABLE 5.1
Isobars with $A = 99$

Isobar	N	Z	N/Z	Decay Mode(s)	Decay Energy (MeV)
$_{47}$Ag	52	47	1.11	β^+, EC	5.4
$_{46}$Pd	53	46	1.15	β^+, EC	3.39
$_{45}$Rh	54	45	1.20	EC, β^+	2.04
$_{44}$Ru	55	44	1.25	Stable	Stable
$_{43}$Tc	56	43	1.30	β^-	0.294
$_{42}$Mo	57	42	1.36	β^-	1.357
$_{41}$Nb	58	41	1.41	β^-	3.64
$_{40}$Zr	59	40	1.48	β^-	4.56

The neutron number, atomic number, and N/Z ratio are tabulated for each isobar. The optimal value for the N/Z ratio is likely close to 1.25 for $A = 99$, since ^{99}Ru is the only stable isobar. The decay mode(s) are given in the next column in Table 5.1. Notice that the isobars above ^{99}Ru have lower N/Z ratios (are proton-rich), and all undergo positron and electron capture decay. The more commonly observed decay mode of the two is listed first for each nuclide. The apparently even mix of the two decay modes is to be expected for the relatively moderate Z values for these nuclides. The isobars listed below ^{99}Ru all have higher N/Z ratios and decay via β^- emission.

Finally, the total energy of decay is listed for each isobar in Table 5.1. Notice that the value for ^{99}Tc is rather small, especially when compared to ^{99}Rh. This suggests that the optimal N/Z ratio for $A = 99$ is probably between 1.25 and 1.30 (the values for ^{99}Ru and ^{99}Tc respectively), although it is likely closer to 1.25. Notice also that the energy values increase the further the isobar is from ^{99}Ru. We'll soon see that this is not true for all mass number values, but it is often true when A is odd. We can illustrate all the isobaric decays listed in Table 5.1 in a simplified decay diagram, such as Figure 5.14.

The decays represented in Figure 5.14 are simplified such that only the total energy of decay is represented. In other words, if excited daughter states are formed in some of these decays, we aren't drawing them; we're only looking at the ground states for each isobar. Since ^{99}Ru is the only stable isobar for $A = 99$, any other (radioactive) isobars formed will eventually decay to ^{99}Ru. Because of the increasing magnitude of the decay energies as the isobars move away from ^{99}Ru, the energy changes for whole set of isobars can be approximated with a parabola. This mathematical model is known as the **semi-empirical mass equation** and was (remarkably) developed by C. F. von Weizsäcker in 1935. He used the word "mass" because of the simple direct relationship between mass and energy ($E = mc^2$). We could've just as easily graphed mass on the y-axis of Figure 5.14, and it would look just the same.

Energy (or mass) is really the third dimension to the chart of the nuclides (the other two are N and Z—Figure 1.9). If we look at the chart as a whole in this way, we'll see that the stable nuclides are all located at the bottom of an energy valley. The belt of stability is really a **valley of stability**, and Figure 5.14 is a cross-sectional slice of one part of this valley. Figure 5.15 displays this valley for all known nuclides with mass numbers of 46 to 71.

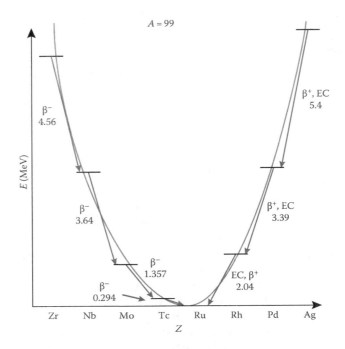

FIGURE 5.14 A simplified decay diagram for some of the $A = 99$ isobars.

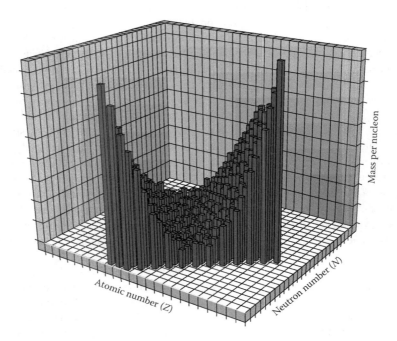

FIGURE 5.15 A 3-D chart of the nuclides from $A = 46$ to $A = 71$.

The walls of this valley become increasingly steep as atomic number (Z) decreases. This means that the energy changes become more dramatic for lower Z isobars. This is often cited as a reason why proton-rich, low-Z nuclides tend to decay via positron emission rather than by electron capture. Apparently, positron decay tends to be favored when the energy differences are great. This can be observed (anecdotally) in Table 5.1. ^{99}Rh has the lowest total energy of decay of all the proton-rich nuclides (high N/Z), and it is the only one to decay via electron capture more often than by positron emission. Don't count on this always being true.

As we scan down through the isobars in Table 5.1, they alternate between even-odd (*eo*) and odd-even nuclides (*oe*). With either N or Z odd (mass number is odd), it is not surprising that there is only one stable nuclide for $A = 99$. It turns out this is always true when the mass number is odd. For all sets of known isobars with an odd mass number, there is only one stable nuclide per set of isobars. Check it out for yourself; it is uncanny! Okay, maybe it's not so strange. Perhaps one stable nuclide for each value of A is intuitively expected. It shouldn't be. When A is even, there can be one, two, or three stable nuclides per set of isobars. Check out Table 5.2 for $A = 104$.

When the mass number is even, then the individual isobars must be odd-odd (*oo*) or even-even (*ee*). With both N and Z odd, it is very unlikely a nuclide will be stable. In fact, it'll definitely be unstable for isobars with $A = 104$ (because the exceptions are all low A). On the other hand, an *ee* nuclide would be expected to exhibit greater stability relative to nearby nuclides with odd numbers of N and/or Z. Scan down through Table 5.2 and the N and Z values. Notice that they alternate between *oo* and *ee*. It's as if there are really two sets of isobars in Table 5.2, those with extra (relative) stability (*ee*) and those with extra instability (*oo*).

The dualistic nature of even isobars is even more apparent when a simplified decay diagram is drawn (Figure 5.16). There are clearly two parabolas that converge at higher energy values. Even mass numbered isobars are also approximately modeled by von Weizsäcker's equations; we just need two parabolas instead of one. Notice also that the two stable nuclides in this set of isobars do not have exactly the same energy. Even among stable nuclides, stability is apparently relative. What is the

TABLE 5.2
Isobars with $A = 104$

Isobar	N	Z	N/Z	Decay Mode(s)	Decay Energy (MeV)
$_{49}$In	55	49	1.12	β^+, EC	7.9
$_{48}$Cd	56	48	1.17	EC, β^+	1.14
$_{47}$Ag	57	47	1.21	EC, β^+	4.28
$_{46}$Pd	58	46	1.26	Stable	Stable
$_{45}$Rh	59	45	1.31	β^-, EC	2.44, 1.14
$_{44}$Ru	60	44	1.36	Stable	Stable
$_{43}$Tc	61	43	1.42	β^-	5.60
$_{42}$Mo	62	42	1.48	β^-	2.16
$_{41}$Nb	63	41	1.54	β^-	8.1

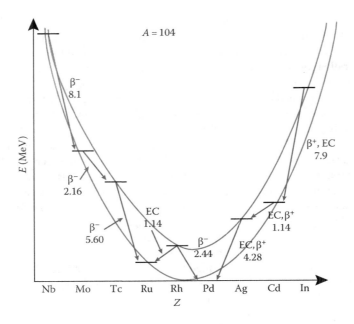

FIGURE 5.16   A simplified decay diagram for some of the $A = 104$ isobars.

optimal N/Z value for $A = 104$? It is between the values for the two stable nuclides: 1.26 (^{104}Pd) and 1.36 (^{104}Ru). It is likely closer to 1.26, as ^{104}Pd is lower in energy than ^{104}Ru. As expected, the optimal N/Z ratio for $A = 104$ is slightly higher than for $A = 99$.

As mentioned earlier, sets of isobars with even mass numbers can include one, two, or three stable nuclides. Generic, simplified decay diagrams illustrate all the observed pathways in Figure 5.17.

There are two ways a set of even isobars can have just one stable nuclide: one with a stable oo nuclide (Figure 5.17a) and the other with a stable ee nuclide (Figure 5.17b). A stable oo nuclide should set off an alarm in your head. Remember there are (only) four stable oo nuclides ($^{2}_{1}$H$_{1}$, $^{6}_{3}$Li$_{3}$, $^{10}_{5}$B$_{5}$, and $^{14}_{7}$N$_{7}$). In every case, they are low Z and are the only known stable nuclide for their set of isobars. The anomalous nature of their stability can be rationalized as due to the narrowness of the valley of stability for low Z nuclides. These nuclides (the oo exceptions) clearly have the best N/Z ratios for their set of isobars, and the adjacent ee nuclides are simply too high (too far up the steep walls of the valley) in energy to be stable.

$A = 104$ is an example of the most popular isobaric decay pathways when A is even. Eighty-three sets of isobars are known with two stable ee nuclides (Figure 5.17c). A close second are 78 sets with one stable ee nuclide (Figure 5.17b), and a distant last are three sets with three stable ee nuclides (Figure 5.17d). Therefore, odds are pretty good that when A is even, there'll be either one or two stable ee nuclides. Close examination of the two and three stable isobar cases illustrated in Figure 5.17c and Figure 5.17d reveals that the stable ee nuclides are always separated by one unstable oo nuclide. Looking over the last couple of pages suggests that two

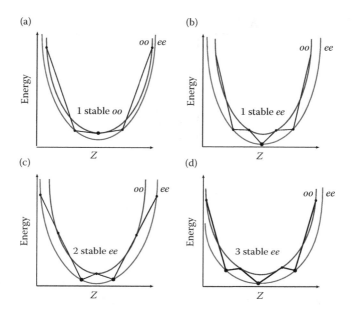

FIGURE 5.17 Simplified decay diagrams representing all possible pathways when mass number (*A*) is even: (a) 1 stable *oo* (b) 1 stable *ee* (c) 2 stable *ee* (d) 3 stable *ee*.

stable isobars will never be adjacent to each other. This is known as the **Mattauch Rule**. It is often stated as "stable, neighboring isobars do not exist." Look at a two-dimensional printed chart of the nuclides, and see if you can find any exceptions.

5.7 ISOMERIC TRANSITIONS

Many nuclear science textbooks define isomers as only those excited nuclear states with measurable lifetimes. This is a rather arbitrary and somewhat inconsistent definition. "Measurable" has changed over the years, and, therefore, what was once considered only an excited state is now considered an isomer. It likely stems from the pre-Internet era when the chart of the nuclides was only available in printed form. It is not possible to show every excited state for every nuclide in a printed format. The 17th (2009) printed edition of KAPL's chart generally provides information only on excited states with half-lives greater than 1 second. To avoid confusion, this chart refers to these as "metastable" states, or simply as "metastates." Recall that this is indicated in the nuclide symbol by adding an "m" at the end of the mass number, for example ^{99m}Tc is a metastate of ^{99}Tc.

We'll use a more inclusive definition for isomer. As stated in Chapter 1, isomers are two nuclides with the same number of protons and neutrons, but they are in different energy states. An **isomeric transition** is, therefore, one isomer changing into another through a nuclear change.

Gamma decay is one type of isomeric transition. We'll have to hold off for a bit before we see the other flavor. In the meantime, let's learn a bit about gamma emission. Since the energy levels of isomeric states are quantized, the energy of the gamma

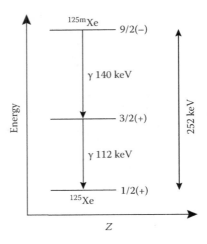

FIGURE 5.18 Isomeric transitions of ^{125m}Xe to ^{125}Xe.

photon emitted is exactly equal to the difference in energy of the two nuclidic states. Like alpha particles, gamma photons are monoenergetic. Unfortunately, many isomeric states are often possible, which can lead to quite a few different gamma photons being emitted as part of a single decay. Also recall that gamma rays have no charge or mass and are strongly penetrating due to their low probability of interacting with matter.

^{125m}Xe decays according to the diagram in Figure 5.18. First, it transitions to another excited state of ^{125}Xe, which is 140 keV in energy below ^{125m}Xe and 112 keV above ground state ^{125}Xe. The second transition is distinct from the first, resulting in the emission of a 112 keV photon. Curiously, a gamma ray with an energy of 252 keV (metastate directly to ground state) is not observed. Why not? It is not "allowed" because of the rather dramatic nuclear changes that would have to accompany such a transition. It has to do with the fractions and the $+$ and $-$ signs written next to each isomeric state in Figure 5.18.

The numbers give the nuclear spin of each state. Each nucleon has a spin of one-half, and when they are combined together in a nucleus, the nucleus will also have a spin. When A is odd, the ground state nuclear spin is always $x/2$, where $x = 1, 2, 3, \ldots$ When A is even, the ground state nuclear spin is 0 or a positive integral value. Nucleons have a tendency to pair up with others of their own kind in the nucleus. When this happens, their spins cancel each other out. As a result, the ground states of most nuclides have spin values that are fairly low. In fact, all *ee* nuclides have a 0 spin in the ground state.

As is apparent in Figure 5.18, different isomers have different spin states. Differences in spin are the main difference between isomers. To transition from one isomer to another, the nucleus needs to change its spin. Generally speaking, the greater the change in spin state, the less likely the transition will take place. For ^{125m}Xe to decay directly to the ground state, the spin state must change from 9/2 to 1/2. Unlikely. Instead it first changes to 3/2, then from 3/2 to 1/2.

The $+$ and $-$ symbols in Figure 5.18 refer to **parity**. Nuclear parity is like your right and left hands; they look the same, but they're really mirror images of each other.

Likewise, a nucleus with + (even) parity looks like a mirror image of an isomer with − (odd) parity. Parity also indicates the direction of the spin; nuclei with even parity spin one way, while those with odd parity spin the other way. Parity plays a role in determining whether a particular isomeric transition can take place, but this is beyond the scope of this text. We will limit ourselves to observing which transitions do not take place and rationalizing them simply based on the magnitude of change in spin.

When the nucleus changes from one isomeric state to another, energy is released. We've already seen that it can be emitted in the form of a high-energy (gamma) photon. In addition to gamma emission, the atom can also kick out one of its electrons to release the energy. This is called **internal conversion**, and this is why this section is titled isomeric transitions rather than gamma decay. When a nucleus transforms from one isomer to another, it has two options to rid itself of the energy it just created.

The emitted electron is called a **conversion electron**. Conversion electrons usually come from the K shell (K int cov), as these electrons have the highest probability of being close to the nucleus. The kinetic energy of the conversion electron is simply the transition energy minus the electron binding energy. The lowest energy excited state for ^{125}Te is 35.5 keV above the ground state. It undergoes an isomeric transition to the ground state according to the probabilities given in Table 5.3.

The transition energy (total energy of decay) is equal to the energy of the gamma photon (35.5 keV) because it ends in the ground state. The energies of the conversion electrons are all lower than the transition energy by an amount equal to the electron binding energy for that shell. Using the numbers in Table 5.3, we can calculate those energies:

K shell electron binding energy = 35.5 keV − 3.7 keV = 31.8 keV

L shell electron binding energy = 35.5 keV − 31.0 keV = 4.5 keV

M shell electron binding energy = 35.5 keV − 34.5 keV = 1.0 keV

The probabilities of each type of emission following the ^{125}Te* → ^{125}Te transition are also presented in Table 5.3. Notice that internal conversion is much more likely than gamma emission. This is partially expected since tellurium has a moderately high atomic number ($Z = 52$). Also note that emission of a K shell electron is most probable, but occasionally electrons from the L and M shell are booted.

The transition used here is rather low in energy. If the transition energy is lower than the binding energy of the electrons, then internal conversion for that shell is not

TABLE 5.3
^{125}Te* Isomeric Transitions

Emission	Probability (%)	Energy (keV)
γ	6.66	35.5
K int cov	80.00	3.7
L int cov	11.42	31.0
M int cov	1.90	34.5

possible. Isomeric transitions that are of greater energy can emit fairly high-energy electrons. Once emitted, they look an awful lot like a beta particle. The only difference between conversion electrons and beta particles is their origin: conversion electrons come from the atom's supply of orbital electrons and result from internal conversion, while beta particles are created in the nucleus by a nuclide undergoing beta decay.

The emission probabilities vary by element *and* by the particular transition. For example, there's another excited state for ^{125}Te that is 145 keV above the ground state. It has a half-life of 57.4 days and is, therefore, a metastable state and can be designated as ^{125m}Te. ^{125m}Te has two options for decay: straight to the ground state, or to the other excited state (the one we've been talking about here and have symbolized as ^{125}Te*). Both of those transitions will have different probabilities for emission than those listed in Table 5.3.

The relative probability of internal conversion to gamma emission is given by the **conversion coefficient** (α_T). It is simply the ratio of ejected electrons to gamma photons:

$$\alpha_T = \frac{I_e}{I_\gamma} \tag{5.4}$$

Instead of counting electrons and photons, we can use the probabilities in Table 5.3 to calculate the conversion coefficient for the ^{125}Te* $\rightarrow$ ^{125}Te transition:

$$\alpha_T = \frac{I_e}{I_\gamma} = \frac{80.00 + 11.42 + 1.90}{6.66} = 14.0$$

This tells us that conversion electrons are 14 times more likely than gamma photons to be emitted in this decay. Additionally, we could calculate the conversion coefficients for the individual electron shells:

$$\alpha_K = \frac{I_e}{I_\gamma} = \frac{80.00}{6.66} = 12.0 \quad \alpha_L = \frac{I_e}{I_\gamma} = \frac{11.42}{6.66} = 1.71 \quad \alpha_M = \frac{I_e}{I_\gamma} = \frac{1.90}{6.66} = 0.285$$

Where α_K is the conversion coefficient for internal conversion of K shell electrons, α_L is for the L shell, and α_M is for the M shell.

5.8   OTHER DECAY MODES

We've covered the major forms of decay: alpha, all three types of beta, and the two types of isomeric transitions. What follows are short descriptions of more unusual decay modes.

5.8.1   SPONTANEOUS FISSION

Spontaneous fission is when a nuclide splits into two large chunks (more or less the same size) and some neutrons without any prompting. This is a significant mode of decay for some of the very heavy nuclides ($A > 230$) and will be discussed in more detail in Chapter 10.

5.8.2 Cluster Decay

Cluster decay happens when very heavy nuclides blow chunks bigger than an alpha particle but not quite big enough to qualify as spontaneous fission. ^{12}C, ^{14}C, ^{20}O, *inter alia* are (very rarely) emitted from large nuclides in a desperate attempt to quickly form a stable nuclide.

Superficially, one might think the spontaneous fission and cluster decay should be much more common than they are. Imagine for a moment that you are a really big nuclide, such as ^{234}U. Stability is a long way away (Pb), and you'd like to get there quickly. Rather than go through the dozen (or so) alpha and beta decays that make up the normal decay chain, ^{234}U can also undergo cluster decay, spitting out ^{28}Mg to form ^{206}Hg:

$$^{234}_{92}U \rightarrow \, ^{206}_{80}Hg + \, ^{28}_{12}Mg$$

^{234}U can also undergo spontaneous fission forming a variety of products, e.g., ^{93}Rb and ^{138}Cs along with three neutrons:

$$^{234}_{92}U \rightarrow \, ^{93}_{37}Rb + \, ^{138}_{55}Cs + 3\, ^{1}_{0}n$$

Yet the branch ratios for ^{234}U defy intuition. It decays via alpha very close to 100% of the time. Spontaneous fission happens only 1.6×10^{-9}% of the time, and cluster decay occurs in roughly 10^{-11}% of all decays. Strange. All three of these decay modes involve basically the same thing, the splitting of the nucleus up into two parts, yet alpha decay is by far the most commonly observed. We'll discuss this in more detail in Chapter 10, but for now we can realize that there must be an energy barrier to all these processes (since they all release energy—do the math!), and this barrier is much more significant for spontaneous fission and cluster decay. A huge advantage that alpha decay has over these two is that it is a relatively small chunk of the nucleus; therefore, it has the smallest barrier to departure. It also forms the doubly magic ^{4}He, which has an unusually high binding energy per nucleon. These two factors strongly favor alpha decay over spitting out larger chunks.

As you might imagine, daughter recoil is a big deal in cluster decay. We can use the same equations (Eqations 5.1 and 5.2) to calculate recoil as we used when we looked at alpha decay.

Example 5.11

^{224}Ra undergoes cluster decay by emitting ^{14}C in 6×10^{-8}% of its decays. Write out a balanced decay equation, and calculate the daughter's recoil energy.

Following Examples 5.1 and 5.2:

$$^{224}_{88}Ra \rightarrow \, ^{210}_{82}Pb + \, ^{14}_{6}C$$

$$[224.020212\,u - (209.984173\,u + 14.003242\,u)] \times \frac{931.5\,\text{MeV}}{u} = 30.55\,\text{MeV}$$

$$K_F = 30.55\,\text{MeV}\left(\frac{14}{210 + 14}\right) = 1.91\,\text{MeV}$$

Notice that recoil is a lot more significant with cluster decay than alpha, just like recoil for a cannon would be larger than for a pistol.

5.8.3 Proton/Neutron Emission

Can particles smaller than an alpha particle be emitted? Certainly, but these types of decay are generally restricted to very proton-rich nuclides (proton emission) and very neutron-rich nuclides (neutron emission). Examples of each are:

$$^{147}_{69}\text{Tm} \rightarrow {}^{146}_{68}\text{Er} + {}^{1}_{1}\text{H}$$

$$^{4}_{1}\text{H} \rightarrow {}^{3}_{1}\text{H} + {}^{1}\text{n}$$

Thulium-147 ($N/Z = 1.13$) can decay by spitting out one of its protons (branch ratio = $\sim$10%), forming erbium-146 ($N/Z = 1.15$, still way too low for this mass number, but better); and hydrogen-4 ($N/Z = 3.00$) ejects a neutron to make tritium (also known as hydrogen-3, $N/Z = 2.00$). As might be expected, these types of decay tend to occur at the fringes of the chart of the nuclides, that is, only when the N/Z ratio is seriously out of whack. While both modes of decay are unusual, proton emission is much more common than neutron emission. This can be understood in terms of Coulomb repulsion. It's easier to kick a proton out of a nucleus because all the protons are pushing each other away. There's not much of a repulsive force acting on neutrons, so they are only rarely ejected as an independent decay mode. In fact, neutron emission is generally observed only for a few highly neutron-rich low Z nuclides.

Multiple protons or neutrons can also be emitted from especially extremist nuclides—those with particularly poor N/Z ratios. These decay modes are exceptionally rare, but an example would be the release of two protons by ^{12}O ($N/Z = 0.50!$) to form ^{10}C ($N/Z = 0.67$):

$$^{12}_{8}\text{O} \rightarrow {}^{10}_{6}\text{C} + 2{}^{1}_{1}\text{H}$$

Once again, recoil will be significant for these decay modes. Equations 5.1 and 5.2 can still be used.

5.8.4 Delayed Particle Emission

Like proton and neutron emissions, **delayed particle emissions** are also sometimes observed for nuclides at the extremes of the valley of stability, that is, up on a high ledge overlooking the valley (what a great view they must have!). In fact, delayed emissions of protons and neutrons are much more common than direct emission of either. The difference is that a delayed particle emission means a proton or neutron is emitted *after* the parent nuclide has already undergone some flavor of beta decay. This tends to happen when the daughter is formed in a particularly high-energy excited state and is an alternative to an isomeric transition to a lower excited state or to the ground state. In this case, an isobar is not formed as the product. For example, almost 7% of all ^{137}I decays emit a beta particle then a neutron (93% of the time it just emits a beta particle). The minor branch (beta plus neutron decay) can be represented as:

$$^{137}_{53}\text{I} \rightarrow\ ^{136}_{54}\text{Xe} + ^{0}_{-1}\text{e} + ^{1}_{0}\text{n} + \bar{\nu}$$

A better representation would include the fleeting formation of the excited daughter, followed by neutron emission from the excited daughter:

$$^{137}_{53}\text{I} \rightarrow\ ^{137}_{54}\text{Xe}^* + ^{0}_{-1}\text{e} + \bar{\nu}$$
$$^{137}_{54}\text{Xe}^* \rightarrow\ ^{136}_{54}\text{Xe} + ^{1}_{0}\text{n}$$

By undergoing beta plus neutron decay, ^{137}I lowers its high N/Z ratio (1.58) two ways, first by converting a neutron to a proton (beta) and then by spitting out another neutron. The end result is formation of a stable, *ee* daughter (^{136}Xe, $N/Z = 1.52$).

Example 5.12

^{19}N undergoes beta plus neutron in approximately 55% of its decays. Assuming a maximum beta energy of 6.12 MeV and a neutron energy of 2.32 MeV, sketch out a decay diagram for this decay mode only.

The balanced decay equation is:

$$^{19}_{7}\text{N} \rightarrow\ ^{18}_{8}\text{O} + ^{0}_{-1}\text{e} + ^{1}\text{n}$$

The total energy of decay is:

$$(19.01702\,\text{u} - (17.999161\,\text{u} + 1.008665\,\text{u})) \times \frac{931.5\,\text{MeV}}{\text{u}} = 8.56\,\text{MeV}$$

Before we draw, we should realize that daughter recoil will be small but significant. We'd better use Equations 5.1 and 5.2 to calculate the recoil for the neutron emission:

$$K_T = 2.32\,\text{MeV} \times \frac{18+1}{18} = 2.45\,\text{MeV} \quad K_F = 2.45\,\text{MeV} \times \frac{1}{18+1} = 0.129\,\text{MeV}$$

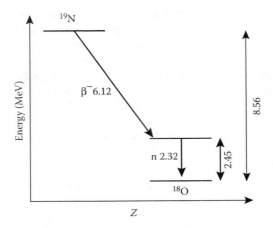

FIGURE 5.19   Beta plus neutron decay of ^{19}N.

The decay diagram is shown in Figure 5.19. Notice the gap left in the neutron emission to indicate daughter recoil.

Like direct emission of these particles, they can also be emitted in bunches. Two, three, or four neutrons can follow beta decay and, as you may have guessed, tend to be emitted in multiples when the N/Z ratio and decay energy are especially high. Likewise, a couple of protons can be emitted following positron decay when the energy of transition is great, and the N/Z ratio is very low. Delayed alpha particles or tritium nuclei can also observed for some nuclides. As we'll see in Chapter 10, emission of delayed neutrons is an important safety feature in nuclear reactors.

5.8.5   Double Beta Decay

Double beta is a rather unusual form of decay and can occur with a neutron-rich, *ee* nuclide. An example is the decay of ^{82}Se to ^{82}Kr by the simultaneous emission of two beta particles. Note that two antineutrinos are also emitted to satisfy the conservation laws discussed earlier in this chapter:

$$^{82}_{34}\text{Se}_{48} \rightarrow\, ^{82}_{36}\text{Kr}_{46} + 2\,^{0}_{-1}\text{e} + 2\bar{\nu}$$

Why does double beta decay happen? ^{82}Se is neutron-rich ($N/Z = 1.41$), so it needs to decay via beta. If that happened, it would form $^{82}_{35}\text{Br}_{47}$ (note it is *oo*). Using the masses of these nuclides, let's calculate the energetics of this putative decay:

$$^{82}_{34}\text{Se} \rightarrow\, ^{82}_{35}\text{Br} + _{-1}\text{e}?$$

$$(81.916699\ \text{u} - 81.916805\ \text{u}) \times \frac{931.5\ \text{MeV}}{\text{u}} = -0.0987\ \text{MeV}$$

This doesn't make sense! The total energy of decay is always positive. Careful examination of the math reveals that it is done correctly. The strange thing is that the mass

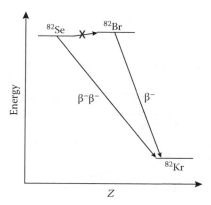

FIGURE 5.20 Simplified decay diagram for ^{82}Se and ^{82}Br.

of the daughter (^{82}Br) is *greater than* the mass of the parent (^{82}Se). Mass should be lost during a decay process, producing energy—not the other way around. In other words, ^{82}Se *cannot* undergo beta decay to ^{82}Br. ^{82}Se is still neutron-rich and must undergo decay, so it emits two beta particles at once and decays directly to ^{82}Kr.

The energetics for the decays of ^{82}Se and ^{82}Br are illustrated in a simplified decay diagram in Figure 5.20. If more $A = 82$ isobars were included, this figure would look like Figure 5.17b with ^{82}Kr as the lone, stable isobar. Notice that ^{82}Br lies higher in energy (has more mass) than ^{82}Se, making that decay pathway impossible.

Nuclides observed to undergo double beta decay typically have very long half-lives. For example, the half-life for ^{82}Se is 9×10^{19} years. The long half-lives are attributed to the fact that creating two particles and two antiparticles all at once in a nucleus is a difficult thing and, therefore, of low probability.

At the time of this writing, double beta decay has been observed for 10 neutron-rich nuclides: ^{48}Ca, ^{76}Ge, ^{82}Se, ^{96}Zr, ^{100}Mo, ^{116}Cd, ^{128}Te, ^{130}Te, ^{150}Nd, and ^{238}U. However, equivalent processes for proton-rich nuclides have only been observed in ^{130}Ba. Theoretically, double positron, electron capture/positron, and double electron capture decays are all possible for certain proton-rich *ee* nuclides. ^{130}Ba decays via double electron capture with a half-life of 2×10^{21} a:

$$^{130}_{56}\mathrm{Ba} + 2\,_{-1}\mathrm{e} \rightarrow\, ^{130}_{54}\mathrm{Xe} + 2\nu$$

QUESTIONS

5.1 Define the following: recoil energy, antineutrino, characteristic X-ray, conversion coefficient, and fluorescence yield.

5.2 If the total energy of decay for ^{245}Cm is 5.623 MeV, what is the recoil energy of the daughter of the decay?

5.3 Calculate the atomic mass of ^{240}U from the primary mode of decay for ^{244}Pu. Write out the decay equation, and use 4.589 MeV as the kinetic energy of the most energetic alpha particle possible for this decay. What is the recoil energy of ^{240}U?

5.4 Using information from Appendix A, draw a decay diagram for ^{268}Mt. Calculate the recoil energy of the daughter. For whom are the parent and daughter elements named?

5.5 List and briefly explain the three conservation laws that are satisfied in beta decay. What must also be emitted for these to hold true?

5.6 Calculate the atomic mass of ^{198}Au using only the information given here. It decays via beta, the atomic mass of its daughter is 197.966769 u, and the total energy of decay is 1.372 MeV.

5.7 The most common beta particles observed in the decay of ^{59}Fe have E_{max} values of 0.466 and 0.274 MeV. The two most prominent gamma rays have energies of 1099 and 1292 keV. If the total energy of decay is 1.565 MeV, and the atomic mass of the daughter is 58.933195 u, what is the atomic mass of ^{59}Fe? Draw a decay diagram consistent with the information given here.

5.8 Explain why ^{37}Ar might be expected to undergo positron decay but doesn't.

5.9 ^{54}Mn (53.940363 u) decays to ^{54}Cr (53.938880 u) via electron capture. In almost every decay, a gamma photon with an energy of 835 keV is emitted. Write out the decay equation, and draw a decay diagram using only the information given here. Give two possible reasons why a gamma photon is *not* seen in a very small percentage (<0.1%) of these decays.

5.10 The binding energies for electrons in element X are given in Table 5.4. Calculate the energy of the $M_2 \rightarrow K$ X-ray, and the $M_2M_3M_5$ Auger electron. What is the energy of the KL_2L_3 Auger electron? What kind of Auger electron would have an energy of 12.315 keV?

5.11 ^{64}Cu undergoes all three forms of beta decay. Draw a decay diagram consistent with the information given here. The total energy of decay for β^- is 0.579 MeV, and for β^+ and EC, it is 1.675 MeV. The only beta particle observed has an E_{max} of 0.578 MeV, and the only positron has an E_{max} of 0.651 MeV. A 1346 keV gamma photon is observed but only in a small fraction of the electron capture decays.

TABLE 5.4
Electron Binding Energies for Element X

Subshell	Binding Energy (keV)
M_5	3.332
M_4	3.491
M_3	4.046
M_2	4.831
M_1	5.182
L_3	16.300
L_2	19.693
L_1	20.472
K	109.651

5.12 A nuclide decays giving off two different positrons and a single gamma ray of 677 keV. If the total energy of decay is 6.138 MeV, what are the likely energies of the two positrons emitted?

5.13 Draw the possible decay diagram for the nuclide in the previous problem.

5.14 Calculate the energy resulting from the reaction of 0.01 moles of positrons with an equal number of electrons.

5.15 The photon torpedoes in *Star Trek* use antimatter as an explosive. Calculate the mass of antimatter necessary to equal the explosive yield of the bomb dropped on Hiroshima. Comment on whether these weapons could someday become a reality. Why is the word "photon" particularly appropriate to these weapons?

5.16 Complete the first four (empty) columns in Table 5.5 without any additional information, and then check your work using Appendix A. Use the appendix to fill in the final column. Finally, draw a simplified decay diagram for this set of isobars, showing only the major mode of decay between nuclides.

5.17 Using the table from the previous problem, what are some conclusions that can be made about a set of isobars?

5.18 Draw a simplified decay diagram for $A = 168$ from tantalum to terbium showing only the major mode of decay between nuclides.

5.19 Why is Mattauch's Rule always true?

5.20 Explain the m in ^{60m}Co. What is meant by the fact that its I_e/I_γ ratio is 41?

5.21 Using only the information given in Appendix A, produce your best guess at a complete decay diagram for ^{97m}Tc.

5.22 94.40% of all ^{137}Cs decays result in the formation of ^{137m}Ba, while the remaining 5.60% produce ground state ^{137}Ba directly. If only 85.1% of all ^{137}Cs decays produce a gamma photon (the remainder produce a conversion electron), what is the conversion coefficient for ^{137m}Ba decay?

5.23 What is the energy of a conversion electron produced from the K shell during the decay of ^{137}Cs?

5.24 What are the differences between conversion coefficient and fluorescence yield?

TABLE 5.5
Data for Some $A = 25$ Isobars

Isobar	N	Z	N/Z	Mode of Decay	Atomic Mass (u)
$^{25}_{9}$F					
$^{25}_{10}$Ne					
$^{25}_{11}$Na					
$^{25}_{12}$Mg					
$^{25}_{13}$Al					
$^{25}_{14}$Si					

5.25 Using mass differences, calculate the amount of energy released by all three decay modes given for ^{234}U in Section 5.8.2. Use the fission equation given in this chapter. Why is ^{234}U naturally occurring?

5.26 Calculate the kinetic energies of the daughter and the emitted nucleus in the cluster decay of ^{234}U.

5.27 Using information in Appendix A, draw a decay diagram for ^{5}He, and estimate the kinetic energies of the products.

5.28 ^{14}B can decay via beta with a delayed neutron. Using information in Appendix A, write out a balanced decay equation, and calculate the total energy of decay for this pathway.

5.29 Using Appendix A, draw a decay diagram for ^{130}Te.

5.30 Explain why double beta decay is only observed for nuclides with an even mass number.

5.31 Sketch out a simplified decay diagram for the $A = 130$ isobars from ^{130}Sn to ^{130}Ce. Note anything unusual.

6 Interactions of Ionizing Radiation with Matter

Most of the previous chapters have established a firm understanding of radioactive decay. This chapter now turns to the ramifications of ionizing radiation. Understanding how ionizing radiation interacts with matter is essential to understanding how it can be detected, shielded, and how it affects living systems. We'll start with a more detailed definition of ionizing radiation and then look at the various ways energetic particles and photons can interact with matter.

6.1 IONIZING RADIATION

Ionizing radiation refers to radiation with enough energy to knock an electron loose from an atom. Alpha particles, beta particles, positrons, gamma rays, and X-rays are all forms of ionizing radiation. Remember that X-rays and gamma rays are located on the high-energy end of the electromagnetic spectrum (Section 1.1) and that the difference between the two is how they are generated. As explained in Section 5.4, X-rays come from electrons (slowing down or transitioning between certain energy levels), and gamma rays from nuclei (IT).

Just below X-rays and gamma rays on the electromagnetic spectrum is UV light. The line between nonionizing UV and ionizing X- and γ-rays is somewhat fuzzy but is generally drawn at 100 eV. A threshold of 100 eV for ionization may seem artificially high, especially since the electron binding energy of the outermost electrons (valence) for most atoms is less than 15 eV, and most chemical bonds (sharing valence electrons) have energies of 1–5 eV. It seems like it should require much less than 100 eV to knock electrons loose from atoms and molecules, but we know that UV is not ionizing. As we will see in this chapter, only a fraction of the photon's (or particle's) energy can be transferred to an electron in an atom or molecule, and a minimum of 100 eV is necessary to consistently knock 'em loose.

6.2 CHARGED PARTICLES

Interactions between charged particles (e.g., alpha or beta particles) and matter can be thought of as the transfer of energy from the particle to the matter. As a particle travels through matter, energy is transferred from it until it has about the same energy as the surrounding matter. It might be strange to think of, but all matter has energy, unless it is at absolute zero! As the particle transfers its energy, it slows down—its energy is kinetic energy and kinetic energy is equal to one-half times mass times velocity squared:

$$\text{KE} = \frac{1}{2}mv^2 \tag{6.1}$$

As kinetic energy decreases, then velocity must also decrease, since mass remains constant. These interactions can also be thought of as collisions between the energetic particles and atoms, although it is not strictly correct. Even so, collisions are often used to describe the transfer of energy from particle to atom. Perhaps it is because that is how we often observe energy transfer in our macroscopic world (e.g., a bowling ball hitting the pins, a billiard ball hitting other billiard balls, etc.). Please keep in mind that the subatomic world is a very different place. Solids and liquids may appear to completely occupy their space, but on the atomic level, they are mostly empty space. An alpha particle traveling through matter is a bit like coasting on a bicycle into a flock of small birds. Each impact on a bird causes the bike rider to slow down a little bit as kinetic energy is transferred from the bicyclist to a bird. If the bicycle and the rider represent an alpha particle, then the birds represent electrons. Interactions with electrons are much more likely than an interaction with the nucleus, as the electrons roam most of the space inside an atom. If the bike rider is especially unlucky, it will strike an object inside the flock with a mass greater than or equal to it (the nucleus). If we continue to let the bike and the rider represent an alpha particle, then the mass of two SUVs would represent a gold nucleus (ouchies!).

An **elastic** collision is one where kinetic energy of the particle is cleanly transferred as kinetic energy to the electron. The total kinetic energy of the particle and the electron it hits remains the same. An **inelastic** collision is one where some of the particle's kinetic energy is converted to some other form of energy (such as an electron moving to a higher shell). Inelastic collisions are very common between a large charged particle (proton, alpha, etc.) and an electron. When the electron is promoted to a higher energy level, the interaction is an **excitation**. As shown in Figure 6.1a, this process is reversible—the electron will eventually (usually right away) de-excite, emitting electromagnetic radiation such as visible light or X-rays.

If the amount of energy transferred by the particle is greater than the electron's binding energy, the electron is removed from the atom (Figure 6.1b). This is called **ionization** because it results in the formation of a cation and a free electron (an **ion pair**). The creation of an ion pair by ionizing radiation is called a primary ionization. This distinguishes it from **secondary ionizations**, which can subsequently be caused by the freed electron. Because the freed electron can have a significant amount of kinetic energy, it can cause its own ionizations. The freed electron is often called a **secondary electron**, or, more collectively, as **delta rays**. The latter term is somewhat archaic and somewhat misleading, as the term "rays" refers to photons, not particles. As indicated in Figure 6.1b, ionizations are also reversible, although it is rarely the same electron that returns to the cation re-forming the neutral atom.

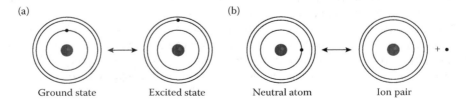

(a)   Ground state   Excited state   (b)   Neutral atom   Ion pair   + •

FIGURE 6.1   (a) Electron excitation and (b) atomic ionization.

Excitations and ionizations are two very common results of the interaction of ionizing radiation with matter. What are their relative odds? Electron excitations are roughly two times more common than ionizations in air.

Example 6.1

A 7.6869 MeV α particle emitted during the decay of ^{214}Po undergoes a collision with an M shell electron in an atom of holmium. If the α particle now has only 7.6830 MeV, will the electron be removed from the atom, and if so, how much kinetic energy will it have?

Remember from Section 5.4 that the electron binding energy is the amount of energy required to completely remove an electron from an atom. Thankfully, the energy of an M shell electron in holmium is given in Figure 5.10 (1.71 keV). The energy of the electron booted out of Ho will be the energy transferred minus the binding energy.
 The energy transferred is:

$$7.6869 \text{ MeV} - 7.6830 \text{ MeV} = 0.0039 \text{ MeV} = 3.9 \text{ keV}$$

This is greater than the binding energy of an M-shell electron in Ho, so the atom will be ionized. The energy of the freed (secondary) electron is:

$$3.9 \text{ keV} - 1.71 \text{ keV} = 2.2 \text{ keV}$$

Note that the electron has very little energy because it appears that very little energy is transferred. In fact, the amount transferred in this problem represents the maximum amount of energy that could be transferred. Remember that the alpha particle is very massive (you and your bike) compared to an electron (a small bird). At most, an alpha particle can transfer only ~0.05% of its total energy to an electron. This is also true with you and the bird—colliding with a bird isn't going to slow you down very much.
 Let's take a closer look at the relationship between kinetic energy and velocity. Earlier in this chapter the following facile formula was quoted (Equation 6.1):

$$\text{KE} = \tfrac{1}{2}mv^2$$

Applying this relationship to the alpha particle in Example 6.1 *should* allow calculation of its velocity. A couple of conversion factors are necessary to get conventional velocity units. Kinetic energy needs to be in units of joules (J). Remember that a joule is the same as a kilogram meter squared per second squared:

$$J = \frac{\text{kg} \cdot \text{m}^2}{\text{s}^2}$$

Converting the energy of the alpha particle emitted by ^{214}Po from MeV to J:

$$7.6869 \, \text{MeV} \times \frac{10^6 \, \text{eV}}{\text{MeV}} \times \frac{1.602 \times 10^{-19} \, \text{J}}{\text{eV}} = 1.231 \times 10^{-12} \, \text{J}$$

Since energy has units of joules, the mass of the alpha particle (electrons not included!) needs to be expressed in kilograms:

$$4.0015 \, u \times \frac{1.66054 \times 10^{-27} \, kg}{u} = 6.6446 \times 10^{-27} \, kg$$

Plugging energy and mass into Equation 6.1:

$$1.231 \times 10^{-12} \, J = \frac{1}{2}(6.6446 \times 10^{-27} \, kg) \times v^2$$
$$v = 1.925 \times 10^7 \, m/s$$

That looks like a pretty zippy little particle! The speed of light is 2.998×10^8 m/s, and no particle (matter) may travel at or above the speed of light. Our alpha particle isn't all that far from this ultimate speed limit. According to Einstein's theory of relativity, increasing amounts of energy are required to continue accelerating a particle whose velocity is approaching the speed of light. More and more energy is required for smaller and smaller increases in velocity, and the simple relationship between kinetic energy and velocity (Equation 6.1—often referred to as the "classical equation") no longer holds true. In this case, it might be a good idea to calculate velocity using the more complex "relativistic equation" (Equation 6.2):

$$KE = \left(\frac{1}{\sqrt{1 - \dfrac{v^2}{c^2}}} - 1 \right) mc^2 \qquad (6.2)$$

where c is the speed of light. This equation is only necessary for particles moving near light speed. As written, it can be a little awkward to solve for velocity. Solving for velocity (not trivial!),[1] we get:

$$v = c\sqrt{1 - \left(\frac{mc^2}{KE + mc^2} \right)^2} \qquad (6.3)$$

If our alpha particle's velocity is any different using this equation, we'll know that relativity needs to be taken into account. If it ends up with the same velocity as the classical equation, we'll know that relativity doesn't apply here and that we've been wasting our time. Let's see!

$$v = 2.998 \times 10^8 \, \frac{m}{s} \times \sqrt{1 - \left(\frac{(6.6445 \times 10^{-27} \, kg) \times \left(2.998 \times 10^8 \, \frac{m}{s}\right)^2}{1.231 \times 10^{-12} \, J + (6.6445 \times 10^{-27} \, kg) \times \left(2.998 \times 10^8 \, \frac{m}{s}\right)^2} \right)^2}$$

$$v = 1.922 \times 10^7 \, \frac{m}{s}$$

---

[1] My thanks to Dr. Craig Tainter for this derivation—JCB.

Once again, it is important to use compatible units. Try the math yourself—there are a lot of buttons to press on this one, and it's important to get them in the right order. The result for velocity is nearly the same as with the classical equation. Had we not carried everything out to at least four significant figures, we might not have seen any difference. Despite the small difference, using the relativistic equation is the correct approach for this problem—because of the high level of precision. If the velocity were higher, or if the problem involved a less massive particle (like a proton or an electron), the difference in velocities calculated by the classical and relativistic equations would be more dramatic.

Velocities near the speed of light are often expressed as a fraction of the speed of light (v/c). For the alpha particle from ^{214}Po, this fraction would be:

$$\frac{v}{c} = \frac{1.922 \times 10^7 \, \frac{m}{s}}{2.998 \times 10^8 \, \frac{m}{s}} = 0.06412$$

Therefore, the alpha particle could be said to be traveling at 6.412% of the speed of light. Despite our earlier amazement at its velocity, this is not really all that fast for a subatomic particle. Notice also that calculating v/c marginally simplifies the relativistic equation (Equation 6.3):

$$\frac{v}{c} = \sqrt{1 - \left(\frac{mc^2}{KE + mc^2} \right)^2} \tag{6.4}$$

The faster a particle is moving, the more energy it has to deposit in the matter it travels through. If the particle is traveling through air, an average of 33.85 eV is required for each ionization of an air molecule. Remember that air is ~78% nitrogen (N_2) and ~21% oxygen (O_2), so 33.85 eV/IP is reflective of this particular gas mixture. The average energy required to produce an ion pair in any medium is called its **W-quantity**. This energy cost includes the electron binding energy and is, therefore, the total energy lost by the ionizing radiation as it travels through the medium and is not the energy of the secondary electron.

As a charged particle travels through matter, it creates a certain number of ion pairs per unit length. This is its **specific ionization** (SI), which depends on the energy of the particle and the nature (density) of the matter it is traveling through. The specific ionization of alpha particles traveling through air typically varies from 3 to 7 million ion pairs per meter (IP/m)—that's a lot of ions!

If we know how many ion pairs a charged particle creates per unit length (SI), and we know the average cost of creating each ion pair (W-quantity), we can also determine the average energy lost by the charged particle per unit length. This is called the particle's **linear energy transfer** (LET), and it is obtained by multiplying the specific ionization by the W-quantity:

$$LET = SI \times W \tag{6.5}$$

Example 6.2

Alpha particles of a certain energy have a specific ionization of 5.05×10^6 IP/m. Calculate the LET value for these particles in air.

Since $W = 33.85$ eV/IP in air:

$$LET = \frac{5.05 \times 10^6 \, IP}{m} \times \frac{33.85 \, ev}{IP} = 1.71 \times 10^8 \frac{eV}{m} = 171 \frac{MeV}{m}$$

Wow! That's a lot of energy for every meter traveled in air. The alpha particle will need to be very energetic to travel a full meter, or it will only travel some fraction of a meter before it runs out of gas. **Range** is defined as the average distance a charged particle will travel before being stopped. For alpha and other "heavy," charged particles (nuclei), range can be calculated by dividing the energy of the alpha particle by its average linear energy transfer (LET). Caution: This equation only applies to alpha particles and other nuclei (not electrons!):

$$\text{Range} = R = \frac{E}{LET} \tag{6.6}$$

Remember that specific ionization varies with the energy of the particle, so the amount of energy transferred per unit length (LET) will also vary. Since particles slow down (lose energy) as they travel through matter, their SI and LET values will also vary. Therefore, many calculations, such as Equation 6.6 need to be done with SI and LET values that are based on averages for the particle's entire journey through matter.

Alpha particle range (cm) in air can also be estimated solely from its energy using the following equation:

$$R_{\alpha\text{-air}} \cong 0.31 E_\alpha^{3/2} \tag{6.7}$$

Example 6.3

^{241}Am is used in smoke detectors because the LET of air is significantly lower than the LET of smoky air. Estimate the range and the average SI and LET in air for the 5.4857 MeV alpha particles emitted by ^{241}Am.

$$R_{\alpha\text{-air}} \cong 0.31 E_\alpha^{3/2} = 0.31 \times (5.4857 \, MeV)^{3/2} = 4.0 \, cm$$

$$LET = \frac{E}{R} = \frac{5.4857 \, MeV}{4.0 \, cm} = 1.4 \frac{MeV}{cm}$$

$$SI = \frac{LET}{W} = \frac{1.4 \dfrac{MeV}{cm}}{33.85 \dfrac{eV}{IP} \times \dfrac{MeV}{10^6 \, eV}} = 4.1 \times 10^4 \frac{IP}{cm}$$

These alpha particles are completely stopped by only 4 centimeters of air! The smoke detector works by detecting the alpha particles after they've passed through ∼1–2 cm of air. If smoke gets between the [241]Am source and the detector, they absorb more of the alpha particles; the detector does not see as many alpha particles per second and sounds the alarm. [241]Am is a pretty good source for this application as it has a 433 year half-life, so its activity is unlikely to change over the lifetime of the detector, yet it is short enough to only require a small amount (∼1 µCi).

As mentioned previously, specific ionization and linear energy transfer vary with the energy of the particle. As a particle moves through matter, it loses energy (slows down). As it loses energy, it transfers increasing amounts of energy and creates more ion pairs per centimeter traveled. Figure 6.2 generically illustrates this phenomenon for a charged particle moving through matter.

Notice that the specific ionization does not change much initially and then increases dramatically after it has penetrated some distance into the matter. This means that the particle transfers energy to the matter at a relatively low and constant rate until it reaches some low-energy threshold. After that point, it begins to transfer energy to the matter at ever-increasing rates until, suddenly, it is out of gas and stops.

The behavior illustrated in Figure 6.2 is common to all charged particles (alpha, beta, protons) in all media (air, water, humans, etc.). The peak observed in Figure 6.2 is called the **Bragg peak** and represents the fact that charged particles deposit much of their energy after penetrating matter some distance. One way to think about this is in terms of the probability of interaction. While the particle is moving quickly (higher energy), it spends only a very short time by each atom it passes, decreasing the probability of interaction. At some point, it is traveling slowly enough (lower energy) that the time spent near an atom increases the probability of interaction. From that point forward, the probability of interaction increases dramatically as the particle continues to decelerate. It's a bit like skipping a flat rock over a still lake. The first couple of skips are far apart, and relatively little energy is transferred per unit distance. When the rock slows, it starts to skip often over short distances, transferring a fair bit of energy over a relatively short distance before coming to a stop and sinking into the lake.

What happens when the particle stops? If it's an alpha particle, it will steal a couple electrons and form a helium-4 atom. If it's a positron, it'll annihilate with an

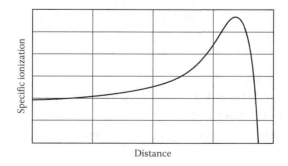

Distance

FIGURE 6.2 Generic representation of the change in specific ionization as a charged particle penetrates matter.

electron, etc. Keep in mind that the particle's energy does not go to zero, so it doesn't really "stop"—this is only possible at 0 K. The particle will have translational and rotational energy that reflect the temperature of the matter it is now part of. Instead of thinking that the particle comes to a full stop, it is better to understand that it transfers energy to its surroundings until its own energy matches that of its environment.

So far, we've focused on the interactions of charged particles with air. What about their interactions with denser media, such as liquids and solids? Solids and liquids have a lot more matter packed into less space. Therefore, they have greater densities, and the probability of interaction per centimeter traveled will increase dramatically. Specific ionization will increase (more energy deposited per cm), and range will decrease. One way to compare the shielding effectiveness of different materials is to determine relative stopping power (RSP):

$$RSP = \frac{R_{air}}{R_{abs}} \tag{6.8}$$

where R_{air} is the range in air, and R_{abs} is the range in another *absorbing* material. Larger values for RSP mean that that material is more strongly absorbing (smaller R_{abs}). Table 6.1 provides some RSP and range values for a 7.0 MeV alpha particle in different media.

As the density increases, so does relative stopping power, and the range drops dramatically. The bottom line here is that alpha particles are not very penetrating— something we already knew to be true, but now we've got some serious numbers to back up the claim. Since we are made up mostly of water, the numbers in the water row can be approximated for us. Skin averages between 1 and 2 mm thick (that's 0.1–0.2 cm), so we can feel pretty safe around alpha radiation. So long as it remains external to our bodies, the outer layers of our skin will stop alpha radiation.

Much of the discussion in this section has focused on large charged particles, such as alpha particles, protons, and other nuclei. What about smaller charged particles such as electrons? They could be beta particles, Auger, or conversion electrons emitted during nuclear decay, or they could be from an artificially produced electron beam. We should also consider positrons here as they have the same mass and unit charge. As a result, they tend to interact with matter in much the same way as electrons.

Like the heavier charged particles, electrons can cause excitations and ionizations as they pass through matter. Additionally, the probability of interaction increases as

TABLE 6.1
RSP and Range Values for a 7.0 MeV Alpha Particle in Various Media

Absorber	Density (g/cm³)	RSP	Range (cm)
Air	0.0012	1	6.1
Water	1.0	970	0.0063
Aluminum	2.7	1700	0.0035

TABLE 6.2

Comparisons of LET and Range Values for an Alpha Particle, a Proton, and an Electron

Particle	LET (water, MeV/cm)	Range (air, cm)	Range (water, cm)
10 MeV alpha	920	10	0.011
10 MeV proton	83	120	0.12
10 MeV electron	1.9	4200	5.2

the velocity (KE) of the **incident electron** decreases. The incident electron is the electron that is transferring the radiation to the matter, that is, the radiation, the one that's causing the whole *incident* (ionizations) to take place. The nature of the interaction and the energy of any secondary electrons produced from these interactions depends on how much energy is transferred from the incident electron.

The obvious difference between electrons and larger, charged particles is mass. Electrons are thousands of times less massive than even the smallest nuclei. This has two major implications. First, if an electron and a proton have the same energy, the electron will be moving a whole lot faster. Electrons will, therefore, require use of the relativistic kinetic energy equation at much lower energy values (>3 keV) than protons (>6 MeV) or alpha particles (>25 MeV).[2] Since they are moving faster, they will spend less time passing atoms and, therefore, have a lower probability of interaction. This means that electrons will be more penetrating than alpha particles or protons with equivalent energy values. Lower probabilities of interaction mean fewer ion pairs created per centimeter and longer ranges, as illustrated in Table 6.2.

Specific ionization values for *electrons* in air can be calculated if their velocity (energy!) is known using Equation 6.9:

$$SI = \frac{4500\,IP/m}{v^2/c^2} \tag{6.9}$$

where v is the electron's velocity, and c is the speed of light. Note that this equation works only for electrons and positrons in air.

Example 6.4

Calculate the SI and LET in air for the most probable energy beta particle emitted by ^{32}P.

Remember that beta particles share their energy with the antineutrino, and that the most probable energy is $0.3E_{max}$ (see Figure 5.3). For ^{32}P, $E_{max} = 1.709$ MeV, and therefore $0.3E_{max} = 0.51$ MeV (assuming two significant figures). Using the relativistic equation (since beta energy > 3 keV), $v/c = 0.87$ for a 0.51 MeV electron (check it yourself!).

---

[2] The energy thresholds given here are when relativistic values begin to differ from classical by more than 0.5%. Threshold values will vary if a different level of precision is used.

$$SI = \frac{4500\,\text{IP/m}}{(0.87)^2} = 6000\,\text{IP/m} = 60\,\text{IP/m}$$

$$LET = SI \times W = \frac{6000\,\text{IP}}{\text{m}} \times \frac{33.85\,\text{eV}}{\text{IP}} \times \frac{\text{MeV}}{10^6\,\text{eV}} = 0.20\,\text{MeV/m}$$

Note that the same average energy per ion pair (W) is used here as was used for alpha particles. The source of energy doesn't matter; the average energy it takes to ionize air molecules remains the same. Note also that a lower energy beta particle would have a higher SI and LET. Lower energy means the particle is moving slower, has a higher probability for interaction, is less penetrating, and creates more ion pairs per meter.

The second major difference between electrons and nuclei as projectiles is that electrons are much more likely to be deflected (scattered) as they pass through matter. If most of the interactions between a charged particle and matter involve the electrons, this makes perfect sense. If alpha particles are like a human cyclist coasting into a flock of birds, then beta particles are simply a (high-speed) bird flying into a flock of equally sized birds. Both are slowed by collisions with the flock, but the cyclist's direction is unlikely to change much, but the bird will likely bounce around like a cue ball, slamming hard into a bunch of other billiard balls. Because of scattering, proper calculation of the average range for a particular energy beta particle is complex; however, electron range in human tissue (R_{e^-}) can be estimated by dividing the electron energy (in MeV) by 2.0 MeV/cm (Equation 6.10). This relationship becomes more important when we discuss electron beam therapy in Chapter 14.

$$R_{e^-} = \frac{E}{2.0\,\text{MeV/cm}} \tag{6.10}$$

Scattering of incident electrons doesn't really change the interactions from the perspective of the matter the particle is traveling through. Electrons that are part of the matter will still experience excitations or ionizations as energy is transferred from the incident electron to the matter. The big difference is that electron projectiles can also be scattered by flying near a nucleus. The most important such interaction is called **inelastic scattering by nuclei**. If the projectile electron passes close enough to a nucleus, it will feel an attractive (positrons will be repelled) Coulomb force. The electron slows and curves around the nucleus, as shown in Figure 6.3.

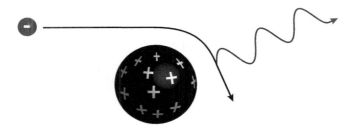

FIGURE 6.3   Inelastic scattering of an electron by a nucleus.

Because the electron slows down (loses kinetic energy) and energy needs to be conserved, the lost energy is emitted as a photon (electromagnetic radiation—the squiggly line in Figure 6.3), typically an X-ray. The level of interaction varies depending on how close the electron travels to the nucleus and how many protons (Z) are in the nucleus. If the electron passes very close to the nucleus, it will be scattered through a higher angle, slow more, and emit a higher energy photon than if it were to pass further away. An infinite number of these kinds of interactions are possible, as the distance between the electron and nucleus can be varied by infinitesimally small increments. The result is that a wide variety of scattering angles can occur corresponding to a wide variety of energies for the emitted X-rays. An energy spectrum of X-rays produced from the beta particles emitted during the decay of ^{90}Sr and ^{90}Y interacting with aluminum is illustrated in Figure 6.4.

X-rays produced by inelastic scattering of electrons by nuclei are often referred to as **bremsstrahlung**, which is a German word roughly translated as "braking radiation." Because they have a broad energy distribution, they are also referred to as **continuous X-rays**. Notice in Figure 6.4 that not all energies have equal probabilities. In general, the probability decreases as the energy increases. This is because lower-energy X-rays are produced when the electron is further from the nucleus, and there are more points in space at a greater distance. Prove this to yourself by drawing two concentric circles. If the nucleus is at the center, and each point on the larger circle represents the probability of the lower energy interaction, and each point on the smaller circle represents the probability of the higher-energy interaction, which is more likely? In other words, which circle has more points? The larger circle, of course! This means that there are more points further away from the nucleus that will produce a lower-energy X-ray.

Now, let's mess with the matter the incident electron is traversing. The effect of increasing the atomic number (Z) of the matter the electron is traveling through can be thought of in two ways. As mentioned previously, the intensity of the interaction will increase with Z at the same distance from the nucleus. This is simply due to the

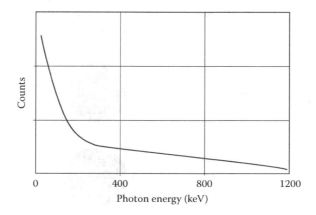

FIGURE 6.4   X-ray spectrum produced by beta emissions from ^{90}Sr and ^{90}Y interacting with Al.

increased positive charge of the nucleus. The other effect of increasing Z is that the nucleus will have greater reach. A larger positive charge means that the electron can interact with it from further away. Overall, as Z increases, the probability of inelastic scattering also increases. This should make intuitive sense; as atomic number increases, so (generally) does density. Denser matter makes better shielding because ionizing radiation has more opportunities to interact as it is trying to pass through.

Elastic scattering by nuclei happens when the projectile electron hits the nucleus dead on and kinetic energy is conserved. No excitations or ionizations occur, at least not initially, and no photons are created. This type of scattering is much less probable than inelastic scattering. The nucleus is really small, and Coulomb attraction can act over reasonably long distances. As illustrated in Figure 6.5a, we can think of the electron initially in motion, heading right toward a nucleus, which is not moving. The collision itself is too violent to be depicted here, but Figure 6.5b shows the aftermath. The electron has reversed direction and has lost some velocity (the arrow is smaller), and the nucleus is now moving very slowly in the opposite direction. Only some of the electron's kinetic energy is transferred to the nucleus, and, because the nucleus is so much more massive, it has very little velocity. Elastic scattering of electrons by nuclei is also known as **backscattering** because the electron ends up going back along the path it came from.

Cherenkov radiation is an obscure form of radiation that results from particles (typically electrons) traveling from one medium into another. The speed of light depends (slightly!) on the medium the particle or photon is traveling though. For example, light travels 1.33 times slower in water than in a vacuum. This means that beta particles traveling near the speed of light in one medium that pass into another medium, where light travels more slowly, could now be exceeding the ultimate speed limit. Since no object may move faster than light, the electron must slow down. In doing so, it must release energy. This effect isn't really important, but it explains why nuclear reactor cores (Chapter 11) have a cool blue glow. High-energy beta particles are emitted by some of the products formed by the fission reactions and travel from the solid fuel pellet, through the solid fuel rod, and into the liquid (usually water) coolant. The energy released by the speeding electrons into the water is converted into blue light. Cherenkov radiation can also be observed by placing higher-energy, beta-emitting nuclides such as ^{90}Y in water and turning out the lights.

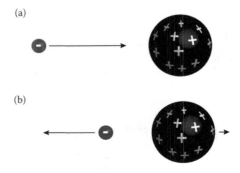

FIGURE 6.5 Elastic scattering of an electron by a nucleus: (a) prior to scattering (b) after scattering.

6.3 PHOTONS

The transfer of energy from X-ray and gamma ray photons to the matter they traverse is a bit more complicated than it is from particles. Nine different interactions are possible, but we will only look at the three that are generally the most probable and have the most significance for medical applications: Compton scattering, the photoelectric effect, and pair production. These three interactions all transfer energy from high-energy photons to matter through fairly complex mechanisms. We'll discuss them in fairly simple terms, but please keep in mind that they are not as simple as they are described here.

6.3.1 COMPTON SCATTERING

Compton scattering can be imagined as a collision between an incident (incoming) photon and an unsuspecting outer shell (valence) electron in an atom of the matter the photon is travelling through. This is represented in Figure 6.6. In reality, it is better to think of it as a transfer of some energy from the incident photon to a valence electron. There's not much electron binding energy to overcome here (valence electrons are typically bound by less than 50 eV), so the electron gets booted away from its atom. The ejected electron is called the Compton (or recoil)[3] electron. Energy is conserved in this process; therefore, the energy of the incident photon (E_0) equals the energy of the scattered photon (E_{SC}) plus the energy of the Compton electron (E_{CE}) plus the binding energy of the electron (B.E.):

$$E_0 = E_{SC} + E_{CE} + \text{B.E.}$$

Since the binding energy of a valence electron is really small compared to the other energies, we can neglect it. This equation simplifies to:

$$E_0 = E_{SC} + E_{CE} \qquad (6.11)$$

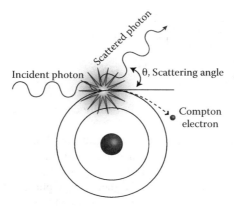

FIGURE 6.6 Compton scattering.

---

[3] An unfortunate term, as it has nothing to do with daughter recoil—the momentum of the daughter following nuclear (particularly alpha) decay.

The photon is scattered, meaning its direction has changed. The angle through which the photon is scattered is called the scattering angle (θ). The energy of the scattered photon is related to its scattering angle and its original energy:

$$E_{SC} = \frac{E_0}{1 + \left(\dfrac{E_0}{0.511}\right) \times (1 - \cos\theta)} \tag{6.12}$$

All energies in the Equation 6.12 *must* be in units of MeV. Basically, this equation tells us that as scattering angle increases, the energy of the scattered photon (E_{SC}) decreases. This makes some sense—if the incident photon "hits" the valence electron dead center, maximum energy is transferred, and the scattered photon returns along the same path of the incident photon, traveling in the opposite direction (backscattering, θ = 180°). Anything other than dead center transfers less energy and deflects the photon less. At the other end of the spectrum, a glancing blow gives a very small scattering angle and transfers only a tiny amount of energy.

Example 6.5

Calculate the energies of the 140.5 keV gamma photons that are produced during the decay of ^{99m}Tc when they are scattered through angles of 25° and 180°. Also determine the energies of the Compton electrons that are produced.

Apply Equations 6.12 and 6.11 to each.

$$E_{SC} = \frac{0.1405\,\text{MeV}}{1 + \left(\dfrac{0.1405\,\text{MeV}}{0.511}\right) \times (1 - \cos 25°)} = 0.1370\,\text{MeV or } 137.0\,\text{keV}$$

$E_{CE} = 140.5\,\text{keV} - 137.0\,\text{keV} = 3.5\,\text{keV}$

$$E_{SC} = \frac{0.1405\,\text{MeV}}{1 + \left(\dfrac{0.1405\,\text{MeV}}{0.511}\right) \times (1 - \cos 180°)} = 0.0906\,\text{MeV or } 90.6\,\text{keV}$$

$E_{CE} = 140.5\,\text{keV} - 90.6\,\text{keV} = 49.8\,\text{keV}$

Note that higher-angle scattering transfers more energy from the incident photon to the Compton electron.

Table 6.3 lists three nuclides that emit gamma photons of different energies. Notice that the percentage of energy that can be transferred from the incident photon to a Compton electron increases with the energy of the photon. All calculated values are for θ = 180° (backscattering). Higher-energy photons transfer a higher percentage of their energy to the Compton electron. Note also that a photon can never transfer all its energy to an electron via Compton scattering. The maximum amount is transferred when the photon undergoes backscattering. In Example 6.5 (and in the second row of Table 6.3), it is when the incident photon transfers 49.8 keV to the Compton electron but retains 90.6 keV.

TABLE 6.3

Energies of Scattered Photons and Compton Electrons when $\theta = 180°$

Nuclide	E_0 (keV)	E_{SC} (keV)	E_{CE} (keV)	% Transferred to Electron
^{125}I	35.5	31.2	4.3	12
^{99m}Tc	140.5	90.6	49.8	35
^{60}Co	1330	214	1116	84

6.3.2 THE PHOTOELECTRIC EFFECT

When a high-energy photon interacts with an atom via the **photoelectric effect**, we can imagine that the energy of the incident photon is completely absorbed by an *inner* shell electron, as illustrated in Figure 6.7. So long as the incident photon has enough energy (greater than the electron binding energy), odds are good it will be a K shell electron that gets booted. This time the ejected electron is called a **photoelectron**, and there is no scattered photon. The energy of the incident photon, therefore, equals the energy of the photoelectron (E_{PE}) plus the electron binding energy:

$$E_0 = E_{PE} + \text{B.E.} \qquad (6.13)$$

Since the binding energy of a K shell electron can be significant compared to the energy of a gamma photon, we can't neglect it. Kicking out an inner shell electron creates an electron vacancy or hole in that shell. This is exactly the same situation that follows electron capture decay (Section 5.4). An electron in a higher-energy shell drops down to fill that vacancy, and a characteristic X-ray or an Auger electron is then emitted.

6.3.3 PAIR PRODUCTION

The least important of the three photon-matter interactions we will study is **pair production**. Like the photoelectric effect, the atom absorbs the entire energy of the incident photon. The difference is that the incident photon is absorbed by the

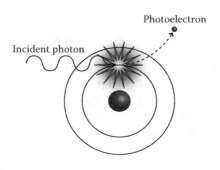

FIGURE 6.7 The photoelectric effect.

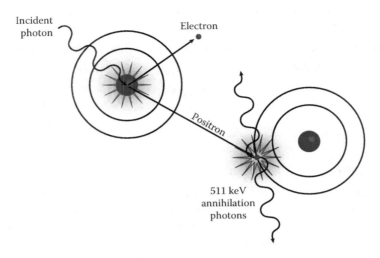

FIGURE 6.8   Pair production.

nucleus, and an electron and a positron are created, as shown in Figure 6.8. They are the "pair" that is produced.

The incident photon needs to be at least 1.022 MeV for pair production to take place. This is the amount of energy needed to create the amount of matter contained in an electron and a positron. If the incident photon has more than 1.022 MeV of energy, the remainder goes into the kinetic energy of the electron and the positron. In this case, the energy of the incident photon is equal to the energy of the emitted particles plus 1.022 MeV:

$$E_0 = E_{\beta^+} + E_{\beta^-} + 1.022 \, \text{MeV} \qquad (6.14)$$

This extra energy is not evenly divided between the pair, as anywhere from 20% to 80% of the excess energy can be deposited in either particle. The remainder goes into the other one.

The positron transfers its kinetic energy to the nearby matter, likely via ionizations and excitations. When it runs out of gas, it will annihilate with an electron to form two 0.511 MeV annihilation photons (Figure 6.8). Note that, theoretically, we could start with a 1.022 MeV incident photon and end up with two 0.511 MeV photons!

Pair production is the least important of the three photon-matter interactions discussed here because it is the least commonly observed for incident gamma photons with low-to-moderate energies. What are the relative probabilities of these three interactions? The relative probability of each depends on both the energy of the photon and the atomic number (Z) of the atom it interacts with.

Figure 6.9 illustrates how the relative probabilities of Compton scattering (solid black lines), the photoelectric effect (dashed lines), and pair production (solid gray line) vary with photon energy. The probability of all three are added together and graphed as "total" (dotted line). Photon energy increases from left to right, and probability increases from bottom to top on each graph. Graph (a) is for photon interactions with water, (b) with iron (Fe), and (c) with lead (Pb). For water (Figure 6.9a), Compton scattering is most likely for all photon energies graphed, except at the highest energies where pair production becomes the dominant interaction. For Fe (Figure 6.9b), Compton scattering still dominates for most energy values, but the relative odds of the photoelectric effect for low-energy photons, and the relative odds of pair production at higher energies, are both increased compared to water. This trend continues as we move further down the periodic table. The photoelectric effect is dominant for low-energy photons (less than 500 keV) traveling through lead (Figure 6.9c), and pair production becomes the most significant interaction above 5 MeV.

Notice also that the values for the probability (y-axis) scale increase dramatically with atomic number for the three graphs in Figure 6.9. It is generally true that the overall probability of photon interaction with matter increases as the atomic number increases. This can be understood of in terms of density. The more matter you pack into the same amount of space, the more likely you'll get high-energy photons to interact within that space. Density generally increases with atomic number.

Finally, it is worth noting that the overall probability of interaction generally decreases with increasing energy of the photon. The overall probability that a photon will interact is the sum of the probabilities for Compton scattering, photoelectric effect,

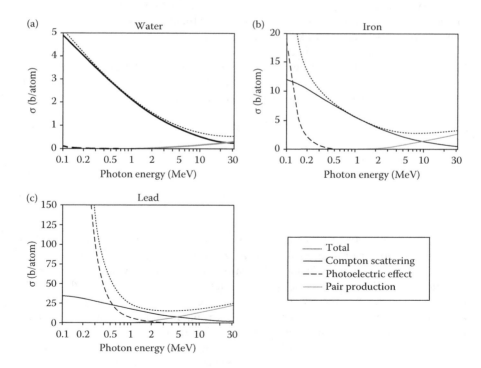

FIGURE 6.9   Relative probabilities (b/atom) of photon-matter interactions with: (a) H_2O, (b) Fe, and (c) Pb. (Produced using data from XCOM https://www.nist.gov)

and pair production. For all three graphs in Figure 6.9, the overall probability of interaction (dotted lines) generally decreases as the energy of the photon increases. This is normally true of photon-matter interactions. Like particles, higher-energy photons generally have a lower probability of interacting with matter. In this case, we *cannot* rationalize this trend in terms of the time spent near the atom because all photons travel at the same velocity. As you might guess, it's complicated and is related to the wavelength of the photon, which does change with photon energy. A shorter wavelength (higher energy) generally makes it more difficult for a photon to interact with matter.

6.4   ATTENUATION OF GAMMA AND X-RADIATION

When gamma and X-ray photons run through matter, three things can happen. They can be absorbed (photoelectric effect or pair production), scattered (Compton scattering), or they can travel straight through without interacting. If a photon is scattered or absorbed, we say it has been attenuated. For example, if 1000 photons hit a slab of matter (a half-pound cheeseburger? mmm, delicious!) and 200 are scattered, 100 absorbed, and 700 transmitted, then we'd say that 300 (30%) were attenuated.

What if a photon in the middle of a broad beam is scattered through a small angle? Would we notice? Only if our detector is far enough away—far enough that the scattered photon would get itself out of the beam and miss the detector. What if the beam is made up of photons of various energies (polyenergetic)? As we can see in Figure 6.9, low-energy photons are always more likely to interact with matter than their high-energy buddies. It can get complicated.

Let's keep it simple for now and assume that we're working only with a narrow beam of monoenergetic (all having the same energy) photons, all traveling in the same direction. This situation is often referred to as "good geometry," and anything else would be "bad geometry." When we have good geometry, the following equation holds:

$$I = I_0 e^{-\mu x} \tag{6.15}$$

which can be re-written as:

$$\ln \frac{I}{I_0} = -\mu x \tag{6.16}$$

where I is the number of photons making it through the matter (transmitted), I_0 is the original number of photons in the beam, x is the thickness of the matter, and μ is the linear attenuation coefficient. The ratio of I/I_0 is also known as the **transmission factor**—which makes sense; it is the fraction of photons that make it straight through the matter. Conversely, the attenuation coefficient is a measure of how much scattering and absorption takes place, that is, it is based on how many photons don't make it straight through. The attenuation coefficient depends on the material and the energy of the photon beam. It should always be the same for a particular material and a specific photon energy. The units on the linear attenuation coefficient are usually reciprocal centimeters (cm^{-1}) requiring x to have units of cm. Other units of length can be used but should cancel when used in the same calculation. With this equation, it should be apparent that we could never fully attenuate a photon beam. There is always a finite probability that some photons in any beam will travel through any matter. Test this for yourself with Equation 6.15 by making the thickness (x) bigger and bigger. Use arbitrary, but consistent, values for I_0 (such as 1 million) and μ (1.0 cm^{-1}).

Example 6.6

A narrow beam of 2000 monoenergetic X-ray photons is reduced to 1000 photons by 1.0 cm thick piece of copper foil. What is the attenuation coefficient of the Cu foil?

$$\ln\frac{1000}{2000} = -\mu \times 1.0\,\text{cm}$$

$$\mu = 0.69 \text{ cm}^{-1}$$

In Example 6.6, the beam is attenuated to half its original intensity; therefore, 1.0 cm is the **half-value layer** (HVL)—sometimes called the half-value thickness (HVT). Since the half-value layer will always have a I/I_0 of 1/2, HVL can be mathematically defined with Equation 6.17:

$$\text{HVL} = \frac{\ln 2}{\mu} \tag{6.17}$$

Some half-value layer values in water for various photon energies are listed in Table 6.4. Notice that HVL increases with the photon energy. As the photon's energy increases, it takes more water to attenuate half of the beam, and the photons become more penetrating.

Example 6.7

What would the beam intensity be in Example 6.6 if the foil's thickness is tripled?

$$I = I_0 e^{-\mu x} = 2000\,\text{photons} \times e^{-0.69 \text{ cm}^{-1} \times 3.0 \text{ cm}} = 250\,\text{photons}$$

This makes sense! For every HVL, half of the beam is cut out. We now have three HVLs in place. The first centimeter cuts the beam from 2000 to 1000, the second from 1000 to 500, and the third from 500 to 250. This is exponential change, just like radioactive decay!

A **tenth-value layer** (TVL) is the thickness required to attenuate a photon beam to one-tenth of its original value. It is calculated in a manner very similar to HVL:

$$\text{TVL} = \frac{\ln 10}{\mu} \tag{6.18}$$

The attenuation coefficient can take other forms and units. Some look pretty crazy. The mass attenuation coefficient (μ_m) is defined as the linear attenuation coefficient divided by the density (ρ) of the absorbing material:

TABLE 6.4

Half-Value Layers in Water for Different Photon Energies

Photon Energy (MeV)	HVL in Water (cm)
0.10	4.0
0.50	7.2
1.0	9.8
5.0	23
10	31
20	38

$$\mu_m = \frac{\mu}{\rho} \tag{6.19}$$

If density has units of grams per cubic centimeter, and the linear attenuation coefficient (μ) is per centimeter, then μ_m would have units of cm²/g.

The atomic attenuation coefficient (μ_a) is defined as the linear attenuation coefficient multiplied by the average atomic mass of the absorbing material, divided by its density times Avogadro's number:

$$\mu_a = \frac{\mu \times \text{atomic mass}}{\rho \times \text{Avogadro's number}} \tag{6.20}$$

Believe it or not, μ_a has units of square centimeters per atom (cm²/atom).

Example 6.8

Calculate the mass and atomic attenuation coefficients for Example 6.6.

The density of copper is 8.96 g/cm³, and its average atomic mass is 63.55 g/mol.

$$\mu_m = \frac{0.69\,\text{cm}^{-1}}{8.96\,\dfrac{\text{g}}{\text{cm}^3}} = 0.077\,\frac{\text{cm}^2}{\text{g}}$$

$$\mu_a = \frac{0.69\,\text{cm}^{-1} \times 63.55\,\dfrac{\text{g}}{\text{mol}}}{8.96\,\dfrac{\text{g}}{\text{cm}^3} \times \left(6.02 \times 10^{23}\,\dfrac{\text{atom}}{\text{mol}}\right)} = 8.2 \times 10^{-24}\,\frac{\text{cm}^2}{\text{atom}}$$

Remember, all of this is for good geometry (monoenergetic, narrow beam). With a broad beam, percent transmittance doesn't drop off as rapidly because some photons can be scattered through small angles, and they might not escape the beam by the time we detect them.

Polyenergetic beams are also problematic because the attenuation coefficient varies with the energy of the photon. As implied by Figure 6.9, low-energy gamma and X-ray photons are more likely to interact with all types of matter. Therefore, a polyenergetic beam will have a higher energy distribution after it passes through some attenuating matter. This is called **filtration**. Higher-energy photons are called "hard," and lower-energy photons "soft." A beam that has been filtered would then be "harder." Filtration is commonly done with medical X-rays. Most medical X-ray machines produce a polyenergetic beam that is filtered before it hits the patient (usually with an aluminum disk—take a look the next time you go to the dentist!). This is a good thing, as the soft X-rays are more likely to be absorbed by the patient and unnecessarily increase their dose during the procedure.

For polyenergetic beams, an *effective* attenuation coefficient can be determined using Equation 6.21:

$$\mu_{eff} = \frac{\ln\dfrac{I_0}{I}}{x} = \frac{\ln 2}{HVL} \tag{6.21}$$

It is based on measurements and is specific to the X-ray source, as different sources will produce different spectra (energy distributions). Calculation of other flavors (analogous to μ_m and μ_a) of the effective attenuation coefficient are performed in the same way as Equations 6.19 and 6.20.

Example 6.9

An X-ray beam has an HVL of 1.5 mm for Cu. What are the effective linear and mass attenuation coefficients?

$$\mu_{eff} = \frac{\ln 2}{0.15\,cm} = 4.6\,cm^{-1}$$

$$\mu_{meff} = \frac{\mu_{eff}}{\rho} = \frac{4.6\,cm^{-1}}{8.96\dfrac{g}{cm^3}} = 0.52\frac{cm^2}{g}$$

QUESTIONS

6.1 Calculate the wavelength and the frequency of a photon with 100 eV of energy.

6.2 Why is a gamma ray considered ionizing radiation, but photons of purple light are not?

6.3 Define the following: ionizing radiation, linear energy transfer, bremsstrahlung, Compton scattering, photoelectric effect, pair production, and half-value layer.

TABLE 6.5
Electron Binding Energies for Argon

K	L_1	L_2	L_3	M_1	M_2	M_3
3.20 keV	327 eV	250 eV	248 eV	27 eV	14 eV	14 eV

6.4   An alpha particle emitted during the decay of ^{222}Rn (5.4895 MeV) transfers 0.0010% of its energy to an argon atom. Is this enough energy to ionize the argon atom by booting any of its electrons? Table 6.5 provides electron binding energies for all occupied subshells in an argon atom. If it can eject electrons from any of the subshells listed, calculate the energy of the secondary electron that is produced.

6.5   How many ion pairs will be formed by a single 5.00 MeV alpha particle traveling through air? Assume all the particle's energy goes into ion pair formation.

6.6   Would two beta particles, with different energies, have different ranges in air? Briefly explain.

6.7   If the range of a charged particle in dry air (at standard temperature and pressure) is known, what other data must be obtained to calculate the range of the same particle in titanium?

6.8   Calculate v/c for a 7.6868 MeV proton. Repeat for an electron (beta particle) of the same energy. Use both the classical and relativistic equations for each particle. Which gives more accurate results?

6.9   Calculate the kinetic energy (MeV) of an electron traveling at a velocity exactly equal to one-half the speed of light. Repeat the calculation for a proton and an alpha particle.

6.10 What is the maximum velocity for a beta particle emitted during the decay of ^{6}He? Why must this be considered a "maximum" velocity?

6.11 Which of the following are true?
  a.   Compton scattering is the most likely interaction for any X-ray photon energy.
  b.   Pair production is the least likely of the three interactions studied.
  c.   The probability of all interactions generally increases with increasing Z.
  d.   Pair production is never the dominant form of interaction.
  e.   The greater the energy of the photon correlates to a greater probability of interaction.

6.12 Calculate the mass of 1.0 μCi of ^{241}Am. AmO_2 is its chemical form in smoke detectors. What is its specific activity? Do ^{241}Am-containing smoke detectors pose a radiation hazard for people who have them in their homes?

6.13 Former KGB agent Alexander Litvinenko was poisoned with ^{210}Po in 2006. ^{210}Po emits an alpha particle with an energy of 5.3044 MeV. What is the specific ionization of this alpha particle in air? On average, how many ion pairs will be created in air by each alpha particle emitted by ^{210}Po? Why is ^{210}Po a particularly clever poison?

6.14 How would you expect protons to interact with matter? Compare to alpha particles and electrons of equal energy.

6.15 How many times more massive are alpha particles compared to electrons?

6.16 Define the following: tenth-value layer, attenuation, photoelectron, Compton electron, and Cherenkov radiation.

6.17 An elastic collision takes place between a 2.51 MeV electron and a stationary ^{12}C nucleus. If exactly 25% of the electron's energy is transferred to the ^{12}C nucleus, calculate the resulting velocities of both.

6.18 An electron initially traveling at 2.997×10^8 m/s slows as it transfers to another medium, emitting a photon of blue light (wavelength = 455 nm). Calculate the electron's new velocity. What is the blue light known as?

6.19 Calculate the energies of the backscatter peak and the Compton electron for the 662 keV gamma ray observed in the decay of ^{137}Cs.

6.20 Calculate the energies of the scattered photon and the Compton electron when incident gamma radiation of 167 keV (from ^{201}Tl) is scattered through an angle of 23°. Assume the Compton electron originated from the L_1 subshell of oxygen (binding energy = 37.3 eV).

6.21 What is the linear energy transfer (LET) of beta particles emitted by ^{90}Y in air? Use the most probable energy of a beta particle emitted by ^{90}Y. Estimate the range in air of an alpha particle with the same energy, and compare to the average range of the ^{90}Y beta particle (11 mm) in human tissue.

6.22 Humans can be considered as mostly water. If a 25 cm thick human is exposed to gamma radiation from ^{137}Cs, estimate the percentage of the gamma ray intensity attenuated. The mass attenuation coefficient for water exposed to ^{137}Cs is 0.088 cm^2/g. What is the most likely interaction between a photon and a water molecule?

6.23 Copper has a density of 8.9 g/cm^3, and its total atomic attenuation coefficient is 8.8×10^{-24} cm^2/atom for 500 keV photons. What thickness of copper is required to attenuate a 500 keV photon beam to half of its original intensity?

6.24 The mass attenuation coefficient for iron exposed to 1.5 MeV X-rays is 0.047 cm^2/g. Calculate the half- and tenth-value layers.

6.25 Using information from Problem 6.24, calculate the atomic attenuation coefficient. Also calculate the intensity of the original beam if the measured intensity is 5800 photons/s on the other side of a 70 mm thick piece of iron.

6.26 Calculate the kinetic energy of a photoelectron released from the K shell (binding energy = 36.0 keV) as a photon with a frequency of 9.66×10^{18} s^{-1} interacts with a chunk of cesium. What part of the electromagnetic spectrum does this photon belong to?

7 Detection of Ionizing Radiation

Perhaps the main reason that most people fear ionizing radiation is that humans have no way to directly sense it. We need instruments to convert ionizing radiation into signals that we can observe and understand. Remember that ionizing radiation can interact with matter by ionization (creation of an ion pair) or excitation. We can take advantage of both in our quest to detect ionizing radiation. This chapter is all about how these instruments work. Special emphasis will be placed on Geiger-Müller tubes (also known as Geiger counters), as they are so commonly used, and on gamma spectroscopy because of its importance to nuclear medicine.

7.1 GAS-FILLED DETECTORS

The basic design and function of any gas-filled detector is illustrated in Figure 7.1. Ionizing radiation passes through some gas in a container, creating ion pairs from the gas molecules. The negatively charged electrons then begin to migrate toward a positively charged piece of metal (anode), and the positively charged cations move toward a negatively charged cathode. When the ions hit the metal, they create an electrical signal (current), which can be read. Many different electrode configurations are possible, but the most common is a wire anode running down the center of a cylindrical cathode.

Remember from Chapters 1 and 6 that X-rays and gamma rays have a relatively low probability of interacting with matter. Since gases have very low densities, gas-filled detectors don't have a lot of matter available to interact and are, therefore, quite inefficient in detecting photons. Typically, less than 1% of all photons entering the tube will interact with the gas inside the tube. That means that more than 99% fly right on through undetected. However, gas-filled detectors are very efficient at detecting beta and alpha radiation as they have much higher probabilities of interacting with matter.

The major difference between the three flavors of gas-filled detectors discussed in this section is the voltage applied to the electrodes. We will look at them in order of increasing voltage.

7.1.1 IONIZATION CHAMBERS

Ionization chambers are the lowest-voltage form of gas-filled detector, typically running at 50–300 V. The idea is to have just enough voltage to keep the ions moving toward the electrodes without accelerating them. If they are accelerated, there's a chance they may cause other ionizations as they travel through the chamber. If we can keep their velocity more or less constant, then the only thing detected are the ions created.

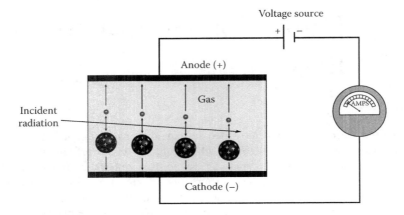

FIGURE 7.1   Schematic of a gas-filled detector.

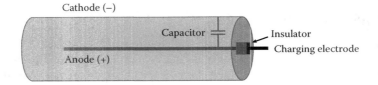

FIGURE 7.2   Schematic drawing of a pocket dosimeter.

Ionization chambers are usually filled with air. Remember that, on average, it only takes about 34 eV to create an ion pair in air, so many ion pairs *can* be expected from every particle or photon of ionizing radiation that enters the chamber. Even so, ionization chambers cannot always detect individual events because the electrical signal from these events can be too weak. Fortunately, most radioactive sources emit radiation at a pretty high rate, and the chamber can measure the overall current created, which can be calibrated to a count rate.

One application of ionization chambers is the pocket dosimeter. A schematic of one is pictured in Figure 7.2. The inner wire (anode) and the outer case (cathode) are given a charge, which is stored in a capacitor. The user can then look through a window on one end and read the level of charge on a gauge. As ionizing radiation interacts with the gas inside the dosimeter, the charge is gradually dissipated. The extent of charge dissipation is a measure of the amount of radiation the user was exposed to. These dosimeters are commonly used when the possibility of a significant dose in a relatively short time exists. The accuracy of these devices is poor, only about 20%.

7.1.2   PROPORTIONAL COUNTERS

One disadvantage of ionization chambers is that they are not very sensitive. They are designed to detect radiation in bulk, not as individual photons and particles. They can have trouble detecting a single ionizing particle because the signal generated is sometimes too weak. Fortunately, we can make a couple simple changes to amplify the signal.

First, we can crank up the voltage. A higher potential means that the electrons will now be accelerated as they approach the anode. If they accelerate enough, they can begin to cause additional ionizations, which will generate more electrons, which then accelerate, causing more ionizations ... and, when all these electrons hit the anode, a stronger signal is generated. This process of accelerating the electrons to create more ion pairs is called a cascade or avalanche. Typically, 10^6 times as many ion pairs are generated in the cascade as were originally generated by the ionizing radiation.

Secondly, we can change the gas. Air contains oxygen, which has a bad tendency to react with free electrons, preventing them from getting to the anode and dampening the signal in ionization chambers. **Proportional counters** typically use Ar or Xe gas inside the detector, in addition to running at higher voltages than ionization chambers.

The signal is now strong enough that individual counts can be made. Another bonus is that the total charge produced is *proportional* to the energy deposited by the ionizing radiation. Therefore, both count rate and energy information on the incident radiation can be obtained. This can be helpful in discerning alpha from beta radiation and in identifying alpha emitting nuclides.

7.1.3 GEIGER-MÜLLER TUBES

Geiger-Müller (G-M) tubes are designed for maximum signal amplification (Figure 7.3). To do this, the voltage is usually cranked up to around 800 V. The detector gas is almost always Ar. As a result, electrons are accelerated like crazy toward the anode. Typically, 10^{10} times more ion pairs are created. There's no missing that signal!

Lots of atomic excitations also happen in the mad rush of electrons to the anode. When these atoms de-excite, electromagnetic radiation (visible light, UV, and X-rays) is given off—*lots* of it. Some of these photons have enough energy to knock electrons loose from the metal electrodes (Einstein's photoelectric effect). Some of the photons also cause additional ionizations in the detector gas.

This cascade could easily get out of hand, so what shuts it down? The slow-moving cations! The cations are much more massive than the electrons and, therefore, move much slower under the same conditions. So many ionizations are created near

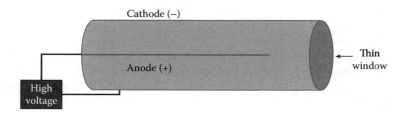

FIGURE 7.3   Schematic of a Geiger-Müller detector.

the anode by all the electrons accelerating toward it that a cylindrical tube of cations forms around the anode. Eventually, all gas inside this tube is ionized, and the cascade shuts down.

As the cations finally approach the cathode (~300 μs later), electrons come off the wall (the cathode) to combine with them. As these electrons accelerate toward the vast cation cloud, they excite electrons in the Ar atoms they meet along the way, and more electromagnetic radiation is generated. This new burst of photons threatens to start another cascade. To shut this down (**quench**), chlorine is added to the Ar gas inside the G-M tube.

$$Cl_2 \xrightarrow{\text{photons}} 2Cl^{\cdot} \rightarrow Cl_2$$

Chlorine efficiently absorbs the higher energy photons, which symmetrically break the Cl–Cl bond, forming two chlorine radicals. These radicals then recombine, dissipating the released energy as heat.

Because of the huge gas amplification (acceleration toward the electrodes), G-M tubes detect individual events quite well. However, the strength of the observed signal is constant, regardless of the amount of energy deposited in the tube by the ionizing radiation. Therefore, only count information can be recorded with G-M tubes, not energy.

For a certain amount of time after ionization takes place in a G-M tube, it is unable to distinguish another ionization of the Ar gas. Even after the cascade has been quenched, a little more time must pass before we can be sure another event will be recorded. Several terms are applied to this time interval:

Dead time is the time interval after a pulse has occurred during which the counter is insensitive to further ionizing events.

Resolving time is the minimum time interval by which two pulses must be separated to be detected as two individual pulses.

Recovery time is the time interval that must elapse after a pulse has occurred before another full-sized pulse can occur.

The differences between these three terms are fairly subtle, but the implication of this required delay between counts is clear. At higher count rates, G-M tubes will begin to miss some counts. For example, if two beta particles enter a G-M tube at about the same time (**coincidence**), only one count will be recorded. What kind of time interval are we talking about here? It depends a bit on the tube, but typically, it is a few hundred microseconds. Because of the possibility of additional ionizations occurring within this dead time increases with the count rate, higher count rates mean more counts will be missed!

Fortunately, it is easy to estimate the true number of counts by experimentally determining how many counts are being missed at various count rates and plotting the results in a graph, such as shown in Figure 7.4. Correcting for the missing counts is usually called a **dead time correction**. The straight line shows the entirely theoretical case when true counts are always equal to observed counts. The curved line shows the reality of observed counts being less than true counts at higher count rates.

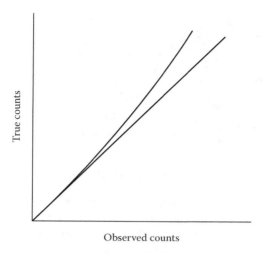

True counts

Observed counts

FIGURE 7.4   Dead time correction graph for a G-M tube.

In the real world, count rates can also be affected by variances in line voltage, background, sample geometry (how the sample is positioned relative to the detector), atmospheric conditions, and movement of wires. As mentioned in Chapter 2, the percentage of decays detected relative to all the source's decays in the same time interval is called the percent efficiency. As is clear here, the percent efficiency can be affected by many factors and would, therefore, be expected to vary somewhat from detector to detector.

7.2   SCINTILLATION DETECTORS

Gas-filled detectors work by converting ionizing radiation into an electrical signal via the formation of ion pairs. Another way to detect ionizing radiation is to convert it to visible light. This is exactly what scintillation detectors do, and it is accomplished through electron excitations (albeit indirectly). The visible light is observed by converting it to an electrical signal. This section begins by looking at the conversion of light to electronic signal (an electronic eyeball), then describes the two most common types of scintillators, inorganic and organic.

7.2.1   PHOTOMULTIPLIER TUBES

When ion pairs recombine, or excited electrons drop back to the ground state, visible light can be given off. **Scintillators** are materials that do this particularly well. Only a few thousand photons are usually produced when ionizing radiation travels through a scintillator. This is not very bright and, therefore, very difficult to detect, so we will need a way to amplify the visible light signal. **Photomultiplier tubes** (PMTs) do this by converting the visible light photons to electrons then amplifying the signal. This process is illustrated in Figure 7.5.

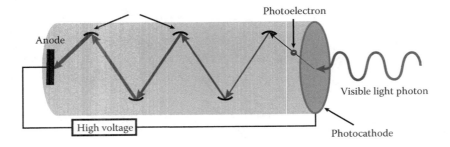

FIGURE 7.5   Schematic of a photomultiplier tube (PMT).

When the visible light photons leave the scintillator, they hit the photocathode. The photocathode converts the photons into electrons (photoelectrons). These electrons now represent the signal generated by the ionizing radiation. The electrons are accelerated using electrical charge toward a series of electrodes (dynodes) arranged in a zigzag pattern. The dynode is made from a material that ejects more electrons than hit it at any given moment. By the time the electrons are ejected by the last dynode, there are a whole lot more electrons than started out. This is how the signal is amplified. Finally, the electrons hit the anode on the far end of the PMT and register an electrical signal. The energy of the original signal is preserved through the PMT, as it is proportional to the number of photoelectrons produced at the photocathode. This means that we can plot an energy spectrum (count rate versus energy) to see how many counts are observed at specific energy values.

7.2.2 Inorganic Scintillators

Inorganic scintillators are salts (inorganic compounds) that light up when hit by ionizing radiation. The most commonly used are crystals of NaI (sodium iodide), CsI (cesium iodide), and $LaBr_3$ (lanthanum bromide). In all three cases, the crystals light up a lot better if they are doped with a tiny amount of either thallium (Tl) or cerium (Ce). The doped crystalline materials are represented as NaI(Tl), CsI(Tl), and $LaBr_3$(Ce).

Notice that these materials all contain a fair amount of high atomic number (Z) elements. Higher Z means higher probabilities of interacting with the elusive X-ray and gamma ray photons, especially through the photoelectric effect (see Figure 6.9). These materials are also very dense when compared to the gases used in gas-filled detectors. In other words, they're packing a lot more matter, increasing the probability of interacting with ionizing radiation. As a result, inorganic scintillators are much more efficient at detecting high-energy photons.

Figure 7.6 demonstrates how a typical NaI(Tl) detector works. The crystal is cylindrically shaped and sealed inside of a shiny metal case. The case is reasonably thin, so that X-rays and gamma rays can easily penetrate it. Once inside the crystal, the photon interacts, most likely with the iodide inside the crystal,[1] causing ionizations and excitations, eventually creating some visible (scintillation) light. The nature

---

[1] There's not much Tl, and the Na is much lower Z.

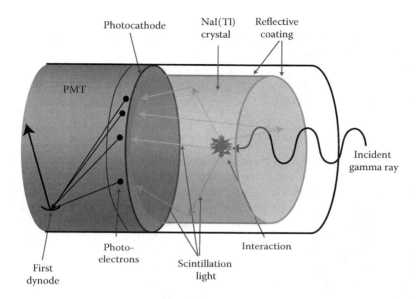

FIGURE 7.6 A NaI(Tl) detector at work.

of these interactions is discussed in more detail in Section 7.4. The light then goes directly toward the photocathode, or it is reflected off the shiny metal enclosure into the photocathode.

After the PMT amplifies the signal, it is run through a multichannel analyzer (MCA), which sorts the signal by energy (actually, it is by channel, with each channel covering a short energy range). Both count and energy data are obtained allowing the production of a spectral plot (photon energy versus count rate). Such spectra will be indicative of its source.

NaI(Tl) detectors are relatively inexpensive and are commonly used with gamma-emitting nuclides. However, they are shock-sensitive (crystals are easy to crack or break if dropped), rather bulky (because they require a PMT), and suffer from poor energy resolution. Gamma photons are monoenergetic and, therefore, should be observed in a spectrum as having a single energy value. Instead, a distribution of energies is observed centered on the actual value. The wider this distribution, the poorer the energy resolution, that is, the harder it would be to see (resolve) two different gamma energies that just happen to have energy values close to each other.

Inorganic scintillators suffer from poor energy resolution because the signal generated at the photocathode is limited. First, not all the incident photon energy is converted to visible light photons; some is converted to heat (displaced electrons traveling through matter will generate some heat). Second, not all the visible light photons will make it to the photocathode *and* generate photoelectrons. This is when the signal is the weakest. As with any measurement, greater relative error is experienced with smaller numbers being measured.

There is also random error associated with the number of photoelectrons produced, which are then amplified with the signal in the PMT. The poor energy resolution of inorganic scintillators also results from the fact that it takes so many steps (conversions of energy) from the incident photon to the final electrical signal. There are random errors introduced at each step, leading to significant random error in the final signal. Fortunately, most gamma energies in a single spectrum are well enough separated to be observed using inorganic scintillators. Their resolution may be poor, but they are still useful.

7.2.3 ORGANIC SCINTILLATORS

All of the detection methods mentioned so far suffer from the fact that they can only detect radiation that enters the detector (and interacts). Since radiation is typically emitted in all directions equally, less the 50% of all emitted radiation can be expected to enter the types of detectors we've seen so far. If we have a low-activity sample, especially if it emits low-energy radiation, it would be better if we could place the sample inside our detector so that we could detect all decays. This can be done with an **organic scintillator**, as shown in Figure 7.7.

Organic scintillation works by dissolving the radioactive material in a "scintillation cocktail." This cocktail is made up of organic (low Z) materials: a solvent, a scintillator, and sometimes other stuff to help make it glow brighter or with the right wavelength. Low Z materials are used to minimize bremsstrahlung that can occur when beta particles are emitted during decay. This solution is placed in a glass or plastic vial, which is put in a dark place (inside an instrument) with PMTs all around it to watch for visible light photons and record data.

How does this work? First, the decay takes place. Since the cocktail is mostly solvent, it is most likely the solvent that absorbs energy from the ionizing radiation. The solvent then (eventually) transfers this energy to a scintillator molecule, which becomes excited. The scintillator then de-excites, emitting a photon. The photon hits a PMT, which amplifies the signal and records it. The more PMTs we can pack around the scintillation vial, the more photons we'll see and the higher the counting efficiency.

As mentioned previously, this is a great way to measure samples emitting low-energy radiation, such as weak beta-emitting nuclides. It is not very efficient for hard X-rays or gamma rays, as these photons are unlikely to interact with the low Z materials inside the vial.

Another drawback is **quenching**. Quenching here is a different issue than for gas-filled detectors. In terms of scintillation counting, it refers to anything that reduces light output. It can be a fingerprint or other smudge on the outside of the vial.

FIGURE 7.7 Organic scintillation gives off photons of visible light.

A highly colored sample (such as blood) could also be a problem. If the sample volume is significant compared to the volume of the cocktail, the photon output will be reduced. Finally, O_2 can inhibit the transfer of energy from the solvent to the scintillator and is best purged from the solution.

7.3 OTHER DETECTORS

7.3.1 SEMICONDUCTOR DETECTORS

Given the poor energy resolution of inorganic scintillation detectors, it might be nice to have a detector with higher resolution when it is important to resolve signals from two gamma photons with similar energy values. As you remember from Section 7.2.2, the Achilles heel of a scintillation detector is the number of conversions that take place from incident radiation to electrical signal and the fact that signal can be weakened with many of these conversions. We need the detector made from a solid, so it has a reasonable chance of interacting with gamma photons, but it'd be nice if it worked like the gas-filled detectors, which convert ionizing radiation more directly to an electronic signal.

The answer would be a detector made from semiconductors. Semiconductors are those elements on the periodic table that have chemical and physical properties that are intermediate between the metals and the non-metals (such as Si, Ge, and Te). These elements are sometimes called metalloids, but most important for us is the fact that they are neither conductors of electricity, nor are they insulators—they are something in between, semiconductors. This means that sandwiching a crystal between two electrodes can allow electrons knocked loose by ionizing radiation to migrate to an electrode and produce an electrical signal.

If you take a slab of silicon (Si) or germanium (Ge) and sandwich it between a couple electrodes, it makes a **semiconductor detector**. These detectors are very similar to gas-filled detectors, as depicted in Figure 7.1, except they are filled with a Si or Ge crystal instead of a gas. The main disadvantages of semiconductor detectors are that they are expensive to make (very high purities are normally necessary, although they can also be doped with certain elements), and they typically need to be cooled to ~100 K (liquid nitrogen!) in order to work. The main advantage is that their energy resolution is excellent, as illustrated in Figure 7.8.

The improvement in energy resolution is largely a reflection of the greater simplicity of the radiation to signal process. Incident gamma photons interact with semiconductors by knocking loose an electron and creating an electron "hole." The hole can be imagined as simply the place where the electron is supposed to be. Instead, the electron has been given extra energy and now resides in what is called the conduction band and is free to move about the semiconductor (toward the anode). Previously, it had been lodged in the valence band where it can't move. The hole also "moves" but toward the cathode. This really isn't physical movement of the hole so much as an electron between the hole and the cathode being pushed into the hole, resulting in a new hole closer to the cathode.

As mentioned previously, two drawbacks of Si- and Ge-based detectors are their expense and their need to be very cold to work properly. A third drawback is that

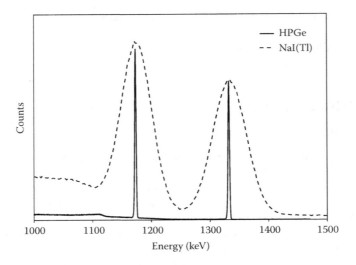

FIGURE 7.8 Comparison of the gamma spectrum (^{60}Co) energy resolution of an NaI(Tl) (dashed line) and high-purity germanium (solid line) detectors.

they have relatively low atomic numbers ($Z = 14$ and 32, respectively). Being made of lower Z material means that detection efficiency will suffer. More recently, new materials have been found that do not share all these drawbacks. In particular, CdTe and $Cd_{1-x}Zn_xTe$ ($0.04 \leq x \leq 0.20$) have shown significant promise. The latter is more commonly known as "CZT." Neither require special cooling; in fact, both function well at normal room temperatures. They also incorporate higher Z material ($Zn = 30$, $Cd = 48$, and $Te = 52$) and, therefore, are more efficient in detecting gamma and X-ray photons. Detectors made from these materials are still expensive relative to NaI(Tl)-based detectors, but they are finding application in medicine, research, and national security where their compact size, good energy resolution, higher counting efficiencies, and convenience (no or less cooling necessary) are valued.

7.3.2 THERMOLUMINESCENT DOSIMETERS

Certain salts (inorganic compounds) can be used to store information on exposure to ionizing radiation. Good materials for this are lithium fluoride (LiF) and calcium fluoride doped with a little manganese, CaF_2(Mn). When electrons in crystals of these salts are excited by interaction with ionizing radiation, they get moved up from the valence band to the conduction band but then get stuck (trapped) in intermediate energy levels. They can stay trapped for months or even years! When the crystal is heated up, the electrons are freed from their traps and return to the ground state. Along the way they emit visible light, which can be recorded by a PMT. The intensity of the light emitted is proportional to the amount of radiation the crystal was exposed to. This process (thermoluminescence) is illustrated in Figure 7.9.

These crystals are typically built into badges worn by personnel working with radioactive material or radiation producing devices to estimate their occupational

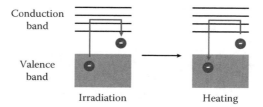

FIGURE 7.9 Thermoluminescence.

dose. These badges are called thermoluminescent dosimeters (TLDs), and can be collected on a monthly or quarterly basis to monitor exposure. Typically, three or four crystals are placed inside a single badge, with different levels of shielding. This way the penetrating ability of the radiation the worker is exposed to can also be estimated.

7.4   GAMMA SPECTROSCOPY

Gamma spectroscopy is the detection of both count and energy information from gamma rays. By plotting energy versus counts, we can see different energy gamma rays at the same time. It is sometimes referred to as pulse-height or energy spectrometry. Regardless of the name, it is a wonderful tool for detecting and discerning various radioactive nuclides. Unfortunately, the information is sometimes complicated by what appear to be odd features in the spectrum. Fortunately, these features can all be understood in terms of the interactions of high-energy photons with matter, discussed in Chapter 6.

As you already know, gamma rays are very penetrating, that is, they have a low probability of interacting with matter. We can boost this probability by making them try to pass through relatively dense matter. That's why lead (Pb) is commonly used for shielding high-energy photons. If we want to detect them, we'll also have to use fairly high Z material. As discussed previously, that's why inorganic scintillators or semiconductor detectors are commonly used. Remember that these detection methods give count *and* energy information from gamma-emitting sources. Instead of just getting count data, we get count data spread out over the gamma energy spectrum, like Figure 7.10 for ^{137}Cs.

But wait a minute! We know that the decay diagram for ^{137}Cs (Figure 7.11) only shows one gamma ray being emitted, with an energy of 662 keV. Since nuclear energy states are quantized, we should *only* see counts for 662 keV. Why is the peak at 662 keV so broad, and what's all the other stuff in the spectrum? Why is there a broad continuum of counts below ~500 keV? What is the bump around 200 keV and the really big peak at relatively low-energy? The answers to these important questions (and maybe some others) have to do with how gamma rays interact with matter because that's how they are detected.

Figure 6.9 shows us that if we build a detector out of low or medium Z matter, most of the gamma photons that interact with it will do so via the Compton effect. If we make it with high Z materials, the photoelectric effect will dominate, at least for

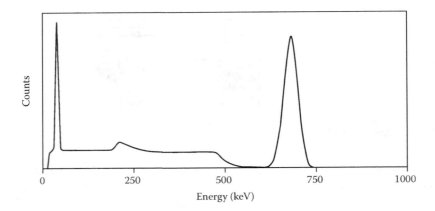

FIGURE 7.10   Gamma spectrum for ^{137}Cs using a NaI(Tl) detector.

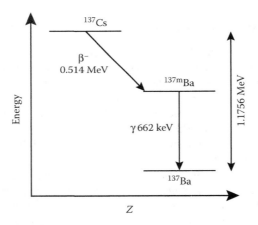

FIGURE 7.11   Decay diagram for ^{137}Cs.

low- to medium-energy gamma rays. Figure 7.12 shows the relative probabilities of the photoelectric effect, Compton scattering, and pair production with iodine. Notice that the photoelectric effect is dominant for photon energies below ~300 keV. Above 300 keV, Compton scattering is most likely.

Does it make a difference whether the photon interacts with the iodide in the detector via the photoelectric or Compton effect? You betcha. When a photon interacts via the photoelectric effect, all its energy is dumped into the detector, giving us a signal corresponding to the energy of that photon. In the gamma ray spectrum, this is called the **photopeak** (or the total energy peak, or the gamma peak). This is what we want to happen; therefore, we prefer to make gamma ray detectors out of high Z materials. As noted in Chapter 6, high Z also means a higher overall probability of interaction, meaning fewer gamma photons will pass completely through the detector without interacting.

If the photoelectric effect gives the energy of the gamma photon, why is the photopeak in Figure 7.10 so broad? As explained in Section 7.2.2, it is because the

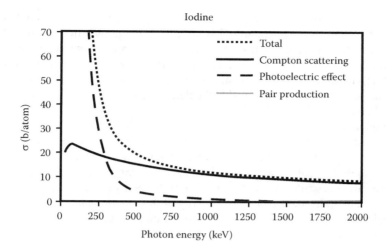

FIGURE 7.12   Relative interaction probabilities of high-energy photons with iodine.

energy of the incident photon undergoes so many conversions that these types of detectors have larger random error.

When the gamma photon interacts with the detector via the Compton effect, the photon *could* deposit only some of its energy in the detector then leave, taking with it some of its original energy as the scattered photon. Since the amount of energy transferred by the photon undergoing Compton scattering is variable, then different amounts of energy will be deposited in the detector for each photon undergoing the Compton effect then escaping the detector. This will give rise to a broad continuum of counts called the **Compton continuum**. This is the broad plateau of counts below (lower in energy) the photopeak in Figure 7.10. One way to decrease the number of counts under the Compton continuum relative to the photopeak is to increase the size of the detector. A larger detector will give the scattered photon more opportunities to interact on its way out, leaving all the incident photon's energy inside the detector.

What's the maximum energy Compton scattering can leave in the detector? Assuming the gamma photon only undergoes one Compton scattering interaction inside the detector, it would be when it is scattered through 180°. This energy can be calculated based on the original energy of the photon using Equation 7.1:

$$E_{ED} = \frac{E_0^2}{E_0 + 0.2555} \tag{7.1}$$

where E_0 is the energy of the incident photon, and E_{ED} is the "maximum energy" that can be deposited in the detector. Both energy values must have units of MeV for this equation to be true. E_{ED} is also known as the **Compton edge**. The Compton edge for the 662 keV photopeak observed in the ^{137}Cs spectrum is 478 keV:

$$E_{ED} = \frac{0.662^2}{0.662 + 0.2555} = 0.478\,\text{MeV} = 478\,\text{keV}$$

Careful examination of Figure 7.10 shows that the Compton edge does not define a strict upper limit to the continuum, but rather it is the point where counts begin to decrease with increasing energy. Why are some Compton continuum counts still observed above the Compton edge? This is partly due to the poor energy resolution of NaI(Tl) detectors. It can also result from a gamma photon having more than one Compton scattering interaction inside the detector and exiting leaving slightly more energy behind than is calculated for E_{ED}.

Compton scattering can also take place *outside* the detector. A gamma photon originally traveling away from the detector could be scattered off an air molecule, you, your lab partner, or some other matter. If it is scattered back into the detector, it shows up in the gamma spectrum with a specific energy. The energy has a certain value because the scattering angle must be 180° (or pretty close to it) for the scattered photon to end up in the detector. The energy of this small peak (E_{BS}, also known as the **backscatter** peak) can be calculated using Equation 7.2, where all energy values are again given in MeV:

$$E_{BS} = \frac{E_0}{1 + (3.91 \times E_0)} \tag{7.2}$$

Even though they originate from different phenomena, there's a simple mathematical relationship between E_{BS} and E_{ED}; they add up to energy of the incident photon!

$$E_0 = E_{BS} + E_{ED} \tag{7.3}$$

This equation is really the same as Equation 6.11, repeated here for your convenience:

$$E_0 = E_{SC} + E_{CE}$$

where E_{SC} is the energy of a scattered photon, and E_{CE} is the energy of the Compton electron. Equation 7.3 represents a special case of Equation 6.11—when backscattering takes place, then E_{SC} is equal to E_{BS}, and E_{CE} is equal to E_{ED}.

^{137}Cs emits a 0.662 MeV gamma ray; therefore, its backscatter peak should be found at 184 keV:

$$E_{BS} = \frac{0.662}{1 + 3.91(0.662)} = 0.184 \, \text{MeV} = 184 \, \text{keV}$$

We now understand most of what we see in the gamma spectrum for ^{137}Cs. These features are labeled in Figure 7.13.

What's up with the low-energy peak, and why is it labeled "Ba X-ray"? The ^{137m}Ba isomer (a metastable state) that is formed after ^{137}Cs emits a beta particle (see Figure 7.11) can also decay to the ground state by spitting out a conversion electron. The conversion coefficient (α_K) is 0.093, which means that ~1 in 10 decays produce a conversion electron instead of the 662 keV gamma photon. The resulting hole in the K shell of the ^{137}Ba daughter can be filled by an electron from an L shell, which could generate an X-ray with an energy of 32 keV. These X-ray photons are also

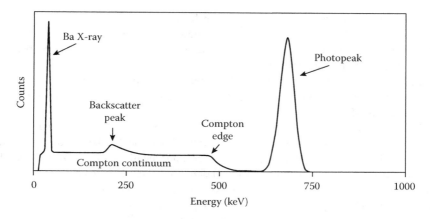

FIGURE 7.13 Labeled gamma spectrum for ^{137}Cs.

detected by the NaI crystal, producing the low-energy peak observed in the gamma spectrum of ^{137}Cs.

Equipped with a solid understanding of the interactions of gamma rays with matter (Section 6.3), let's see if we can understand what other odd features we might come across in gamma spectra.

The relative intensities of gamma peaks depend on a few factors. As we already know, higher-energy photons have a lower probability of interacting with matter. Therefore, we would expect to see smaller peaks for higher-energy photons even if the source is emitting equal numbers of low- and high-energy photons. Intensities of two different peaks may also vary if they are part of two different decay branches, or if they have significantly different conversion coefficients. Finally, relative intensities can vary due to the relative probability that each photon will be emitted as part of the decay.

A great example is ^{57}Co. As illustrated in Figure 7.14, it decays via electron capture to an excited state of ^{57}Fe with a 5/2 spin state. This excited state has two options.

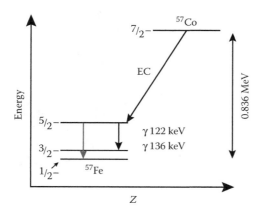

FIGURE 7.14 Decay diagram for ^{57}Co.

It can decay directly to the ground state (spin 1/2), emitting a 136 keV gamma photon (gray), or it can decay to a lower energy excited state (spin 3/2) emitting a 122 keV gamma photon (black). The 122 keV gamma photon is more likely to be observed (more intense in the spectrum) than the 136 keV gamma photon because the change in nuclear spin states is less dramatic. We would also expect the 136 keV peak to be smaller because it is higher in energy than the 122 keV peak. The observed relative intensities are 85.6:10.7.

Using NaI(Tl) detectors, it is not possible to resolve the two ^{57}Co gamma peaks— their energy values are too close to each other. Since the 122 keV peak dominates intensity, the 136 keV peak tends to be buried underneath it. Figure 7.15 shows the gamma spectrum for ^{57}Co using a NaI(Tl) detector. The 122 keV peak is the largest feature on the spectrum, and it appears to have a low shoulder on the right (higher energy) side. This is probably a part of the 136 keV photopeak.

There is another feature in the ^{57}Co spectrum (Figure 7.15) that is often observed in the spectra of low-energy gamma photons. It shows up as a small peak that is 28 keV lower in energy than the photopeak (at ~94 keV in this case). When a low-energy gamma photon enters a NaI(Tl) detector, it will most likely interact with the iodide (I^-) via the photoelectric effect. This creates an electron vacancy in the K shell of the iodide. When an electron from a higher-energy shell fills this vacancy, an X-ray photon is generated. Normally the crystal absorbs the energy of this photon, that is, it interacts before it can get out of the crystal. However, if it manages to find its way out of the detector, the recorded energy will be missing 28 keV (the energy of the iodine K_α X-ray). This is called an **iodine escape peak**. It is only observed with low-energy gamma photons, since they are most likely to interact with I^- via the photoelectric effect. Low-energy photons also have a higher probability of interacting with matter in general. Therefore, odds are better that they will interact close to the surface of the detector. This increases the probability that the K_α X-ray will escape the detector, since it (potentially) has less matter to travel through to get out.

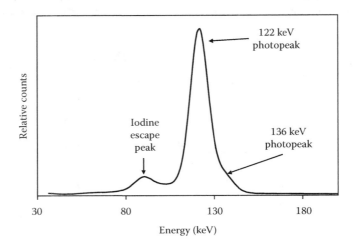

FIGURE 7.15   Gamma spectrum of ^{57}Co.

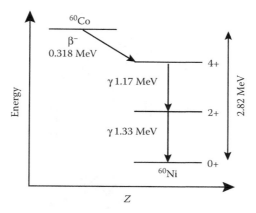

FIGURE 7.16   Decay diagram for ⁶⁰Co.

Figure 7.16 shows the decay diagram for ⁶⁰Co. After beta emission, an excited state of ⁶⁰Ni with a spin state of 4 is formed. Like the ⁵⁷Co example previously, this excited state has two choices: decay to another excited state (spin 2) or go all the way to the ground state (spin 0). In this case, the change in spin state is too dramatic to allow direct decay to the ground state. As a result, photopeaks are observed at 1.17 and 1.33 MeV.

If a sample of ⁶⁰Co is reasonably toasty (high activity), a small peak may also be observed at 2.50 MeV. However, this is not due to a 4+ → 0+ transition but is a **sum peak**. This happens when two different gamma photons hit the detector at the same time (or within the detector's resolving time) and are recorded as a single count. Sum peaks become more probable when a well detector is used. A cutaway diagram of a well detector is shown in Figure 7.17. Basically, it's a NaI(Tl) detector with a hole drilled into it so the source can be placed in the center of the crystal. This dramatically increases the counting efficiency as well as the odds that two photons will enter the scintillation crystal at more or less the same time.

There's also a gamma spectral feature that indicates positron decay. Figure 7.18 shows the decay diagram for ²²Na, which decays by positron emission and electron capture. Both decay modes form the same excited state of the ²²Ne daughter, which emits a 1.275 MeV gamma photon when it decays to the ground state. As expected, its gamma spectrum has a photopeak at 1.275 MeV, but a large peak is also observed at 0.511 MeV. This peak is due to positron annihilation (with an electron) and is called an **annihilation peak**. The positron will most likely annihilate outside of the detector, generating two 511 keV photons. A number of these photons enter the detector, and those that deposit all their energy are recorded at 0.511 MeV. The ²²Na spectrum can also exhibit sum peaks at 1.02 MeV (two annihilation photons entering the detector at the same time) and at 1.786 MeV (1.275 + 0.511 MeV).

If the energy of the gamma photon is greater than 1.022 MeV (as in the last two examples), a peak at 0.511 MeV may be observed because of pair production *outside* the detector crystal. Remember that pair production creates an electron and a positron. If the positron annihilates outside the detector and sends one of the annihilation

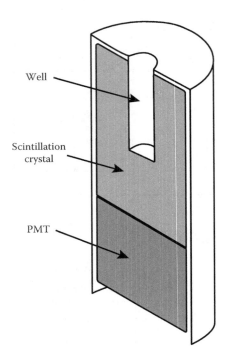

FIGURE 7.17   Cutaway view of a well detector.

photons into the detector, and it deposits all its energy in the detector, a count will be recorded at 0.511 MeV. Remember also that the odds of pair production are pretty low or zero at energies we're likely to see gamma photons (below ~1.5 MeV, Figure 7.12), that is, it's pretty unlikely you'll ever observe this from a gamma source. It would require a relatively high-activity source emitting pretty high-energy gamma photons—something you wouldn't want to be near.

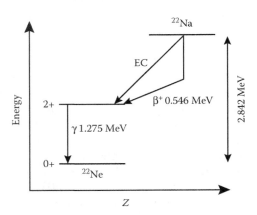

FIGURE 7.18   Decay diagram for ^{22}Na.

What if a high-energy photon gets inside the detector and interacts via pair production? Then the positron and the annihilation photons are created inside the detector. If one or both of the annihilation photons get out of the detector without interacting, then counts will be recorded with an energy of 0.511 (one escapes) or 1.022 MeV (both escape) below the photopeak. The peak that is 511 keV below the photopeak is called a **single escape peak**. The peak 1.022 MeV below the photopeak is called a **double escape peak**. An example of a spectrum exhibiting these escape peaks is shown in Figure 7.19. For these peaks to be observed, the photopeak must have an energy well above the 1.022 MeV threshold for pair production and have a fairly high activity. Again, the odds of most humans seeing this phenomenon in their lifetime is pretty low. Notice also that the intensity of the double escape peak is much lower than the single escape peak. The odds of both annihilation photons escaping is much lower than just one.

If lead is used for shielding or as a collimator, or is just lying around nearby, a peak at ∼75 keV could be observed. If a gamma photon from the radioactive source interacts with Pb via the photoelectric effect, it can create a vacancy in the lead K electron shell. When this vacancy is filled with an electron from a higher-energy shell, a lead characteristic X-ray can be produced. The X-ray can then enter the detector and record a count of its energy, which is called a **lead X-ray** peak.

When *a lot* of beta particles, relative to the number of gamma photons, are emitted from a radioactive source, *bremsstrahlung* is observed. Remember that bremsstrahlung is the production of X-rays as a high-speed electron slows down while interacting with matter. These interactions are highly variable and can produce a broad-energy spectrum of X-ray photons. Bremsstrahlung can occur either inside or outside the detector. All that matters to us is whether the detector records the X-ray. If it does, it'll show up as a count at the low-energy end of the gamma spectrum. Collectively, it is manifested in the far left of the gamma spectrum as a slow rise in the count rate as photon energy decreases. Figure 7.20 gives an example spectrum and the corresponding decay diagram.

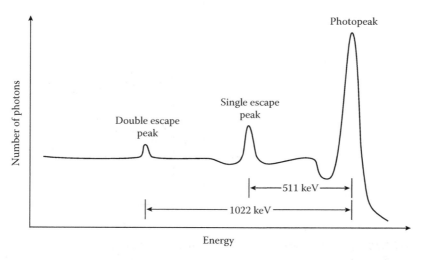

FIGURE 7.19 A gamma spectrum with escape peaks.

The decay diagram for ^{86}Rb (Figure 7.20b) shows that the beta decay branch ratio leading to gamma emission is small. This nuclide is spitting out mostly beta particles and only a few gamma photons. As a result, some bremsstrahlung is also observed in its gamma spectrum (Figure 7.20a).

If something is between the source and the detector, the gamma photons *can* undergo Compton scattering while traveling through the object on their way into the detector. This is known (logically!) as **object scattering**. Since Compton scattering lowers the energy of the photon, object scattering leads to relatively more counts observed under the Compton continuum and relatively fewer counts under the photopeak.

The **peak-to-total ratio** is a measure of detector quality. It is determined by summing the counts for all the channels under the photopeak and dividing by all counts observed in the entire spectrum (Figure 7.21). The peak-to-total ratio will depend on the energy of the gamma ray. The ratio generally decreases with increasing energy of the gamma ray. This makes sense as the *relative* probability of Compton scattering increases with increasing gamma ray energy. Overall, the probability of interaction goes down as gamma energy increases, but the odds of undergoing the photoelectric effect go down a lot faster than the odds of Compton scattering (Figure 7.12); therefore, the relative odds of Compton scattering increase for higher-energy photons. This means that the peak-to-total ratio is expected to be lower for a higher-energy

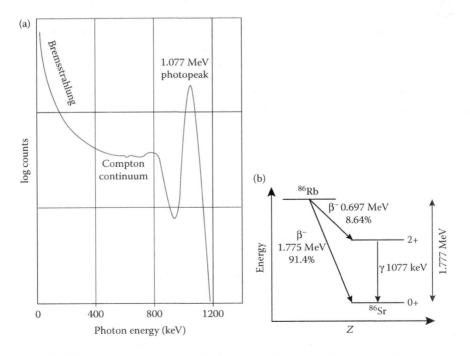

FIGURE 7.20 (a) Gamma spectrum. (Adapted from Heath, R. L. *Scintillation Spectrometry: Gamma-Ray Spectrum Catalogue*, 2nd ed., 1997, Idaho National Laboratory, Idaho Falls, ID.) (b) Decay diagram for ^{86}Rb.

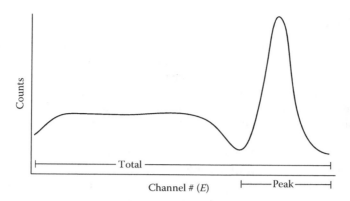

FIGURE 7.21   Peak-to-total ratio.

peak because more of its photons will interact with the detector via Compton scattering rather than by the photoelectric effect. If the detector is damaged, this ratio will increase.

Another measure of detector quality is its ability to resolve two peaks close in energy to each other. As mentioned previously in this chapter, resolution is a measure of how much random error is introduced by the detector and is manifested in the width of the photopeak. A wider peak means more error. Resolution is quantified by measuring the width of the peak halfway to the top (full width at half-maximum, or FWHM) and dividing by the energy of the photopeak (E_0):

$$\% \text{ resolution} = \frac{\text{FWHM}}{E_0} \times 100\%$$

Full width at half-maximum is illustrated in Figure 7.22. NaI(Tl) detectors typically have a resolution of 6%–7% at 662 keV (^{137}Cs), while lanthanum bromide detectors are ~3%, and germanium detectors are a phenomenal 0.3% at the same energy.

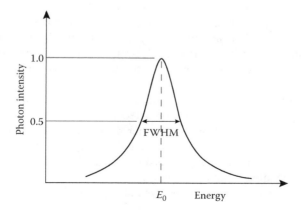

FIGURE 7.22   Resolution for a photopeak.

QUESTIONS

7.1 How could a proportional counter discern between alpha and beta radiation?

7.2 List the three types of gas-filled detectors in order of increasing voltage. Very briefly give a distinguishing characteristic or typical use for each.

7.3 Using information from Problem 6.4, determine which electronic transitions in argon atoms produce X-rays, UV, and visible light.

7.4 Would a Geiger-Müller counter be efficient at detecting neutrinos? Briefly explain.

7.5 A high-energy ($\sim$1.3 MeV) gamma-emitting nuclide is placed about 2 cm below a G-M tube, which records 360 counts over 30 s. A thin ($\sim$0.035 mm) sheet of aluminum is then placed directly over the source, and the detector records 460 counts over 30 s. Such an increase in counts is often observed under these circumstances. Can you explain why?

7.6 What would be the best detector to use if we only wanted approximate count information for an intense, polyenergetic photon beam, and we didn't want to significantly attenuate the beam?

7.7 What would be the best detector to obtain count and energy information from a beta-emitting nuclide? A gamma-emitting nuclide? Assume only one energy (particle or photon) is emitted in each case.

7.8 A nuclear medicine technologist working in radiopharmacy performs swipe tests weekly to look for areas of radioactive contamination. The swipe test is performed by wiping a small piece of paper (such as filter paper) across the surface to be tested, placing it in a glass test tube, then placing the assembly into a gamma spectrometer. If the only radionuclide in use at this pharmacy is ^{99m}Tc, comment on why this might not be an especially good quality control (QC) procedure.

7.9 How is bremsstrahlung avoided in organic scintillators?

7.10 Radiation therapists work with equipment that generates high-energy photons (X-rays). If they wanted to determine the monthly dose from small amounts of radiation "leaking" from their machines (*not* the main beam) into the treatment room, what would be the best way to go about it?

7.11 A 0.100 MeV gamma ray interacts inside a NaI(Tl) detector by the photoelectric effect. What element is most likely involved in this interaction? What is the atomic shell most likely to lose an electron from this element? Briefly explain how an iodine escape peak might be formed from this gamma ray. Calculate the energy of the freed electron.

7.12 In terms of detecting ionizing radiation, what is the fundamental difference between gas-filled and scintillation detectors?

7.13 Calculate the energy of the backscatter peak and the Compton edge for the gamma ray most commonly observed in the decay for ^{41}Ar.

7.14 The energy of a backscatter peak is observed at 213 keV. What is the energy of the corresponding Compton edge?

7.15 Backscatter peaks are often observed at slightly higher energies than calculated. Why might this be?

7.16 Sketch the gamma ray spectrum for ^{99m}Tc labeling the important features.

7.17 During a routine survey of the Crab Nebula, the science staff of the Stargrazer discovers an abundance of one element. Sensor analysis shows it to be Al, P, Cu, Br, or Tm. You obtain a small sample and place it in a neutron beam generated by one of the EPS conduits. Upon removal, you detect a 0.51 MeV photon using a flat-faced NaI(Tl) detector. The same sample shows peaks at 0.51 and 1.02 MeV when placed in a well detector. Using either detector, at least one other peak is observed in the gamma spectrum, but interference from the nebula is preventing you from observing their energies. Your report on this element is due to the ship's science officer in one hour. What can you contribute to the solution of this mystery? What experiment would you suggest be done next?

7.18 Explain the origin of the Compton continuum often observed in gamma spectra.

7.19 Starting with the relevant type of decay, outline the process by which an annihilation peak is observed in a gamma spectrum.

7.20 What two types of decay are likely to eventually produce an X-ray peak in a gamma spectrum.

7.21 Label the features observed in the gamma ray spectrum of ^{65}Zn (Figure 7.23).

7.22 When a lead collimator is used with a gamma camera, a peak at 75 keV is observed. When the collimator is absent, no peak is seen at 75 keV. Briefly explain the origin of this peak.

7.23 Why are Compton scattering and pair production undesirable interactions inside of a gamma detector? How are detectors designed to minimize these interactions?

7.24 A gamma photon enters a NaI(Tl) detector, but a count is not registered under the photopeak. Give four reasons why this might happen.

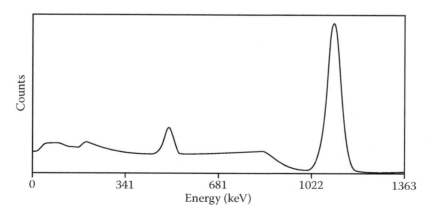

FIGURE 7.23   The gamma spectrum of ^{65}Zn.

7.25 The dashed line in the Figure 7.24 represents that gamma spectrum of a radio-nuclide in air, while the solid line is the spectrum for the same source inside of a patient. Briefly explain the differences between the two spectra. What happens to the peak-to-total ratio as this source is put inside of a patient?

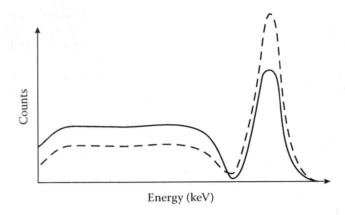

FIGURE 7.24 Gamma spectra taken of a nuclide outside (dashed line) and inside (solid line) a patient.

8 Applications of Nuclear Science II
Medicine and Food

Radioactivity was a newly discovered phenomenon early in the twentieth century. During the early decades of the century, it was commonly believed that deliberate exposure to ionizing radiation had health benefits. There were numerous radiation "treatments." Radium salts (inorganic compounds) were sold in health elixirs. People paid for mine tours to breathe air containing relatively high concentrations of radon or to soak in waters containing radioactive ions. Some of these practices continue today, and a health benefit from low-level exposures to ionizing radiation may be real.

There was also some excitement over consumer products containing radioactive materials at that time. Radium-containing cosmetics and radium-painted clock and watch dials were popular in the early part of the twentieth century because they glowed all night long. For much of the twentieth century, uranium compounds were used to provide colors in the glazes of ceramic products, most famously in orange and ivory Fiesta®[1] dinnerware from 1936 until 1973. Uranium was also used to provide a yellow or yellow green color to transparent glass. Called "uranium," "Vaseline," or "gemstone" glass, they were also manufactured until the early 1970s.

The excitement faded then turned to horror as the adverse health effects from significant doses of ionizing radiation became public. First were the "radium girls." Some of the workers (almost all young women) that applied radium-containing paint to clocks, watches, and aircraft instruments began to get sick and die. As they painted, they used their lips to sharpen the tips of their brushes, ingesting a significant amount of radium. In the middle of the century, images of the adverse health effects experienced by Japanese exposed to significant amounts of ionizing radiation from the Hiroshima and Nagasaki bombings were made public.[2] Shortly after World War II, numerous fictional representations of the possible horrors resulting from exposure to ionizing radiation became a staple of popular fiction that unfortunately persists today.

While fiction took considerable advantage of the possible health effects of exposure to ionizing radiation, scientists sought to quantify it. As discussed in Section 1.6, it is clear that above certain levels, adverse health effects are observed, and the risk increases in a roughly linear fashion with increased dose. At lower doses, it remains less clear whether there is still some risk, no risk, or a health benefit. While

---

[1] Fiesta is a registered trademark of the Homer Laughlin China Co., Newell, West Virginia.

[2] Less publicized were long-term studies of the radium girls and Japanese atomic bomb survivors who received lower doses showing they generally lived long, healthy lives.

scientific evidence supports either no risk or a health benefit to low doses of ionizing radiation, and there is no evidence to support the linear nonthreshold hypothesis, radiation workers still follow the ALARA (keeping doses as low as reasonably achievable) rule. Health physicists also work to lower patient doses whenever possible, assuming that additional exposure *could* present a risk to the patient. What if the additional exposure is also a health benefit? As evidence mounts to support this notion, we may discover that *some* of our horrors over the radiation treatments of the early twentieth century are misplaced.

This chapter deals mostly with the applications of ionizing radiation to health care, primarily in the treatment of cancer (radiation therapy) and the diagnosis of a variety of diseases (radiography and nuclear medicine). These health professions are normally considered part of the medical specialty of radiology, so it will be briefly discussed first. Because of its relationship to radiation therapy, food irradiation is also briefly discussed in this chapter. It may sound like an odd relationship now, but hang in there; it's a pretty good one.

8.1   RADIOLOGY

Radiology is a medical specialty that uses both ionizing (X-rays) and nonionizing (e.g., ultrasound) radiation for the diagnosis and treatment of disease. We'll only be concerned with those modalities that use ionizing radiation. One that is often thought of as involving ionizing radiation is magnetic resonance imaging (MRI). Perhaps it is because it was formerly known as nuclear magnetic resonance (NMR). An MRI takes advantage of the fact that the nuclei in the molecules in our body are all little (very little!) magnets. If placed in a strong magnetic field, they will align with the field. Once aligned, they can be tipped over using a radio wave, and, as they realign with the field, they give off their own radio signals, the exact energies of which depend on their local environment. Thus, muscle looks different from the blood vessels running through it. Radio signals are transmitted by photons, but they are low-energy (see Figure 1.1) and are nonionizing. An MRI simply tickles nuclei rather than violently transforming them.

Radiology departments are very high-tech and include a wide variety of imaging and therapeutic techniques. Perhaps the most familiar part of radiology is radiography—using X-rays to image the body's innards. The technique is pretty simple: shine a beam of X-rays on part of a patient and collect an image on the other side. As discussed in Section 6.4, X-rays are more likely to be absorbed by higher-density materials; therefore, they are generally absorbed by bones and transmitted through soft tissues. Radiography is, therefore, well suited to bone imaging, as bone shows up with the greatest contrast. Soft tissues have slight variations in density, which then appear as subtle differences on the radiograph (X-ray image). When necessary, contrast media can be administered to help light up particular areas. For example, a patient could swallow a $BaSO_4$ suspension,[3] which would then absorb X-rays trying to pass through the gastrointestinal system.

---

[3] Barium sulfate has an extremely low solubility in water and will, therefore, pass through the patient as a solid.

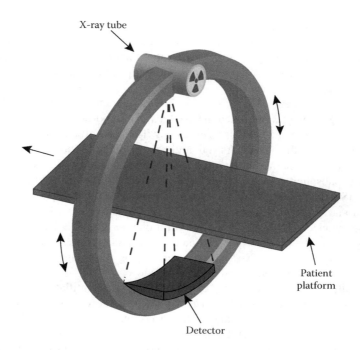

X-ray tube

Patient platform

Detector

FIGURE 8.1   A schematic of a CT scanner.

Conventional radiography is two-dimensional and cannot provide cross-sectional views. The X-ray beam travels through the body in one direction and produces an image of everything it travels through. If there are different organs stacked along the beam's path, they may be difficult to discern. Compton scattering will also take place as the beam passes through the (mostly low Z) patient, which will cause some fogging of the image overall, making it difficult to see some of the subtler differences in the image. If these limitations are important, they can be overcome using computed tomography (CT, also known as a CAT scan). The key difference from conventional radiography is that CT imaging involves the movement of both the X-ray source and detector around the patient, as shown in Figure 8.1. Several X-ray images taken at different angles can be put together to form a three-dimensional image of the patient's anatomy.

Computed tomography scans allow cross-sectional views without having to slice the patient in two (messy!). It also allows separate and complete viewing of organs that could otherwise be in the way of each other on a conventional X-ray. Finally, the errors produced by scattered photons can be minimized using collimators both before the beam enters the patient and after it exits. The collimators are basically honeycombs of lead that absorb any photons not traveling on a straight line from the X-ray source to the detector.

CT scans provide fairly detailed anatomical information but come with a cost. As you might guess, taking lots of X-rays to get enough for a clear 3-D image means higher doses to the patient. In fact, CT scans typically involve the highest patient doses for any diagnostic procedure. Manufacturers of CT machines, physicians, and health physicists have worked diligently to find ways to lower CT doses while still obtaining the information needed. The focus has been on limiting the amount of patient scanned while using the lowest intensity of photons needed to obtain a quality image. Their efforts have largely paid off, and patient doses have been lowered for many CT protocols.

Various detectors are used for CT and 2-D radiography, and most are based on the detector technologies presented in Chapter 7. An important consideration for any X-ray detector used in medicine is its efficiency. The more efficient it is, the lower the dose needs to be for the patient. As you might guess, higher Z solids tend to dominate because they can place more matter in the beam path and are more likely to interact.

Storage phosphor detectors work in a manner similar to thermoluminescent detectors (Section 7.3.2). When the X-rays interact with the phosphor, electrons are promoted and trapped in higher energy states. Instead of heat, laser light is used to release the electron, which then gives off a photon of visible light (at a different wavelength than the laser), which can be read. If made properly, phosphors can be quite efficient in detecting X-rays. However, getting the image is a two-step process—exposing the phosphor, and then reading it.

Also quite efficient are detectors based on inorganic scintillators (Section 7.2.2). These detectors use a scintillator, such as CsI(Tl), to convert the X-ray to visible light. The light then can be read several different ways, but one of the more interesting is by a charge-coupled device (CCD). CCDs are commonly used in digital cameras to snap pictures of you and your friends. These types of detectors produce the image in a single step, instantly displaying on a computer screen after the patient is zapped.

8.2   RADIATION THERAPY

Radiation therapy (RT) is the medical application of ionizing radiation to treat disease. Today, most radiation therapy procedures involve the bombardment of tumors with high-energy X-rays. The remaining RT procedures involve the placement of a sealed radioactive source inside or near diseased tissue with the intent to kill it. The former are generally termed external beam therapy and the latter implant or brachytherapy. The goal in performing radiation therapy is to deliver the greatest dose to the tumor while minimizing dose to the surrounding (healthy) tissue. The ratio of abnormal cells killed to normal cells killed is called the **therapeutic ratio**. These ratios are generally high for tumor treatment because tumor cells are rapidly dividing and, therefore, are much more susceptible to radiation damage from a high dose delivered over a relatively short time. The history of RT is all about increasing the therapeutic ratio.

The first 50 years (1895–1945, or so) of external beam therapy were dominated by treatment using machines that could only generate photons up to 500 keV. Because these instruments generated X-rays in much the same fashion as diagnostic X-ray machines (used to look for a broken bone), they are generally called "conventional" X-ray machines.[4] These treatments were often limited by skin reactions because the

---

[4] When generating X-rays from a tube potential of 150-500 kV, they are known as "orthovoltage" or "deep therapy" machines.

maximum dosage was always delivered to the skin. In fact, treatment time was often determined by observation of the skin under the X-ray beam—when the skin started getting too red, treatment was terminated.

Implant therapy was fairly popular in those days. Back then, radioactive materials were pretty much limited to the members of the naturally occurring decay series. ^{226}Ra was commonly used as it could be separated from uranium ores. As you will recall (Chapter 1), ^{226}Ra is in the middle of the ^{238}U decay series—there are still a dozen or so radioactive nuclides in the series before it stops at ^{206}Pb. This is bad news, as it means that most of these other nuclides will be present in any sample of ^{226}Ra once it has sat around for a while (about a month). All kinds of different ionizing radiation will be emitted. Enough high-energy gamma radiation is emitted that those working with these sources on a daily basis couldn't help but be exposed to significant doses.

Alpha decay is a big part of the decay of ^{226}Ra to ^{206}Pb. Every one of these alpha particles becomes a helium atom inside the sealed source. Pressure will eventually build up from the helium (and the ^{222}Rn produced in the decay chain) and burst the container. This was a big problem with radium implant sources because the radium was in the form of a fine powder—when a sealed source burst, it created quite a mess to clean up.

Finally, it was tough to get these sealed sources to stay put. Our bodies move— even at rest we breathe, circulate blood, and digest. It is very difficult, even for the most disciplined person, to remain motionless for long. When we move, something placed in our tissues will also move—not always in the same way. As a result, it was often difficult to ensure adequate dose distributions from sealed sources placed inside of humans. For all the reasons mentioned above, implant therapy fell out of favor during the first half of the twentieth century.

After World War II, nuclides such as ^{60}Co (emitting 1.17 & 1.33 MeV gamma rays during decay) became available, and radiation therapy instruments using ^{60}Co sources became the mainstay for about 30 years. As a result, some aspects of being a radiation therapist became more complicated. Higher energy photons meant that the maximum dose was now received below the skin, improving the therapeutic ratio but eliminating skin damage as a way to determine treatment times.

It was also during this time that radiation therapists started repositioning the gamma ray beam during or between treatments—firing at a tumor from different angles, as shown in Figure 8.2. This is known as **isocentric** therapy because the tumor is located at the intersection (isocenter) of the various beam paths. Isocentric therapy minimizes the dose to the healthy tissue between the tumor and the source by changing the path the photons take to zap the tumor. This technique greatly increases the therapeutic ratio. Determining variables such as treatment time, optimal beam shape, and angles of entry became quite complex when using isocentric techniques. Fortunately, computers were also developed in this time period, which greatly facilitated treatment planning.

In the late 1960s, researchers at Los Alamos National Laboratory developed compact linear accelerators that were capable of generating X-rays with maximum energies greater than 1 MeV in a relatively small space (such as a hospital room). Once again, higher-energy beams meant the ability to deliver more dose deeper inside the patient, improving the therapeutic ratio. As a result, these **megavoltage** machines (commonly

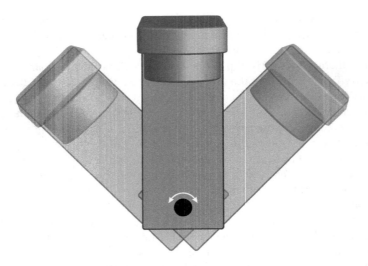

FIGURE 8.2 Isocentric radiation therapy.

called linacs) started replacing ^{60}Co instruments in the 1970s. It took a while, but they are now pretty much the only external beam therapy machines used by radiation therapists (Figure 8.3). With the dawning of the megavoltage era, still greater complexity was introduced. These are pretty high-tech instruments that produce very intense, very high-energy, very sharp beams. They can also be moved with greater flexibility relative to the tumor, and some can even change the shape of the X-ray beam as the beam is moved around the patient (**intensity-modulated radiation therapy**, or IMRT). Imaging of the tumor can also be done at the same time as treatment. After all, you're

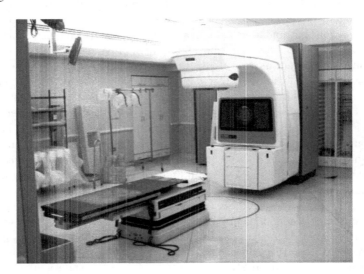

FIGURE 8.3 A modern-day linear accelerator radiation therapy machine. (Courtesy of Kris Saeger.)

running a whole lot of X-rays through a patient; might as well see if you're hitting the target. This technique is called **image-guided radiation therapy** (IGRT), and it allows for modification of the beam shape in real-time with the therapy. Finally, treatment can be synchronized with the patient's breathing (**respiratory-gated therapy**) when the treatment volume moves with respiration. Treatment planning is now heavily dependent on computers, but therapeutic ratios are higher than ever.

A few ^{60}Co radiation therapy treatment machines have survived for specialty applications. A machine known as a **Gamma Knife** arranges a couple hundred ^{60}Co sources in a dome-shaped device. Each source is carefully collimated so only a narrow beam of photons is not absorbed by shielding (these machines are heavy!). The beams are oriented so they all converge at a point. The patient's tumor (typically brain) is placed at the convergence point, and the patient is moved so the entire tumor gets some quality time at this point. A more flexible version of the Gamma Knife is the **CyberKnife**. A CyberKnife machine has a single ^{60}Co source in a shielded container that produces a narrow beam. This container is mounted on a robotic arm that can then position the beam hundreds of different ways for treatment. By maximizing the number of pathways taken by the high-energy photons, the therapeutic ratio can be improved. Treatment with the CyberKnife is best for relatively small tumors with lots of angles of approach. As a type of IGRT, CyberKnife machines collect diagnostic images during therapy to optimize location of the beam.

Implant therapy has experienced a bit of a renaissance in recent years, thanks in part to the use of new nuclides such as ^{192}Ir and ^{103}Pd. Iridium and palladium are chemically inert and, therefore, require minimal or no encapsulation. Other implant sources were typically encapsulated in stainless steel, while ^{192}Ir and ^{103}Pd can be encased in thin plastic. This allows more ionizing radiation (low-energy gamma photons and beta particles) to penetrate the casing. ^{192}Ir and ^{103}Pd both emit lower-energy gamma photons, and ^{192}Ir also emits beta particles. Because of their high specific activities and minimal encapsulation requirements, these sources can be quite small. In fact, they are often referred to as "seeds." Implant sources can be placed permanently inside a patient, but they are usually only temporarily put inside someone. To accomplish this, they can be placed on a wire or plastic line and inserted into the patient through a catheter. To minimize dose to the therapist, these sources can be inserted into and removed from the patient by remote control. This is known as **remote afterloading**.

Electron and proton beams are sometimes used for radiation therapy. Electron beams are readily available from the same X-ray machines used by radiation therapists to generate beams of high-energy photons. As explained in more detail in Chapter 13, X-rays are generated from beams of high-energy electrons hitting a metal target. Simply removing the target allows these electron beams to hit the patient instead. Proton beams are a little harder to come by, as they need a cyclotron to accelerate them, and rather large and powerful magnets are required to bend the beam. While a typical therapy room may not look very big (Figure 8.4), the rest of the equipment for the room is at least as large as the room itself. As a result, proton therapy centers are relatively expensive to construct and maintain, and they are much less common than those using X-ray and electron beams.

Since both electrons and protons are charged particles, they tend to deposit a large proportion of their energy at a specific depth below the surface of the matter (the

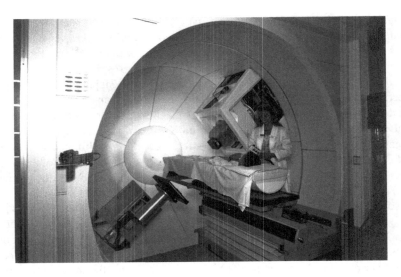

FIGURE 8.4   A proton therapy gantry. (Courtesy aof the National Association for Proton Therapy.)

Bragg peak—see Section 6.2). This depth depends on the energy of the particle: the greater the energy, the greater the penetration. This means that electron and proton beams tend to have great therapeutic ratios. The downside for electron beams is that they can undergo scattering by the nuclei of the matter they penetrate (more from Section 6.2), causing the beam to spread a bit inside the patient. Both electrons and protons also have a much greater probability of interacting with matter than high-energy photons and, therefore, cannot always penetrate as deeply as needed. Despite their drawbacks, both have specific applications where they outperform high-energy photon beams or implants, so both experience some use.

Beams of other particles have been examined for radiation therapy, but none have found a practical application. Beams of neutrons, pions (also known as π-muons), alpha particles, and nuclei of ^{20}Ne and ^{40}Ar have all been examined. None are currently in common use.

As mentioned later in this chapter, nuclear medicine is currently researching molecular radiotherapy—designer molecules with a very specific biodistribution and a radioactive payload. If successful, these therapies might be well suited to conditions that cannot be effectively treated with current beam or implant techniques. One example would be a cancer that has metastasized to several different locations within a patient. Beam and implant therapies work best on tumors that are well localized—if one has spread throughout a vital organ or the body, it is difficult to treat with these techniques without destroying the organ or body.[5] Ionizing radiation delivered on the cellular or molecular level could be much more effective.

---

[5] Some studies suggest that low half-body or full-body X-ray doses (~150 cGy over five weeks) are effective against more distributed cancers and are helpful in preventing recurrence.

8.3   FOOD IRRADIATION

Food irradiation is the use of ionizing radiation to kill biological organisms that are deleterious to foods. It may seem strange to place this topic in the same chapter with radiation therapy, but they have a lot in common. Both typically use high-energy photons to kill the bad stuff (diseased cells or bacteria) while trying to leave the bulk (healthy tissue or edible food) unaffected. Before we get into food irradiation, let's take a quick look at some other uses of high-energy photons.

The U.S. Postal Service uses ionizing radiation to kill anthrax spores or other potentially harmful organisms that could be sent through the mail. High-energy photons are typically used because of their ability to penetrate packaging. Ionizing radiation is also widely used to sterilize medical supplies such as bandages and consumer products such as tampons and condoms. A common public misconception is that such irradiation will make their mail, bandage, or condom become radioactive. We will learn in Chapter 9 that this can't be true—only projectiles such as neutrons, protons, and other nuclei could cause residual radioactivity in the matter they pass through. Most likely, ionizations will be caused, and chemical bonds will be broken—hopefully killing the bacteria and/or bugs while leaving the rest of the irradiated object alone. Just like radiation therapy, we want to kill the bad stuff while minimizing damage to the rest.

Food irradiation uses X-rays, gamma rays, or electron beams to rid food of pests or bacteria so that the food will not be damaged or spoil as quickly. In some cases, it can serve as an effective substitute for chemical pesticides and preservatives. It only works well with certain foods, and is typically used on fruits, vegetables, and spices. It can dramatically increase shelf life of some of these products. Food irradiation can be used with meats and is an effective method to prevent *E. coli* contamination in ground beef. Unfortunately, irradiation of meats can leave it with a bad taste—some say it is like burnt hair, disgusting!

Because very high doses are used, we can't expect that the ionizing radiation will only target the bugs. Some chemical damage is also done to the food. In the case of meat, it clearly alters some of the chemicals that give it flavor. Opponents to food irradiation suggest that irradiation affects the chemical makeup of any food and should not be used. Attempts to detect these chemicals have largely proved fruitless, which is expected since these chemicals could only be formed in very low quantities. The important question to ask is which has the most benefit and the least risk: treatment with chemicals, treatment with ionizing radiation, or no treatment at all. There are risks and benefits to each.

A typical layout for an irradiation facility is shown in Figure 8.5. The stuff to be zapped is loaded on a conveyer belt that takes it into the irradiation room then out again to be distributed. The irradiation room has extra-thick walls and contains a high-activity source (typically ^{60}Co or ^{137}Cs) or a radiation-generating device (for an X-ray or electron beam). The energy and type of the ionizing radiation as well as the size and materials of the packaging must all be considered to ensure that all the package contents receive the required dose. When not in use, the radioactive source can be lowered into a shielded storage area or switched off.

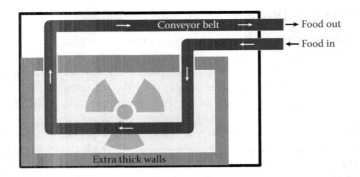

FIGURE 8.5   A schematic for a food irradiation facility.

8.4   NUCLEAR MEDICINE

Nuclear medicine is a healthcare specialty that uses radioactive nuclides to diagnose and treat disease. Nuclear medicine technologists (NMTs) prepare radiolabeled drugs (**radiopharmaceuticals**), which localize in a specific organ, then obtain images that provide information on the structure and function of the target organ. A wide variety of scans are possible, covering every major organ system in the human body. Nuclear medicine scans are even possible for some animals other than humans. This is a growing field, and entire textbooks are devoted to it. This section is an introduction to nuclear medicine from the perspective of the radionuclide.

8.4.1   RADIONUCLIDE PRODUCTION

Radioactive nuclides can be prepared using a variety of nuclear reactions (which we'll explore in depth in the next few chapters) using accelerators or nuclear reactors. Traditionally, most nuclides for use in nuclear medicine came from nuclear reactors. There are two ways reactors can prepare these nuclides: (1) as a product in a fission reaction, and (2) from neutron bombardment of a material placed in or near a reactor. To use a fission product, the nuclear fuel must be removed, and the desired nuclide must be separated from all of the other components in the fuel pellet. Fission products make up only about 3% of the spent fuel, and hundreds of different fission product nuclides are formed during fission. Additionally, the spent fuel is rather toasty (hot and radioactive!), making it difficult to handle. For all these reasons, obtaining medical radionuclides from spent nuclear fuel is usually undesirable.

A more attractive option for obtaining fission products would be to place a relatively pure ^{235}U target in the reactor. As the neutrons bombard it, some of its nuclei will fission, providing a much higher percentage of fission products when compared to starting mass (fuel pellets are mostly ^{238}U). ^{99}Mo, ^{131}I, and ^{133}Xe are all commonly obtained this way:

$$^{235}\text{U} + {}^{1}\text{n} \rightarrow {}^{135}\text{Sn} + {}^{99}\text{Mo} + 2{}^{1}\text{n}$$

$$^{235}U + {}^{1}n \rightarrow {}^{131}I + {}^{102}Y + 3{}^{1}n$$

$$^{235}U + {}^{1}n \rightarrow {}^{133}Xe + {}^{100}Sr + 3{}^{1}n$$

With nuclear fission, hundreds of fission reactions are possible; the three shown here are simply examples that happen to form desired products. Other reactions that produce more neutron-rich isobars can also decay to produce the desired products. For example, if ^{99}Nb is formed by a fission reaction, it will quickly decay ($t_{1/2} = 15.0$ s) to ^{99}Mo:

$$^{99}Nb \rightarrow {}^{99}Mo + {}_{-1}e$$

Using highly enriched uranium (HEU) as a target to make nuclides for radio-pharmaceuticals has become a concern in recent years. If intercepted while being shipped, the HEU could be used to make a nuclear weapon. Alternative methods, most still using fission but with lower enrichments of ^{235}U, are being developed to allay proliferation concerns.

Placing other elements/nuclides as targets in a nuclear reactor to undergo a nuclear reaction is also an attractive option. For example, ^{50}Cr can absorb one of the many neutrons zipping around inside a nuclear reactor and form ^{51}Cr along with a gamma photon:

$$^{50}Cr + {}^{1}n \rightarrow {}^{51}Cr + \gamma$$

Many such reactions are possible, but the main point is that if the target is a pure element or nuclide, then isolating the product should be a lot easier than fishing it out from among all the other stuff present in spent fuel or irradiated ^{235}U. Regardless of how they are prepared, reactor-produced nuclides *tend* to be neutron-rich and undergo beta (minus) decay (note that ^{51}Cr is an exception).

Increasing numbers of medical radionuclides are being prepared with cyclotrons. Cyclotrons are a type of accelerator, which are machines used to get particles (protons, electrons, etc.) moving fast enough to slam them into nuclei to make nuclear reactions happen. Cyclotrons are the accelerator of choice for clinical applications because they are relatively compact (about the size of a large fridge or a really small car) and affordable when compared to larger, more powerful accelerators. Protons, deuterons, and alpha particles are typically the particles accelerated in a clinical environment. For example, ^{18}F can be produced from firing a proton into ^{18}O. This reaction also spits out a neutron:

$$^{18}O + {}^{1}H \rightarrow {}^{18}F + {}^{1}n$$

Accelerators are necessary for nuclear reactions because the projectile (the ^{1}H in the reaction above) usually needs to be moving fast enough to hit the target nucleus (^{18}O). Since both have positive charges, they repel each other; so, if the proton doesn't have enough velocity, it'll simply be deflected by the ^{18}O nuclei. For some reactions, the required velocity might be too high for a cyclotron—an inherent limitation that must be considered. Other practical limitations are that the target nuclide be readily available, and it should be easy to separate the product. Efficient separation of the product can sometimes be of paramount importance because of the short half-lives of some of the cyclotron-produced nuclides. Some are so short that they are administered directly to the patient from the cyclotron facility (e.g., ^{13}N, $t_{1/2} = 9.97$ min). In contrast to reactor-produced nuclides, cyclotron-produced nuclides tend to be proton-rich, typically decaying via positron emission or electron capture.

How can a nuclear medicine department in a hospital, clinic, or private practice be assured of a consistent supply of the radionuclides it needs? Both cyclotrons and nuclear reactors are expensive facilities to have around, so most medical radionuclides are purchased and shipped to where they are needed. The farther your practice is from the production facilities, the more limited the variety of radionuclides that can be shipped. Those with very short half-lives would not survive the journey if it takes too long.

A short half-life is desirable for nuclear medicine (*vide infra*). How can a radiopharmacy assure a continuous supply of short half-life nuclides in the absence of a reactor or a cyclotron? If a longer-lived parent (to the desired nuclide) can be produced at the reactor or cyclotron, it may be packaged and shipped to a remote location. As the parent decays, the daughter is gradually produced. If the two can easily be separated, then the daughter can be used for radiopharmaceutical preparations. Such a system is called a **radionuclide generator**.

The parent needs to stay inside the generator while the daughter flows out. This is usually accomplished with an absorbent or ion exchange material. After the daughter is removed from the generator, time must pass for its activity inside the generator to build back up. Since separation can be repeated many times at regular intervals, the generator is often called a "cow" and the separation procedure "milking." Radionuclide generators are generally examples of transient equilibria, and the formulas discussed in Section 2.6.2 can be used to determine how often to milk the cow.

The most common generator in use today is the ^{99}Mo/^{99m}Tc ("moly-tech") generator. The ^{99}Mo is produced in nuclear reactors, as mentioned earlier in this section. It is not produced to be used directly in a radiopharmaceutical but to generate ^{99m}Tc. Figure 8.6 shows the effects on activity by typical milking of this generator. If enough time is allowed to elapse between milking, daughter (^{99m}Tc) activity can build up to close to or exceeding parent activity. Remember from Section 2.6.2 (Equation 2.15) that maximum ^{99m}Tc (daughter) activity can be calculated from the half-lives of ^{99}Mo (parent, 2.7476 d = 65.942 h) and ^{99m}Tc (6.008 h):

$$
\begin{aligned}
t_{max} &= \left[\frac{1.44 t_{1/2(^{99}\text{Mo})} t_{1/2(^{99m}\text{Tc})}}{\left(t_{1/2(^{99}\text{Mo})} - t_{1/2(^{99m}\text{Tc})} \right)} \right] \times \ln \frac{t_{1/2(^{99}\text{Mo})}}{t_{1/2(^{99m}\text{Tc})}} \\[2mm]
&= \left[\frac{1.44 \times 65.942\,\text{h} \times 6.008\,\text{h}}{65.942\,\text{h} - 6.008\,\text{h}} \right] \times \ln \frac{65.942\,\text{h}}{6.008\,\text{h}} \\[2mm]
&= 22.8\,\text{h}
\end{aligned}
$$

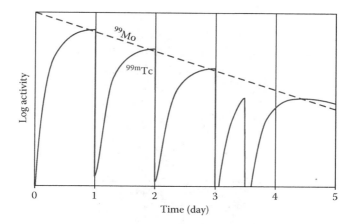

FIGURE 8.6 ^{99}Mo (parent, dashed) and ^{99m}Tc (daughter, solid) activities when ^{99m}Tc is removed at occasional intervals.

This time interval is quite convenient for us humans, as it is very close to one day. In Figure 8.6, ^{99m}Tc is removed every 24 hours for the first three days; shortly after, the ^{99m}Tc activity has peaked. Later in the third day, additional doses were necessary, so the generator was milked again. If insufficient time between milking elapses, daughter activity will still be below parent activity. The generator was not milked on the fourth day, and transient equilibrium will finally be reached late on the fifth day if the generator is not milked again (more than seven daughter half-lives).

Note also that parent (^{99}Mo) activity also decreases with time in Figure 8.6, as expected for transient equilibrium. Eventually (typically a week), the generator will no longer be able to crank out sufficient ^{99m}Tc activity to provide the doses necessary, and a new generator will be purchased.

A cut-away view of the ^{99}Mo/^{99m}Tc generator is shown in Figure 8.7. Overall, the generator is a cylinder, about 25 cm tall. Sterile saline solution is added through a port on the top. This solution flows through an Al_2O_3 (alumina) column with the ^{99}Mo packed in at the top. Before loading it on the alumina, it is oxidized to MoO_4^{2-} (molybdate, probably using H_2O_2). TcO_4^- (pertechnetate) is produced when molybdate undergoes beta decay and does not bind to the alumina as strongly as molybdate. The relative affinity for alumina can be related to ionic charge. Molybdate has a 2− charge, while pertechnetate only has only a 1− charge. Generally speaking, the higher the charge, the more likely it'll stick to alumina. This allows elution of pertechnetate using the saline solution without bringing much, if any, parent along. It is usually eluted directly into a **kit**—a bottle containing chemicals that will react with pertechnetate to form a radiopharmaceutical. Other radionuclide generators used in nuclear medicine include ^{68}Ge/^{68}Ga, ^{82}Sr/^{82}Rb, and ^{113}Sn/^{113m}In.

Nuclides that are produced without the presence of isotopes or isomers (carrier) of the desired nuclide are called **carrier-free**. In reality, most radionuclide preparations will result in the presence of some carrier. Therefore, most radionuclide preparations are termed as **no carrier added** (NCA) to indicate that no isotopes or isomers of the

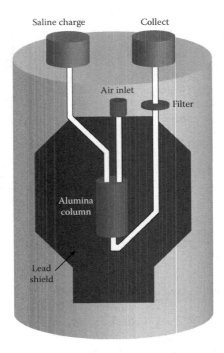

Saline charge   Collect

Air inlet

Filter

Alumina
column

Lead
shield

FIGURE 8.7   Cut-away view of a ^{99}Mo/^{99m}Tc generator.

product were deliberately added. The ^{99m}Tc preparation described previously may initially seem carrier-free, but as soon as it is produced, ^{99m}Tc begins to decay to ^{99}Tc. Some ^{99}Tc (carrier) will be present in any sample of ^{99m}Tc. The presence of carrier is undesirable because isotopes or isomers of the desired nuclide will have different nuclear properties but nearly identical chemical properties. They will, therefore, be unlikely to help in imaging the target organ and may interfere. The term "carrier" generally refers to a stable isotope or isomer, but a radioactive isomer such as ^{99}Tc is also considered a carrier because it is (very nearly) a purely beta-emitting nuclide with a very long half-life (213,000 years!).

8.4.2   RADIOPHARMACEUTICALS

As already mentioned, a radiopharmaceutical is a drug containing a radioactive nuclide used for diagnostic or therapeutic purposes. In reality, $\sim$95% of all radiopharmaceuticals are used for medical diagnoses, and only $\sim$5% are used for therapy. Radiopharmaceuticals can be as simple as an atom (^{133}Xe), ion (^{125}I$^-$ or ^{82}Rb$^+$), or a small organic (^{18}F-glucose) or inorganic (^{99m}Tc(CNR)$_6^+$) compound, or it can be as complex as a large biomolecule (radio-labeled protein, antibody, etc.). The exact nature of the radiopharmaceutical depends on how it will be used.

What are the qualities of a good diagnostic radiopharmaceutical (sometimes called a radiodiagnostic agent)?

1. It should have a good biodistribution. In other words, it should localize primarily in the target organ. It is also helpful if it does not localize in other organs nearby, as this may interfere with obtaining a good image of the target.

2. The radionuclide should emit only gamma rays. Gamma emissions are important, as a nuclide emitting solely alpha or beta particles will not be visible outside of the patient and will give the patient a significant internal dose. If it decays via IT, it should have a low conversion coefficient (α_T). If it decays via EC, it should have a high fluorescent yield (ω_T). Additionally, the gamma rays should be between 50 and 500 keV in energy (roughly speaking). Any lower and the patient will absorb too many, and much higher will make detection too inefficient (see Section 6.2). The amount of radionuclide administered should be enough to provide a good image but no more—minimizing the dose to the patient.

3. The radioactive nuclide should also have a short half-life. The half-life needs to be long enough for the radiopharmaceutical to be prepared, injected into the patient, localize in the target organs, and be imaged. As a practical matter, it should be at least a few minutes. A short half-life will maximize the number of decays that take place during imaging, ensuring good pictures. It will also minimize the dose to the patient (and others nearby) and possible contamination (due to excretion) following the procedure.

4. Finally, the nuclide should be readily available, relatively pure, and decay to a stable, benign daughter. As discussed earlier, the radionuclide needs to be produced on-site or shipped to the site. A radioactive daughter will mean additional patient dose and may interfere with the scan. Benign simply means that it should not be harmful to the patient and the environment—the radionuclide should not decay to an element with high toxicity for humans or that is particularly harmful to the environment. These are not great concerns, as very small amounts of radioactive materials are used in diagnostic procedures.

5. The radiopharmaceutical should be relatively inexpensive and easy to make. If the materials (radionuclide + chemicals) to make a radiopharmaceutical are too costly, it is less likely to be approved by insurance companies (the kiss of death!). If its synthesis is too complex, radiopharmacies are less likely to be able to make it.

6. Once imaging is complete, the radiopharmaceutical should be metabolized and excreted by the patient in a reasonable amount of time. This also minimizes the patient dose. If the patient will excrete a significant amount of the dose soon after the procedure, then a special bathroom should be used. The patient should also be warned if their urine and/or feces represent any radiological hazard.

^{99m}Tc is the most commonly used radionuclide for radiopharmaceutical preparation. Its generator is convenient in that ^{99}Mo is easily prepared in a reactor, and the half-lives of ^{99}Mo and ^{99m}Tc lend themselves well to human schedules (as shown previously). Also noted previously is that the Mo-Tc generator has a useful lifetime of about one week, also convenient for institutions working a five-day week. The only significant disadvantage to ^{99m}Tc is that it does not decay to a stable daughter.

^{99m}Tc generators are also relatively inexpensive,[6] and ^{99m}Tc emits a 140 keV gamma photon in 89% of its decays (low α_T) with a half-life of six hours. Technetium is an element with a great deal of chemical flexibility. It can form a wide variety of chemical compounds—with coordination numbers (number of things attached to the metal) around Tc from 4 to 9, and oxidation numbers (relative number of electrons localized on the metal) from –1 to + 7. ^{99m}Tc is typically produced in the generator as TcO_4^- (pertechnetate ion, coordination number = 4, oxidation number = + 7), and reduction (lowering of the oxidation number) is almost always required. This is accomplished by the tin(II) (Sn^{2+}) ion, which is oxidized to tin(IV) (Sn^{4+}). When the technetium is reduced, other chemicals (**ligands**) must bind to it. For example, the preparation of the myocardial perfusion agent, ^{99m}Tc-sestamibi, can be characterized by the following chemical reaction:

$$TcO_4^- + 6CNR + 3Sn^{2+} + 8H^+ \rightarrow Tc(CNR)_6^+ + 4H_2O + 3Sn^{4+}$$

where CNR is 2-methoxy isobutyl isonitrile, the ligand. Technetium in $Tc(CNR)_6^+$ has a coordination number of 6 and oxidation number of +1.

The chemistry is easy to write down on paper, but there are some challenges when preparing radiopharmaceuticals. First, the solutions are very dilute—$^{99m}TcO_4^-$ has a concentration of about 10^{-9} moles/liter (nanomolar). The reactions must also proceed in high chemical yield, as injection of other radioactive chemical species will likely interfere with imaging of the target organ.

$Tc(CNR)_6^+$ has a rather symmetric structure and appears to the body as a large positively charged sphere. The body confuses it for a potassium ion, so a fair bit of it ends up in the heart, making it a good myocardial imaging agent. Other radiopharmaceuticals used for heart imaging look even more like the potassium ion (K^+); they are $^{82}Rb^+$, $^{13}NH_4^+$, and $^{201}Tl^+$. ^{82}Rb and ^{13}N both decay via positron emission. The positron annihilates in nearby tissue, generating two 511 keV annihilation photons. Because of the higher photon energy (relative to ^{99m}Tc), ^{82}Rb or ^{13}N are sometimes preferred for larger patients. Because they have more tissue for the photons to penetrate, more of ^{99m}Tc's 140 keV photons will be absorbed before they can leave a larger body. Note also that two photons are produced for every positron decay.

Sometimes the physical form of the radiopharmaceutical is more important that its chemical form. Several ^{99m}Tc colloids are used for different imaging applications. A colloid is a bunch of molecules that stick together or a really large molecule. Colloids are still small enough to disperse, but not dissolve, in a solvent, and they are not big enough to precipitate out. Generally speaking, a colloid is the dispersal

---

[6] Atomic Energy of Canada, Ltd. (AECL) produced a significant portion of the ^{99}Mo used for Mo-Tc generators worldwide and did not recover its full production costs. Like the element itself, the cost for a Mo-Tc generator was artificial. This is changing, as AECL plans to end ^{99}Mo production in 2018.

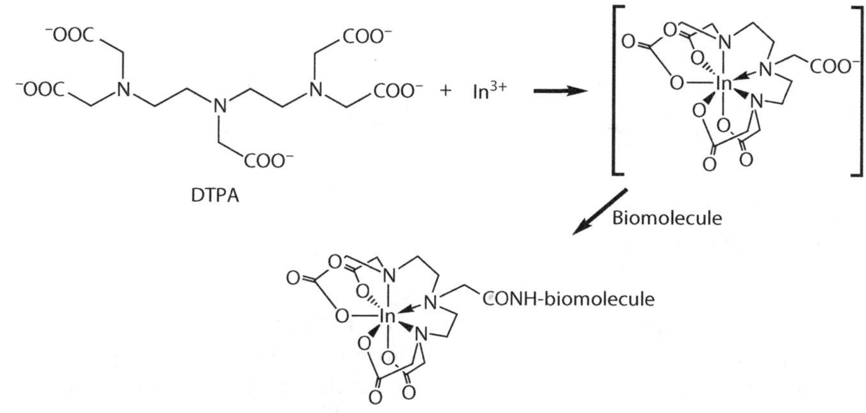

FIGURE 8.8   Preparation of a "tetherball" radiopharmaceutical using indium and DTPA.

of one phase of material in another. Smoke is a solid dispersed in gaseous air and is, therefore, a colloid. Milk is also a suspension of liquid fats in liquid water—the fat is a colloid. The subsequent mixture is called an emulsion because its two components are immiscible. The key to a ^{99m}Tc colloid is the size of the multimolecular aggregate that forms, as the size of the colloid will determine its application.

Most ^{99m}Tc radiopharmaceuticals in current use can be related to ^{99m}Tc-sestamibi in that they started out as a chemical compound that was easily prepared from nanomolar solutions of pertechnetate, was stable under physiological conditions, and had some (typically unanticipated) favorable biodistribution. Most were discovered through trial and error, which can be a time consuming and inefficient way to discover new radiopharmaceuticals. More recently, radiopharmaceutical research has started with biodistribution and then looked for a way to attach a radioactive nuclide. Two successful approaches are discussed next.

The first involves forming a chemical linkage between a molecule with a known biodistribution, such as an antibody, carbohydrate, enzyme, protein, or peptide. In the example shown in Figure 8.8, a diethylenetriamine pentaacetate (DTPA) molecule is wrapped around an indium(III) ion (^{111}In^{3+}). When a single ligand binds to a single metal ion with more than one group, it is called a **chelate**. Chelating ligands such as DTPA can also be used to help sequester radioactive materials that have been inadvertently ingested (chelation therapy). The DTPA wraps itself around the radioactive ion and facilitates its clearance from the body. Back to the reactions in Figure 8.8, notice that one of the five acetate groups ($-COO^-$) remains unbound after the DTPA binds the indium ion and is used to form a chemical bond to the biomolecule. This approach could be called a "tetherball," where the tether is the biomolecule, and the ball is the chelated metal ion.

This approach can fail if the presence of the ball affects the tether's ability to localize in the desirable organ. It could be that the ball binds to a critical part of the tether or that the addition of the ball changes the overall shape of the biomolecule to a significant extent. Synthesis of the radiopharmaceutical can also be more complex if the two steps need to be separated.

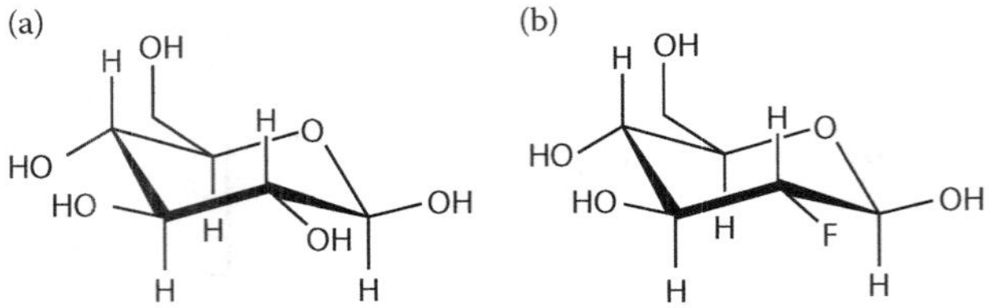

FIGURE 8.9 (a) Glucose and (b) FDG.

The most elegant approach is to incorporate the radioactive nuclide in the biomolecule—replacing one of the atoms with a radioactive isotope, for example, replacing the nitrogen in an amino acid (a basic biochemical building block) with a ^{13}N. The synthetic challenges here are daunting, especially since most of the radioactive nuclides that could be used have rather short half-lives and are cyclotron-produced. ^{18}F-fluorodeoxyglucose (FDG) is probably the most widely used of this type of radiopharmaceutical, although it represents a minor biochemical compromise.

As can be seen in Figure 8.9, the compromise is that FDG replaces one of the hydroxyl (–OH) groups with a fluorine atom (–F). In terms of size and arrangement of electrons, these two groups are similar enough that the body mistakes FDG for glucose. FDG is used to study metabolism in the brain and heart and, most significantly, is used to detect tumors at the cellular level. Tumors are sugar hogs, so FDG will collect in tumors (even pretty small ones) if the body isn't doing much else. As a result, oncology patients injected with FDG are asked to sit or lie quietly lest the FDG start collecting in whatever muscles are in use. The Mayo clinic used to allow FDG patients to watch television while they waited for their scans but found uptake in the right thumb and parts of the neck, especially in male patients. FDG in the right thumb came from continuous channel surfing and the neck from holding the head in a position to watch TV. They removed the TVs from the injection rooms, and the "anomalous" uptakes disappeared. Just goes to show that watching TV is quite a workout!

Like ^{82}Rb and ^{13}N mentioned earlier, ^{18}F is a positron-emitting nuclide; therefore, FDG is used as a positron emission tomography (PET) agent. As you remember from Section 5.3, positrons annihilate with electrons to produce two 511 keV photons that travel on trajectories that are approximately 180° apart. If a patient is injected with some FDG, it will be emitting positrons that travel a short distance before annihilating, generating two photons traveling in opposite directions. To efficiently detect these photons, the patient is placed in the center of a ring of detectors (Figure 8.10).

The chief advantage of PET over using a nuclide such as ^{99m}Tc is that PET generates two photons per decay, whereas ^{99m}Tc only produces a single photon to be detected. Both can provide 3-D images of the target organ provided detectors can (more or less) encircle the patient. When this is done with a single photon nuclide such as ^{99m}Tc, it is called single-photon emission computed tomography (**SPECT**).

PET imaging is growing in importance within nuclear medicine, in part because radiopharmaceuticals such as FDG are examples of **molecular imaging**. That is to say, they illuminate biological processes at the molecular or cellular level. They allow

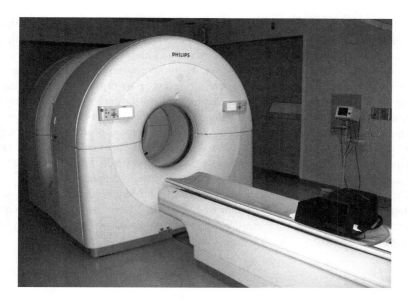

FIGURE 8.10   A Phillips PET/CT camera.

understanding of physiology through the actual biochemical reactions that take place inside of human beings. Since these processes are somewhat unique to each individual, this has the effect of personalizing the diagnosis and treatment of disease. Molecular imaging is so prominent for nuclear medicine that the U.S. professional organization for nuclear medicine, the Society of Nuclear Medicine (SNM), expanded its mission in 2007 to include "advancing molecular imaging and therapy." As our understanding of human beings continues to become more detailed at the biochemical level, our ability to detect and treat disease at this level will become more effective.

Coupled with the increasing abilities of radiopharmaceuticals to detect disease at the cellular and molecular level are **hybrid** (or fusion) technologies. Not to be confused with high-mileage cars, this type of hybrid involves the coupling of nuclear medical imaging with computed tomography (CT) or magnetic resonance imaging (MRI). The nuclear medicine scan provides physiological information, while a CT or an MRI provides anatomical information. When the two scans are put together, it provides medical professionals with detailed information within and around the volume of interest. To accomplish both scans, the immobilized patient is typically placed on an exam table that travels through both types of scanners. For example, a PET/CT camera will pass the patient through a CT scanner, then a PET detector like poking your finger through the holes of two doughnuts sitting up on their sides (Figure 8.10).

So far, we've only discussed diagnostic agents. As mentioned above, nuclear medicine is moving toward "molecular imaging *and therapy*." As molecular imaging advances, it is easy to imagine swapping out a gamma- or positron-emitting nuclide with one that emits alpha or beta particles, Auger or conversion electrons, or low-energy gamma rays. These forms of ionizing radiation deposit a lot of energy (i.e., do a lot of damage!) in nearby tissue. Only specific tissues (such as tumors) are targeted for destruction, so the radiotherapy agent will need an excellent biodistribution. Like radioimaging agents, they should also be readily available and easy to make.

At the time of this writing, molecular radiotherapy remains an active area of research, but only a few agents are currently in use. Part of the problem is that the agents do not localize strongly or quickly enough in the target tissue, leading to a significant dose to healthy tissue. It should also be noted that beta- and low-energy gamma-emitting nuclides are currently showing the most promise. The daughter recoil that follows alpha emission tends to eject the daughter nuclide from the radio-pharmaceutical, and likewise, Auger electron emission tends to be disruptive to the chemical bonds that hold the daughter to the rest of the radiopharmaceutical.

The most common radiotherapy agent in use by nuclear medicine today is a simple ion: $^{131}I^-$. As mentioned earlier in this chapter, ^{131}I can be separated from the other fission products of ^{235}U, but it can also be produced by throwing neutrons at ^{130}Te, producing ^{131}Te, and then letting it decay:

$$^{130}Te + {}^1n \rightarrow {}^{131}Te + \gamma$$

$$^{131}Te \rightarrow {}^{131}I + {}_{-1}e$$

Iodide collects in the thyroid, so $^{131}I^-$ is used to treat hyperthyroidism and thyroid cancer. ^{131}I emits gamma rays as part of its beta decay, so it can also be used for imaging. This is also handy when used for therapy because the biodistribution can be determined as well as the dose received by the thyroid. ^{131}I decays to the stable and chemically inert nuclide, ^{131}Xe.

$$^{131}I \rightarrow {}^{131}Xe + {}_{-1}e$$

Other radioactive nuclides used in therapy include ^{153}Sm, ^{89}Sr, and ^{32}P, which are used for bone pain palliation for patients with bone cancer; ^{90}Y can be used in combination with Zevalin to treat non-Hodgkin's lymphoma; and ^{32}P can also be used to treat polycythemia vera (increased red blood cell counts due to bone marrow hyperactivity) and certain cases of leukemia. $^{223}Ra^{2+}$ is also used for therapy for certain patients with bone cancer. The radium ion chemically mimics the calcium ion, so it is rather efficiently incorporated into parts of the body experiencing rapid bone growth (tumors). Notice that ^{223}Ra decays via alpha but gets around the recoil problem by not being part of a molecule.

Radiopharmaceuticals are sometimes referred to as radiotracers, or simply as tracers. These are more general terms that can apply to experiments using small amounts of radioactive nuclides to "follow" any chemical or biological process. The basic idea is that a radioactive nuclide will exhibit the same chemical and physical behavior as its stable isotopes. By mixing a small percentage of radioactive nuclide into a system under study, a lot can be learned about it without changing it in any way. Examples include a wide variety of analysis, such as determining the solubility product constant (how much dissolves, K_{sp}) of relatively insoluble salts or understanding how CO_2 is assimilated in plants during photosynthesis.

8.4.3 GAMMA CAMERAS

The NMT injects the radiopharmaceutical into the patient; it localizes in the target organ and is busy decaying away, sending out gamma photons in all directions. To "see" the organ being imaged, this radiation needs to be detected but also localized as to its point of origin. To accomplish this, a **gamma camera** (also known as scintillation or Anger camera) is used. Most gamma cameras work like an inorganic scintillator detector, described in Section 7.2.2, that is spread out over two dimensions to allow them to locate the gamma ray interaction within the crystal. In essence, they take a photo of the gamma rays emerging from the patient.

One of the key modifications over a simple detector is the addition of a collimator between the scintillator and the source of the radiation (the patient). Collimators are large slabs of high Z metal[7] with holes drilled through them along the short dimension. These holes can be hexagonal (making it look like a honeycomb), square, triangular, or round—the first two shapes are the most common. As shown in Figure 8.11, this allows photons to travel only on a straight-line path from the patient to the camera. This path is perpendicular to the face of the scintillator crystal.[8]

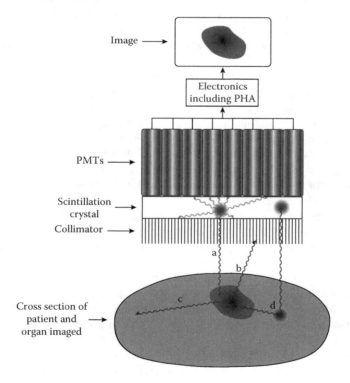

FIGURE 8.11 How a gamma camera works.

---

[7] High atomic number means a higher probability of absorbing the photon. Lead and tungsten are most common.

[8] Collimators can have different configurations than the one shown in Figure 8.11.

To better understand how a gamma camera works, let's start with the various ways a gamma photon can emerge from the patient and examine how they may or may not end up generating a signal in the camera, as shown in Figure 8.11. Path a is the desirable path because these photons travel straight up from the patient to the camera, that is, we know where it came from. Path b is also straight from the patient to the detector but is at an angle. If this generates a signal, we'll think it came from directly below its point of entry in the camera, which would be bad information. The collimator thankfully absorbs this photon. Path c misses the camera altogether and is not recorded. Path d starts out like it will miss the camera but then undergoes Compton scattering within the patient and ends up in the camera. This is a potential problem that will be resolved, but first we need to know a little more about how a gamma camera works.

NaI(Tl) crystals are most commonly used in gamma cameras and vary in size and shape. They are often roughly $38 \times 50\ cm^2$ ($15 \times 20\ in^2$) but are only about 1 cm (~1/2 inch) thick. The thinness is important, as it minimizes multiple Compton scattering within the crystal, which could give erroneous location information. The thinness is also a tradeoff as a thinner crystal will be less efficient in detecting gamma photons, that is, too many photons will traverse the crystal without interacting. Cameras with thicker crystals are used for nuclides that produce higher energy photons (e.g., annihilation photons) and thinner crystal machines for lower energy photons (^{99m}Tc, ^{201}Tl, and ^{123}I).

When a gamma photon deposits its energy in an inorganic scintillator via the photoelectric effect (an "event"), it will produce visible light photons. These photons can then take different paths through the crystal, as shown in Figure 8.11, entering several different photomultiplier tubes (PMTs) covering the top face of the crystal. Modern cameras have 60–120 PMTs packed on top of their scintillator. The tube directly above the event will likely see the most photons, but those around will also see varying amounts. By weighting the signal intensities from nearby PMTs, the exact position of the event (x, y coordinates) within the crystal (and therefore within the patient) can be calculated.

As part of the electronic processing of the signals, all the PMT signals for each event are summed (called the "Z pulse") by the pulse height analyzer (PHA) to determine if the total energy of the gamma photon was deposited in the crystal. Because of the relatively large amount of random error associated with NaI(Tl) detectors, a broad energy window is usually set. This insures inclusion of the entire photopeak. The main purpose of the PHA is to reject lower energy signals that may be due to Compton scattering, such as from path d in Figure 8.11. It will also reject X-ray photons that enter the crystal after a gamma photon interacts with the collimator (path b).

The scintillator and PMTs and some of the electronics constitute what is called the **detector head**. Detector heads on a camera can be held in place relative to the patient and collect a 2-D image. They can also move around the patient collecting multiple images from a variety of angles, like a CT scan, to produce 3-D images. Cameras often have two to three heads to facilitate 3-D imaging. In the case of PET imaging, detectors completely encircle the patient (Figure 8.10).

While the head design described here remains the mainstay for gamma cameras, new designs are available. These designs can provide greater counting efficiency (lowering patient dose and/or counting time) and more accurately locate the origin of the radiation within the patient. The detector heads are also considerably smaller than conventional units. These designs use lots of small scintillator crystals, each with a position-sensitive PMT, which allows location of the event within the crystal. While NaI(Tl) is used in some, others contain CsI(Tl) or LaBr$_3$ (see Section 7.2.2), and others use semiconducting CZT (see Section 7.3.1).

QUESTIONS

8.1 Why is ALARA followed when working with radioactive materials and X-ray machines? How is ALARA potentially harmful when applied to possible ionizing radiation exposures to the general public following nuclear power plant accidents like Chernobyl and Fukushima?

8.2 Why might some working in radiology departments be expected to have greater longevity than the general population?

8.3 What areas of radiology use ionizing radiation? Briefly describe each.

8.4 TLDs can be placed on a patient to measure X-ray exposure during radiography procedures. The badge is often invisible on the image. How can this be?

8.5 What is the difference between a CT scan and a conventional (2-D) radiograph?

8.6 How might CT scans be useful in radiation therapy? In nuclear medicine?

8.7 What is the therapeutic ratio, and how do radiation therapists improve it?

8.8 Why is gas production undesirable for sealed implant sources? Where does this gas come from?

8.9 Calculate the specific activity of ^{192}Ir. Why is high specific activity good for an implant source?

8.10 Why do electrons tend to spread out their dose when used in therapy when photons do not?

8.11 What is the most likely interaction between high-energy photons ($\sim$10 MeV) and the low Z matter that makes up our bodies?

8.12 The dose delivered to radiation therapy patients undergoing external photon beam therapy is due primarily to the photons interacting and what else?

8.13 Describe a simple way that dose at some depth inside a human could be estimated for external beam photon therapy.

8.14 A patient has a small tumor in his neck. It is only 2 cm below his skin, but his spine is not far behind the tumor. The spine is very sensitive to ionizing radiation, so it is important to minimize its dose. Would it be better to treat this tumor with photons or electrons? Briefly explain.

8.15 Calculate the activity for an ^{192}Ir seed that was calibrated 200 days ago at 3.50×10^5 MBq.

8.16 Briefly explain why there might be some convergence of radiation therapy and nuclear medicine in the future.

8.17 How are radiation therapy and food irradiation alike?

8.18 What are the main benefits of food irradiation? What are the main drawbacks?

8.19 Why do reactor-produced nuclides tend to be neutron-rich while accelerator-produced nuclides tend to be proton-rich?

8.20 Why would radioactive nuclides produced by cyclotrons be preferred over the same ones produced in a nuclear reactor? When would those from reactors be preferred?

8.21 A 28 mCi dose of a ^{99m}Tc myocardial agent is injected into a patient. How many atoms of ^{99m}Tc were injected? What mass of ^{99m}Tc is present? This radiopharmaceutical kit also contains 0.050 mg of $SnCl_2 \cdot 2H_2O$. What is the mole ratio of tin to technetium?

8.22 A ^{99}Mo/^{99m}Tc generator is loaded with ^{99}Mo but currently has no ^{99m}Tc activity. How long will it take for ^{99m}Tc activity to peak?

8.23 What would be a simple, safe, and expeditious way to test for ^{99}Mo impurities in a ^{99m}Tc sample recently eluted into a radiopharmaceutical kit from a ^{99}Mo/^{99m}Tc generator?

8.24 What is the purpose of a radionuclide generator? How does it relate to dairy farming?

8.25 Would ^{60m}Co be a good choice for a nuclear medicine diagnostic procedure? Explain.

8.26 Could ^{34m}Cl be used as a nuclear medicine diagnostic agent? Explain.

8.27 What are the differences between a diagnostic and a therapeutic radiopharmaceutical?

8.28 Why are ^{82}Rb$^+$ and ^{201}Tl$^+$ good myocardial imaging agents?

8.29 Why are PET images inherently fuzzy?

8.30 If ^{131}I is produced exactly as illustrated at the end of Section 8.4.2, would it be carrier-free, no carrier added, or neither? Briefly explain.

8.31 How could a lead X-ray be observed by a gamma camera? Why would it not be used to generate the image?

8.32 Why are the new detector materials mentioned at the end of Section 8.4.3 more efficient at detecting gamma photons?

9 Nuclear Reactions

Radioactive decay is not considered a nuclear reaction. Some readers may consider this strange because of the obvious analogy to chemical reactions. One chemical compound breaking up to form two or more compounds is called a decomposition reaction by chemists and considered no less a chemical reaction than an acid-base neutralization reaction or any other type of chemical reaction. So why is nuclear decay not considered a reaction?

Nuclear scientists define nuclear reactions as the collision of two nuclei, or a nucleus and a neutron, resulting in the production of a different nuclide. This definition does not include decay processes, which could be defined as the spontaneous transformation of one nuclide into another, usually accompanied by the nuclear emission of a particle or photon. This may all seem like a relatively pointless semantic debate, but the reader should be aware that it is generally considered improper to refer to nuclear decay as a reaction.

The following is a generic representation of the nuclear reactions we'll study in this chapter:

$$A + x \rightarrow B + y$$

where A is the **target** nuclide, x is the **projectile**, B is the **product** nuclide, and y is an emitted particle or photon.[1] The projectile x and emitted y can be a gamma photon or a wide range of particles, such as an alpha particle (α or ^{4}He), neutron (n), proton (p or ^{1}H), deuterium (d or ^{2}H), or tritium (t or ^{3}H). Curiously enough, something is always spit out; the target and the projectile do not simply combine to form a new nuclide. We can, however, imagine the temporary formation of C from the addition of projectile to target. Inserting C into our generic representation of a nuclear reaction gives:

$$A + x \rightarrow [C]^* \rightarrow B + y$$

The square brackets around C indicate its transient nature, meaning that it doesn't exist for reasonable amounts of time. The asterisk indicates that this nuclide is in an excited state, typically much higher in energy than the ground state. Taken together, [C]* represents what is called a **compound nucleus**.

The following shorthand can also be used to represent a generic nuclear reaction:

$$A(x,y)B$$

where the target and product nuclides are placed outside the parentheses, and the projectile and the emission are placed inside. As a result, these reactions are often

---

[1] Students have suggested the term "emitron" for y.

referred to as **x,y-type reactions**. From this generic point of view, this may not seem like much of a shorthand. Hopefully, its utility will be illustrated in the examples to come.

Neutron capture is a fairly common nuclear reaction. An example would be tossing a neutron at ^{51}V to form ^{52}V. This reaction emits a gamma photon, so it would be called an "n,gamma" reaction. ^{51}V is the target, the neutron is the projectile, and ^{52}V is the product, while a gamma photon is emitted:

$$^{51}_{23}V_{28} + {}^{1}_{0}n \rightarrow {}^{52}_{23}V_{29} + \gamma$$
$$^{51}V(n,\gamma)^{52}V$$

Notice that the numbers of nucleons are conserved in this n,gamma reaction. Just like decay equations, the atomic and mass numbers on both sides of the arrow need to each be equal.

9.1   ENERGETICS

Are mass and energy collectively conserved in nuclear reactions? Let's do the math!

Example 9.1

How much energy is produced, or required, by the n,gamma reaction of ^{52}V? We can do this by looking at the difference in mass between reactants and products, just like we did with decay.

Mass of reactants	^{51}V	50.943960 u
	^{1}n	+1.008665 u
		51.952625 u
Mass of product	^{52}V	−51.944780 u
Reactants − products		0.007845 u

Note that *atomic* masses are used for ^{51}V and ^{52}V just as we did with decay. Also note that mass is lost in this n,gamma reaction; therefore, energy is produced. The variable Q is typically used to designate energy in nuclear reactions, instead of the E that is used for decay. While potentially confusing, it serves to further differentiate these two types of nuclear processes. Let's calculate how much energy is produced by this reaction:

$$Q = \frac{931.5\,\text{MeV}}{u} \times 0.007845u = 7.308\,\text{MeV}$$

When mass is lost in the course of the reaction, then Q is greater than 0, and the reaction is termed **exoergic** (or exothermic). Decay is always exoergic, but that won't be true for nuclear reactions. Occasionally, we will run into reactions where mass is gained, Q is negative, and the reaction is termed **endoergic** (or endothermic).

Dealing with energy in nuclear reactions is much more complex than with decay processes. For example, in the n,γ reaction, the energy produced comes from binding of an additional neutron to a ^{51}V nucleus and is released through the gamma photon and the kinetic energy of the product nuclide. There's also energy on the reactant side that we need to be concerned with; namely, the kinetic energy of the projectile, which is also nonzero. In some reactions, the projectile *must* have a certain amount of energy before a reaction will proceed. How can we sort all of this out? We'll get to all this eventually; first, let's consider the projectile.

The projectile in Example 9.1 is a neutron. Neutrons make great projectiles because they have no electrical charge. As a result, they are neither repelled from nor attracted to an atom's protons or electrons. This means a neutron can approach a nucleus without hindrance. Positively charged projectiles do not have the same advantage. As they approach a nucleus, they will begin to feel Coulomb repulsion of the positively charged nucleus.

The first example of a charged particle being used in a nuclear reaction was discovered by Ernest Rutherford in 1919. It is the reaction of an alpha particle with ^{14}N, forming ^{17}O and a proton:

$$^{14}_{7}N + ^{4}_{2}He \rightarrow ^{17}_{8}O + ^{1}_{1}H$$

In this reaction, ^{14}N is the target, an alpha particle is the projectile, ^{17}O is the product, and a proton is emitted. This is an "alpha,p" reaction and can also be represented as $^{14}N(\alpha,p)^{17}O$. We'll take a detailed look at the energetics of this reaction, starting by calculating the energy released (or required):

Mass of reactants	^{14}N	14.003074 u
	^{4}He	+4.002603 u
		18.005677 u
Mass of products	^{17}O	16.999132 u
	^{1}H	+1.007825 u
		18.006957 u

Reactants − products = 18.005677 − 18.006957 = −0.001280 u

Mass is *gained* in this reaction, so energy needs to be put in—this is an endoergic reaction. Nowadays this is usually accomplished by accelerating the projectile using sophisticated equipment. Rutherford accomplished this in 1919 by using an alpha particle projectile that was already packing a punch. The alpha particles were generated from the decay of ^{214}Po, which have an energy of 7.69 MeV. How much energy does this reaction require?

$$Q = \frac{931.5\,MeV}{u} \times (-0.001280\,u) = -1.19\,MeV$$

The Q value is the minimum energy required by this reaction to manufacture the necessary mass. We also need to be concerned with the energy, or momentum, of

the products. Both mass and momentum are accounted for in the **threshold energy**
(E_{tr}, Equation 9.1):

$$E_{tr} = -Q \times \left(\frac{A_A + A_X}{A_A} \right) \tag{9.1}$$

where A_A is the mass number of the target, and A_X is the mass number of the projectile. Note the negative sign in front of Q. This equation only makes sense when Q is
negative, that is, the reaction is endoergic. For Rutherford's reaction:

$$E_{tr} = 1.19\,\text{MeV} \times \left(\frac{14 + 4}{14} \right) = 1.53\,\text{MeV}$$

This tells us the alpha particle needs a minimum of 1.53 MeV of energy for this
reaction to produce the required mass and momentum of the products. While this
is greater than the 1.19 MeV required solely for matter creation, it is still quite a bit
less than the 7.69 MeV of the ^{214}Po alpha particle. Any excess energy brought in by
the projectile will end up as kinetic and excitation energy of the products. In other
words, the products are movin' and shakin' more than they would otherwise.

Great! We've got Rutherford's reaction under control, right? Not quite. We need to
be concerned about the fact that we're bringing together two positively charged particles. They will try to repel each other (Coulomb repulsion) with increasing force as
they get closer and closer together. The projectile must have a certain amount of energy
to overcome this repulsion. The minimum amount of energy required to overcome this
Coulomb barrier (E_{cb}, in MeV) can be approximated as shown in Equation 9.2:

$$E_{cb} \approx 1.11 \times \left(\frac{Z_A Z_X}{A_A^{1/3} + A_X^{1/3}} \right) \tag{9.2}$$

where Z_A and Z_X are the atomic numbers for the target and the projectile, respectively.
This energy doesn't get soaked up and doesn't enter into our energy accounting for
the reaction; it is simply a barrier to overcome. Just as we did with Q, we must also
be concerned with conservation of momentum here. The minimum energy required
to overcome the Coulomb barrier and provide enough energy for the momentum
of the products is the **effective Coulomb barrier** (E_{ecb}). Its calculation is shown in
Equation 9.3, and it also has units of MeV:

$$E_{ecb} \approx 1.11 \times \left(\frac{A_A + A_X}{A_A} \right) \times \left(\frac{Z_A Z_X}{A_A^{1/3} + A_X^{1/3}} \right) \tag{9.3}$$

Note that Equations 9.2 and 9.3 apply to any nuclear reaction with a positively
charged projectile. For Rutherford's reaction, the effective Coulomb barrier is:

$$E_{ecb} \approx 1.11 \times \left(\frac{14 + 4}{14} \right) \times \left(\frac{7 \times 2}{\sqrt[3]{14} + \sqrt[3]{4}} \right) = 5.00\,\text{MeV}$$

For endoergic reactions, the minimum amount of energy required to make it proceed in good yield is the greater of the threshold energy (E_{tr}) and the effective Coulomb barrier (E_{ecb}), with the effective Coulomb barrier usually being the larger of the two. The alpha particles in this example have ample energy (7.69 MeV) to make it over this hump.

Finally, we should also be concerned about aim. Projectiles don't always hit their targets dead-on. Glancing blows can sometimes be productive, but problems with conserving angular momentum come into play. Fortunately for us, this concern is beyond the scope of this text.

Let's take a step back and look at this reaction as two separate steps. The first step is the formation of the compound nucleus ([^{18}F]*), followed by its decomposition into the ^{17}O product and a proton:

$$^{14}_{7}N + {}^{4}_{2}He \rightarrow \left[{}^{18}_{9}F\right]^* \rightarrow {}^{17}_{8}O + {}^{1}_{1}H$$

The compound nucleus can have two types of energy, excitation and kinetic. Excitation is placement of nucleons in a configuration other than the ground state, and kinetic is the energy of motion. The compound nucleus is moving because the projectile just slammed into the target. Equation 9.4 provides us with a simple way to estimate the kinetic energy of the compound nucleus:

$$E_{KC} \approx E_{KX} \times \left(\frac{A_X}{A_C}\right) \tag{9.4}$$

where E_{KX} is the kinetic energy of the projectile, A_C is the mass number of the compound nucleus, and A_X remains the mass number of the projectile. Note that this equation applies to *any* nuclear reaction (regardless of the sign of Q). Applying this equation to Rutherford's reaction:

$$E_{KC} \approx 7.69\,\text{MeV} \times \left(\frac{4}{18}\right) = 1.71\,\text{MeV}$$

We can use this information to calculate just how far [^{18}F]* is above ground state ^{18}F. Consider the (hypothetical) reaction:

$$^{14}N + {}^{4}He \rightarrow {}^{18}F$$

Let's also assume (even though it is ridiculous!) that all three nuclides are in the ground state. How much energy is produced or consumed? In other words, what is Q?

$$Q = (14.003074\,\text{u} + 4.002603\,\text{u} - 18.000938\,\text{u}) \times \frac{931.5\,\text{MeV}}{\text{u}}$$

$$= +4.41\,\text{MeV}$$

Therefore, when the Rutherford reaction takes its first step and forms a compound nucleus, 4.41 MeV of energy is *produced*. Remember that the alpha particle already has 7.69 MeV of energy. Where does all this energy go? There are only two places it can go—excitation and kinetic energy of ^{18}F. The excitation energy of the $[^{18}$F$]^*$ in the Rutherford reaction is, therefore, the energy of the projectile plus the energy released for the first step of the reaction minus the kinetic energy of the compound nucleus.

$$7.69\,\text{MeV} + 4.41\,\text{MeV} - 1.71\,\text{MeV} = 10.39\,\text{MeV}$$

That's one excited compound nucleus! It's no wonder it can't even hold itself together. If you're clever enough, you can now calculate the energy (excitation plus kinetic) of the products, but this question can be answered without the hassle of calculating the excitation energy of the compound nucleus:

$$7.69\,\text{MeV} - 1.19\,\text{MeV} = 6.50\,\text{MeV}$$

where 7.69 MeV is still the energy that the alpha brings to the reaction, 1.19 MeV is the energy needed to make the mass (Q, from a couple pages back), and 6.50 MeV is the energy (excitation plus kinetic) of the products.

With all this information, we can draw a tidy little diagram that shows the energy changes during the course of this reaction (Figure 9.1). We know that going from the ground state of the reactants to the ground state of the products costs 1.19 MeV of energy (Q), so the products should be slightly higher in energy than the reactants. Note that the products of nuclear reactions are rarely generated in their ground state. They typically have some level of excitation and kinetic energy. Therefore, the line representing the products in Figure 9.1 is somewhat misleading. We can also show the ground state, and actual state of the compound nucleus, since 4.41 MeV

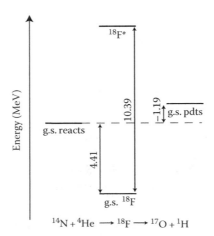

FIGURE 9.1   Energy diagram for Rutherford's reaction.

is released in the formation of the compound nucleus, and it is formed in an excited state that is 10.38 MeV above that ground state.

Example 9.2

What is the minimum energy required to make the following reaction proceed in good yield?

$$^{240}_{94}\text{Pu} + {}^{4}_{2}\text{He} \rightarrow {}^{243}_{96}\text{Cm} + {}^{1}\text{n}$$

This is an alpha,n reaction whose shorthand representation would be: $^{240}\text{Pu}(\alpha,n)$ ^{243}Cm. First find the difference in mass between reactants and products:

Mass of reactants	^{240}Pu	240.053814 u
	^{4}He	+4.002603 u
		244.056417 u
Mass of products	^{243}Cm	243.061389 u
	^{1}n	+1.008665 u
		244.070054 u

Reactants – products $= 244.056417–244.070054 = -0.013637$ u

Once again, mass is gained, therefore this is an endoergic reaction. The energy required to make the mass needed for this reaction (Q) is calculated next:

$$Q = \frac{931.5\,\text{MeV}}{\text{u}} \times (-0.013637\,\text{u}) = -12.7\,\text{MeV}$$

Since it is endoergic, we need to calculate threshold energy (mass + momentum):

$$E_{tr} = 12.7\,\text{MeV} \times \left(\frac{240 + 4}{240}\right) = 12.9\,\text{MeV}$$

Since it involves the collision of two nuclei, we also should calculate the effective Coulomb barrier:

$$E_{ecb} \approx 1.11 \times \left(\frac{240 + 4}{240}\right) \times \left(\frac{94 \times 2}{\sqrt[3]{240} + \sqrt[3]{4}}\right) = 27.2\,\text{MeV}$$

This is a lot higher than the threshold energy and is, therefore, the minimum amount of energy required to make this reaction proceed in good yield. Alpha particles will need to be moving pretty fast to make this reaction work.

In both examples we've looked at so far, the effective Coulomb barrier was higher than the threshold energy and, therefore, determined the minimum energy required for the reaction. The effective Coulomb barrier varies somewhat consistently with the masses of the colliding objects, as illustrated in Figure 9.2. The two lines show how the effective Coulomb barrier changes for ^{1}H and ^{4}He as the projectile while the atomic number of the target nucleus increases. Only data for the most abundant, stable, target nuclides were used to generate this figure.

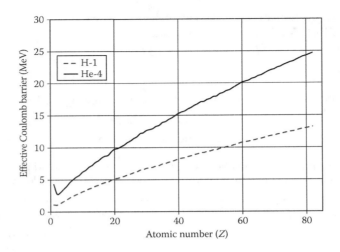

FIGURE 9.2 Variations in the effective Coulomb barrier with projectile and target nucleus.

The upward sloping lines in Figure 9.2 tell us that the bigger the projectile and target are, the harder it is to get them together. This makes sense as the amount of positive charge generally increases with size. This is nicely illustrated in our two examples. In Rutherford's reaction, we slammed an alpha particle (two protons) into a ^{14}N (seven protons) atom and saw an effective Coulomb barrier of ~5 MeV. In the second example, the effective Coulomb barrier jumped to ~27 MeV because we ran an alpha particle into a ^{240}Pu (94 protons) nucleus.

Let's approach nuclear reactions from another perspective. Let's say we wanted to make some ^{18}F (commonly used in PET imaging). What possible reactions could be used? A few possibilities are listed in Table 9.1.

That's a lot of options, but wait, there's more! Figure 9.3 illustrates possible "low-energy" nuclear reactions leading to a particular product. Low-energy simply means these reactions are accessible with conventional accelerators (such as clinical cyclotrons), not huge, high-power atom smashers. Figure 9.3 is laid out just like a chart of

TABLE 9.1

Some Nuclear Reactions Producing ^{18}F

Reaction #	Reaction	Shorthand
1	$^{15}_{7}N + ^{4}_{2}He \rightarrow ^{18}_{9}F + ^{1}n$	$^{15}N(\alpha,n)^{18}F$
2	$^{16}_{8}O + ^{3}_{1}H \rightarrow ^{18}_{9}F + ^{1}n$	$^{16}O(t,n)^{18}F$
3	$^{17}_{8}O + ^{2}_{1}H \rightarrow ^{18}_{9}F + ^{1}n$	$^{17}O(d,n)^{18}F$
4	$^{18}_{8}O + ^{1}_{1}H \rightarrow ^{18}_{9}F + ^{1}n$	$^{18}O(p,n)^{18}F$
5	$^{19}_{9}F + ^{1}_{1}H \rightarrow ^{18}_{9}F + ^{1}_{1}H + ^{1}n$	$^{19}F(p,pn)^{18}F$
6	$^{20}_{10}Ne + ^{2}_{1}H \rightarrow ^{18}_{9}F + ^{4}_{2}He$	$^{20}Ne(d,\alpha)^{18}F$

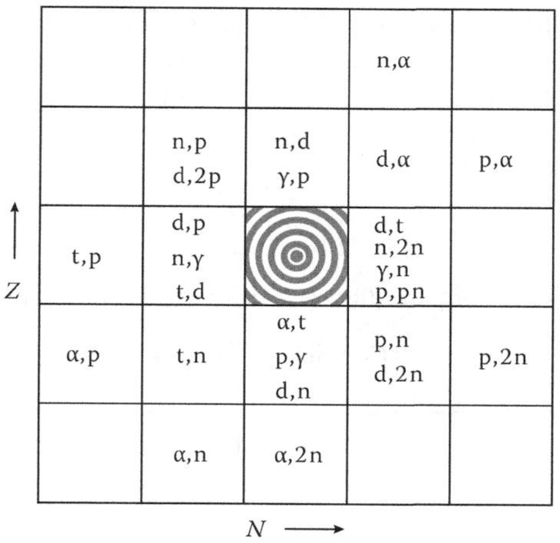

			n,α	
	n,p d,2p	n,d γ,p	d,α	p,α
t,p	d,p n,γ t,d	◎	d,t n,2n γ,n p,pn	
α,p	t,n	α,t p,γ d,n	p,n d,2n	p,2n
	α,n	α,2n		

Z ↑ N ⟶

FIGURE 9.3 Nuclear reactions leading to a particular product (center square).

the nuclides with numbers of neutrons increasing from left to right and atomic number increasing from bottom to top. The square in the center (bull's-eye) represents the desired product nuclide, and the reaction types needed to get from a particular starting nuclide to the product are given in the surrounding squares. For example, we know that an n,gamma reaction produces a nuclide that is one square to the right on the chart. Sure enough, "n,γ" is listed as a possible reaction type one square to the left of the desired product. A deuterium-proton (d,p) reaction accomplishes the same result as an n,gamma reaction and, therefore, appears in the same square. Adding deuterium (^{2}H) increases both N and Z by one; emitting a proton lowers Z by one—the net result is an increase of N by one.

Our list of possible reactions leading to ^{18}F suddenly looks a little short. That's because those in Table 9.1 are limited to a single reaction starting from a naturally occurring nuclide. This is our first criterion in considering potential nuclear reactions—it's usually easier to start with a nuclide that is naturally occurring. Let's see if we can narrow the list a bit more. Reactions that emit more than one particle (such as p,pn) generally require more energy than those that only emit one particle. We can cross off reaction #5.

We also want a high yield of product. Just like chemical reactions, sometimes nuclear reactions work well and make lots of product[2], and other times they don't. We'll discuss this more later. Finally, we'd also like to keep the energy requirements within reason. Of the remaining reactions, numbers 4 and 6 have the highest yields and lowest energy requirements. Deciding between these two may depend on what kind of gun is available—in other words, can you easily accelerate protons (reaction #4) or deuterons (reaction #6)? It may also depend on how easy it is to separate the product. The advantage of reaction #4 is that the ^{18}O can be incorporated into water

---

[2] This is relative to other nuclear reactions. As we'll soon see, overall yields from nuclear reactions are abysmally low compared to chemical reactions.

molecules, and then the ^{18}F will be produced as the fluoride anion, which can easily be separated from the ^{18}O-water using an ion-exchange column. Both reactions 4 and 6 are commonly used to produce ^{18}F for PET imaging. As an example, let's check out the energetics of reaction #4.

Example 9.3

What is the minimum energy required to make reaction #4 proceed in good yield?

$$^{18}_{8}O + ^{1}_{1}H \rightarrow ^{18}_{9}F + ^{1}n$$

Mass of reactants	^{18}O	17.999161 u
	^{1}H	+1.007825 u
		19.006986 u
Mass of products	^{18}F	18.000938 u
	^{1}n	+1.008665 u
		19.009603 u
Reactants − products = 19.006986 u − 19.009603 u = −0.002617 u		

Energy required:

$$Q = \frac{931.5\,\text{MeV}}{\text{u}} \times (-0.002617\,\text{u}) = -2.44\,\text{MeV}$$

Threshold energy:

$$E_{tr} = 2.44 \times \left(\frac{18+1}{18}\right) = 2.57\,\text{MeV}$$

Effective Coulomb barrier:

$$E_{ecb} \approx 1.11 \times \left(\frac{18+1}{18}\right) \times \left(\frac{8 \times 1}{\sqrt[3]{18} + \sqrt[3]{1}}\right) = 2.59\,\text{MeV}$$

For this reaction the threshold energy and the effective Coulomb barrier are about the same, and they are both relatively low. The effective Coulomb barrier is slightly higher, so the minimum energy required for this reaction to proceed in good yield is 2.59 MeV.

9.2   CROSS SECTION

You've probably noticed that we've always qualified the minimum energy requirement with the phrase "in good yield." This suggests that the reaction can proceed at lower energies but in poor yield. This is true. Nuclear reactions sometimes do proceed at energies lower than they should. When this happens, it is called **tunneling**. It is somewhat analogous to a car driving through a hill rather than over it. Driving through will take a lot less energy than driving over. Sadly, tunneling isn't common enough to run nuclear reactions in good yield on the cheap, so it'll still be a good idea to give those projectiles enough energy to ensure success.

Like chemical reactions, you don't always get what you want with nuclear reactions. Various products can be obtained with the same target/projectile combo. For example, hitting a ^{63}Cu target with a proton projectile can result in all the products shown in Figure 9.4. The product ratios will depend on the energy of the projectiles and the probabilities (**cross sections**) for the reactions. With higher energy projectiles, a greater product distribution is generally observed.

Every possible nuclear reaction has a cross section, which is symbolized by the Greek letter sigma (σ). When the projectile is a neutron, a subscript following σ indicates what is emitted along with the formation of product. For instance, σ_γ is the cross section for an n,gamma reaction. While it is best to think of cross section as the probability of success, it is more strictly defined as the *apparent* cross-sectional area of the target nucleus as seen by the projectile. Cross section has units of **barns** (b). One barn is equal to 10^{-24} cm^2 (10^{-28} m^2), which is roughly the projected area of the average nucleus. If nuclear size were the only criterion for the chances of a reaction proceeding, then we'd expect cross section to be low for low Z nuclides and high for high Z nuclides. While this is sometimes true, there are other criteria.

Nuclear cross sections depend not only on Z but also N (and therefore A), the density of the target, the charge, mass (size), and the velocity of the projectile. Figure 9.5 shows how dramatically the probability of all neutron-induced reactions (σ_a) varies with the velocity (energy) of the neutron for the element silver. The general trend is a decreasing probability of reaction as the energy of the neutron increases. This can be rationalized in terms of time (see also Section 6.2). As the neutron moves faster, it spends less time near each nucleus it passes by, thereby decreasing the probability of interaction. Neutrons slowed to that of its surroundings by room temperature water and/or plastic are called **thermal neutrons** and have an energy of about 0.025 eV.

Cold neutrons have even lower energies as they are passed through materials below room temperature. For example, cold neutrons are generated at Oak Ridge National Laboratory by passing them through supercritical hydrogen at 20 K. Fast neutrons are those with energies significantly higher than room temperature.

The "spikes" observed in Figure 9.5 for fast neutrons indicate that, for neutrons with certain energies, the probability of reaction increases dramatically. These spikes are due to **resonance capture**, which is the formation of a compound nucleus

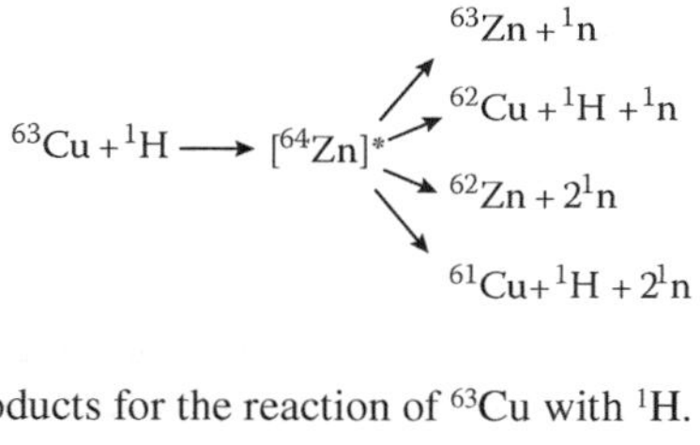

$$^{63}Cu + {}^1H \longrightarrow [{}^{64}Zn]^* \nearrow \begin{array}{l} {}^{63}Zn + {}^1n \\ {}^{62}Cu + {}^1H + {}^1n \\ {}^{62}Zn + 2{}^1n \\ {}^{61}Cu + {}^1H + 2{}^1n \end{array}$$

FIGURE 9.4 Possible products for the reaction of ^{63}Cu with ^1H.

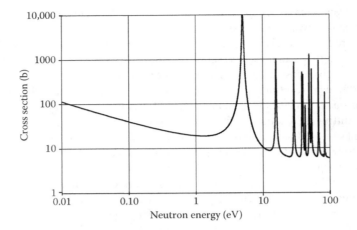

FIGURE 9.5 Probability of a neutron-induced reaction with a silver target as the energy of the neutron increases. (Produced using data from the evaluated nuclear data file (ENDF) at http://www.nndc.bnl.gov/.)

with an especially favorable amount of excitation energy. The energy where they are observed is called the **resonance energy**. They usually occur for 1–100 eV neutrons.

What do the spikes mean? If a particular nuclear reaction is desired, then there may be a specific projectile energy that will greatly increase the probability of reaction. Therefore, it may be desirable to run that reaction with projectiles at the resonance energy. The (enhanced) probability that a reaction will occur at a particular resonance energy is called the **resonance integral**.

The bottom line is that variations in nuclear cross sections are not as simple as comparing sizes of target nuclei. They sometimes vary wildly even for closely related target nuclides. For example, the thermal neutron cross section for the n,gamma reaction involving ^{1}H is 0.332 b but is 0.52 mb (*milli*barns) for ^{2}H.

Not every interaction between projectile and target results in a reaction. Sometimes the projectile is simply scattered by the target. The cross section for all possible interactions can be represented as σ_{tot}, while the cross section for all scattering interactions could be σ_s, and those involving absorption σ_a. The total probability of interaction is equal to the sum of the probability of scattering and the probability of absorption:

$$\sigma_{tot} = \sigma_s + \sigma_a$$

The nature of the product makes a difference because it could be stable or radioactive. If it's hot, it'll be decaying as it is produced, making it a little more challenging to calculate yield.

9.3 YIELD

How can we tell how much product is made? Reaction yield depends on how much target there is, the projectile flux, the reaction cross section, time, and the nature of the product. **Flux** (Φ, the Greek letter phi) is a measure of how many projectiles pass through a certain area per unit time. Our focus is on neutron flux, so it'll have units of neutrons per square centimeter per second (n/cm^2·s).

We'll look at three examples, each with a different product scenario: (1) production of a stable product; (2) production of a radioactive product; and (3) production of a radioactive product with a long half-life. In every scenario we'll assume: (1) the flux does not change with time; (2) the number of target atoms (A) is so large, relative to the number of product atoms (B), it can be considered constant throughout the time of irradiation (t); and (3) the target is thin enough that the flux entering the target is not significantly decreased as it goes through. These assumptions may seem a little perilous, but they are necessary to make the math a bit more tractable, and generally speaking, they are true.

When the product is stable, the amount of product (N_B) formed can be calculated using Equation 9.5:

$$N_B = \sigma \Phi N_A t \qquad (9.5)$$

where N_B is the number of atoms of the product nuclide formed after an irradiation time t with units of seconds, σ is the reaction cross section in square centimeters, Φ is the flux of projectile x in particles per square centimeter per second, and N_A is the number of target atoms A (the target doesn't need to be pure nuclide).

Example 9.4

A sample containing 1.00 g of ^{16}O is placed in a nuclear reactor for 10.0 hours. The reactor has a neutron flux of 6.00×10^{10} n/cm^2·s. Assuming that $^{16}O(n,\gamma)^{17}O$ is the only reaction that takes place, how many ^{17}O atoms are formed?

The neutron cross section (0.19 mb) can be obtained from a chart of the nuclides or a variety of online sources and converted to cm^2. Time also needs to be converted to seconds. Finally, the number of target atoms need to be calculated from their mass:

$$0.19\,mb = 0.19 \times 10^{-27}\,cm^2$$

$$10.0h \times \frac{60\,min}{h} \times \frac{60\,s}{min} = 36000\,s$$

$$N_A = \left(1.00\,g \times \frac{mol}{15.995g} \times \frac{6.022 \times 10^{23}\,atoms}{mol}\right) = 3.76 \times 10^{22}\,atoms$$

$$N_B = \sigma \Phi N_A t$$

$$= (0.19 \times 10^{-27}\,cm^2) \times \left(6.00 \times 10^{10}\,\frac{n}{cm^2 \cdot s}\right) \times (3.76 \times 10^{22}\,atoms) \times (36000\,s)$$

$$= 1.55 \times 10^{10}\,atoms\ of\ ^{17}O\ formed$$

If the rate of formation of product is desired, Equation 9.5 can be modified by dividing both sides by time:

$$\text{rate of formation of B} = \frac{N_B}{t} = \sigma\Phi N_A \tag{9.6}$$

If the product is radioactive, we'll have to worry about it decaying as we make it. The amount of hot product formed after irradiating for time t is given by Equation 9.7:

$$N_B = \frac{\sigma\Phi N_A}{\lambda}(1 - e^{-\lambda t}) \tag{9.7}$$

where λ is the decay constant and should have units of reciprocal seconds. The other variables are the same as we've seen before. Note that N_B can be calculated only for the moment the target is removed from the projectile beam. At any time after that, N_B will be smaller due to its radioactive decay.

Since $A = \lambda N$, Equation 9.7 can also be written as:

$$A_0 = \sigma\Phi N_A(1 - e^{-\lambda t}) \tag{9.8}$$

where A_0 is the activity of the product at the moment the target is removed from the projectile beam and has the units of decays per second (dps or Bq). Activity is, therefore, a function of irradiation time (t) and will eventually **saturate**. Saturation occurs when the rate of formation of product is equal to the rate of decay of product. Sound familiar? It's a bit like nuclear equilibria (Section 2.6). For production of radionuclides, irradiation for seven product half-lives gives 99% of saturation and is a reasonable time to harvest the product nuclide, if maximum yield and minimum time are desired. The path to saturation is illustrated in Figure 9.6. Activity gradually increases as more and more product is made, until it reaches saturation.

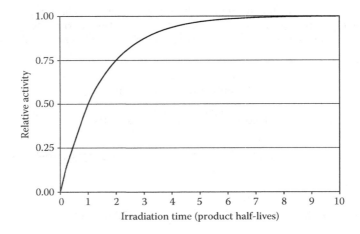

FIGURE 9.6   Activity as a function of irradiation time.

At saturation, $e^{-\lambda t}$ becomes relatively small, and $1 - e^{-\lambda t}$ (sometimes called the saturation term) approaches one. Equation 9.8 then simplifies to:

$$A_{sat} = \sigma \Phi N_A \qquad (9.9)$$

at saturation, where A_{sat} is the activity at saturation.

Equation 9.8 finally gives us something tangible—activity. Counting atoms is possible, but activity is easy to measure. This equation allows easy experimental determinations of flux, cross section, or the amount of a particular nuclide in a sample. The last is known as **neutron activation analysis**—a powerful analytical tool, especially if the target nuclide has a large neutron cross section. It is capable of detecting as little as 10^{-14} g of target per gram of sample.

Example 9.5

A 2.50 g sample of ^{18}O water was placed in a cyclotron beam of protons (flux = 3.01×10^3 p/cm²·s) for 5.00 h, resulting in an activity of 153 dps. Assuming that a $^{18}O(p,n)^{18}F$ is the only reaction that takes place, calculate the cross section for this reaction.

First, calculate the number of ^{18}O atoms in the target, remembering it is $H_2{}^{18}O$:

$$2.50 \text{ g } H_2{}^{18}O \times \frac{\text{mol } H_2{}^{18}O}{20.0 \text{ g } H_2{}^{18}O} \times \frac{1 \text{ mol } {}^{18}O}{1 \text{ mol } H_2{}^{18}O} \times \frac{6.022 \times 10^{23} \text{ atoms}}{\text{mol}}$$

$$= 7.53 \times 10^{22} \text{ atoms}$$

The half-life of ^{18}F is 1.8293 h, therefore:

$$A_0 = \sigma \Phi N_A (1 - e^{-\lambda t})$$

$$153 \text{ dps} = \sigma \times \left(3.01 \times 10^3 \frac{p}{s \cdot cm^2}\right) \times (7.53 \times 10^{22} \text{ atoms}) \left(1 - e^{-\frac{\ln 2}{1.8293 \text{ h}} \times 5.00 \text{ h}}\right)$$

$$\sigma = 7.95 \times 10^{-25} \text{ cm}^2 = 0.795 \text{b}$$

Note that this reaction had not yet reached saturation. We can see this by dividing the half-life into the irradiation time:

$$\frac{5.00 \text{ h}}{1.8293 \text{ h}} = 2.73 \text{ product half-lives}$$

At least seven product half-lives need to pass before saturation is reached. The other way to see that saturation has not been reached is to calculate the saturation term separately. If it is very close to one, then saturation has been reached:

$$(1 - e^{-\lambda t}) = \left(1 - e^{-\frac{\ln 2}{1.8293 \text{ h}} \times 5.00 \text{ h}}\right) = 0.850$$

Not close enough.

When the radioactive product has a really long half-life, the math also simplifies. When $t_{1/2} \gg t$, then $(1 - e^{-\lambda t}) \approx \lambda t$. Therefore, our equation for the activity at the time the sample is removed from the beam simplifies to:

$$A_0 = \sigma \Phi N_A \lambda t \tag{9.10}$$

Nuclide production is linear with time! Because of the long half-life, more product is made as more time passes. As a result, the issue of saturation is a bit moot if the half-life of the product is too long. Also, irradiating for seven (or more) product half-lives may not be practical. Finally, the reader should notice that Equation 9.10 is the same as Equation 9.5. In other words, production of a nuclide with a sufficiently long half-life is the same as producing a stable nuclide.

Example 9.6

9.507 g of thallium is irradiated in a neutron howitzer with a flux of 3.67×10^6 n/cm²·s. How long will it take to obtain a ^{204}Tl activity of 2894 dpm?

Thallium is made up of ^{203}Tl (29.52%) and ^{205}Tl (70.48%). We're only interested in the reaction of ^{203}Tl with neutrons, assuming it is an n,gamma reaction:

$$^{203}_{81}\text{Tl} + {}^{1}\text{n} \rightarrow {}^{204}_{81}\text{Tl} + \gamma$$

First, let's figure out how many target atoms we must shoot at:

$$9.507\,\text{g} \times \frac{\text{mol}}{204.4\,\text{g}} \times 0.2952 \times \frac{6.022 \times 10^{23}\,\text{atoms}}{\text{mol}} = 8.268 \times 10^{21}\,\text{atoms}$$

The cross section for this reaction is 11.4b, and the half-life of ^{204}Tl is 3.78 years. Assuming that this will be long relative to the irradiation time, we can use Equation 9.10:

$$\frac{2894\,\text{decays}}{\text{min}} \times \frac{\text{min}}{60\,\text{s}} = (11.4 \times 10^{-24}\,\text{cm}^2)$$

$$\left(3.67 \times 10^6\,\frac{\text{n}}{\text{cm}^2 \cdot \text{s}}\right)\left(8.268 \times 10^{21}\,\text{atoms}\right)\left(\frac{\ln 2}{3.78\,\text{a}}\right)t$$

$$t = 7.60 \times 10^{-4}\,\text{a}$$

Our assumption that irradiation time is relatively short is correct! Let's convert the time value to something more tangible:

$$t = 7.60 \times 10^{-4}\,\text{a} \times \frac{5.259 \times 10^5\,\text{min}}{\text{a}} \times \frac{\text{h}}{60\,\text{min}} = 6.66\,\text{h}$$

If a mixture is being irradiated, and there are short-lived *and* long-lived nuclides being formed, we can take advantage of their difference in time to saturation. If we want the short-lived nuclide(s), we irradiate for a short time—because very little of the long-lived product will be formed. If we want the long-lived nuclide(s), we irradiate for a long time, take the sample out, and wait for the short-lived stuff to decay.

9.4 ACCELERATORS

How do you get energetic projectiles? Probably not from your favorite online super store. Aside from radioactive decay, bombarding a gas with energetic electrons can produce charged particles. The positive gas ions produced are separated by attraction to an electrode with negative voltage. They are then accelerated through electrical potentials. Passing a particle through a 1000 V potential would give it another keV per unit charge. There are a couple of ways to get this done.

A schematic for a traditional linear accelerator is shown in Figure 9.7a. Linear accelerators are often called linacs, and they accelerate charged particles in a straight line through single or multiple stages. Each stage is a cylindrical electrode (also known as drift tube) that has the opposite charge of the particle as the particle moves toward it, then the same charge of the particle as it is moving away. Since opposite charges attract and the same charge repels, the particle accelerates as it approaches *and* as it moves away from two adjacent drift tubes. The tubes have no charge while the particle is moving through them, thus it is coasting (or drifting!) as it passes through the tube. The final energy of the particle beam depends on the voltage applied at each stage and the total number of stages.

While they can accelerate a variety of charged particles, linacs are most commonly used to accelerate electrons. As the electrons are accelerated, the stages get longer; this is done to ensure the electron spends the same amount of time traveling through each tube. This simplifies the electronic switching between positive and

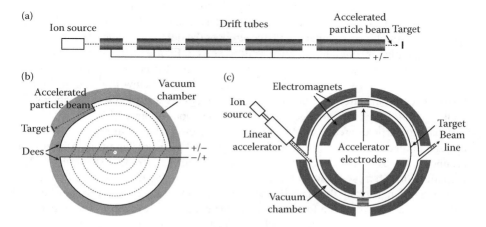

FIGURE 9.7 Schematic representations of (a) linear accelerator, (b) cyclotron and (c) synchrotron.

negative charges on the tubes. To accelerate particles to very high velocities (energies), this type of linac needs to be very large—some are even a couple miles in length. While this is fine for large research centers, more compact designs are necessary in more common settings such as a hospital.

Most **modern-day linear accelerators** use microwaves to accelerate electrons. These microwaves are often produced in much the same way they are for an ordinary microwave oven. Microwaves, like all forms of electromagnetic radiation, can only travel at the speed of light; therefore, the radiation itself cannot be used directly for acceleration of the electrons—the photons are not going to slow down and push the electrons to higher and higher speeds. Instead it is the phase of these waves that is manipulated to accelerate electrons.

Remember that electromagnetic radiation behaves like waves in many respects. We can think of phase in relation to the peaks and troughs of a wave. A peak represents one phase, and the trough represents the opposite phase. The electron is pushed and pulled by these phases, so if we arrange them just right, they will accelerate the electron. It's a bit like a surfer riding a wave as it approaches the beach. The surfer is constantly moving downhill and is being pushed forward by the wave. In our case, the wave is the phase, and it's accelerating.

There's another way to look at the acceleration of charged particles in this kind of linear accelerator. As the microwave moves through a metal tube, it will induce temporary positive and negative electrical charges in the walls of the tube. If these temporary charges move just right, they can be used to accelerate charged particles.

For clinical applications, such as radiation therapy, these accelerators only need to be a couple of feet long, and can accelerate electrons up to 20 MeV. The Stanford linear accelerator (SLAC) uses microwaves to accelerate electrons up to 50 GeV and is nearly two miles long!

Cyclotrons are the most commonly used accelerators for radionuclide production in hospitals and other, more ordinary places. They take advantage of the fact that charged particles float in a circle when placed in a constant magnetic field, and the radius of the circle is directly related to the velocity (energy) of the particle. In other words, the faster the particle is moving, the bigger the circle it makes. The size of the circle also depends on the strength of the magnetic field—the stronger the field, the smaller the circle.

A schematic for a simple cyclotron is illustrated in Figure 9.7b, although the magnets have been removed for clarity. Only the accelerator portion of the cyclotron is shown. Imagine two round magnets sitting above and below the page in Figure 9.7b—just like an Oreo™[3] cookie, where the magnets are the crunchy chocolate cookie, and the particles are accelerated in the delicious creamy filling.

Charged particles are injected into the center of the cyclotron (the small white circle) and immediately begin to move in a circle parallel to the two magnets. As it circles around, it moves in and out of two hollow, semicircular electrodes. These electrodes are called **dees** because they look like the letter D. Imagine making them by cutting a tuna can in half along the can's diameter and taking all the tuna out (somewhere, a cat is purring...). As is true for all particle accelerators, the cyclotron is kept under a vacuum so that the accelerating particles don't run into any matter along the way.

---

[3] Oreo is a registered trademark of Kraft Foods Inc., Northfield, Illinois.

Just like the traditional linear accelerator described previously, acceleration of the charged particle takes place as it travels between the dees. As it leaves one dee, that dee is given the same charge as the particle, pushing it away. At the same time, the dee the particle is approaching is given the opposite charge, attracting the particle to it. As it's accelerated, the particle moves in a larger circle, giving it a kind of spiral path as it is accelerated in the cyclotron. It is not a smooth spiral, as the particle is not accelerated as it travels through the dees and, therefore, moves along a semicircular path with a fixed radius. Eventually, the circular path of the particle is large enough (energy is high enough) to allow it to exit the cyclotron.

Cyclotrons are not very good at accelerating electrons. Because of their low mass, it doesn't take a whole lot of energy ($\sim$100 keV) to get them traveling at velocities approaching the speed of light. When particles approach the speed of light, increasing amounts of energy are required to accelerate the particle by increasingly small amounts (Section 6.2). In a cyclotron, this means that once the particle gets to a certain velocity, it'll be difficult to continue to accelerate it, and it will pretty much stay in a circle (not continuing to spiral). Additionally, there are practical limits to the strengths of the electromagnetic fields within the cyclotron. As a result, cyclotrons tend to have energy limits for various particles. For most particles accelerated in cyclotrons (^{1}H, ^{2}H, ^{4}He, etc.), this limit is 25–50 MeV.

A common nuclide produced by cyclotrons for clinical use is ^{18}F. As we saw in Section 9.1, the following reaction requires a minimum energy of $\sim$2.6 MeV. Protons can easily be accelerated in cyclotrons to this energy:

$$^{18}_{8}\text{O} + {}^{1}_{1}\text{p} \rightarrow {}^{18}_{9}\text{F} + {}^{1}_{0}\text{n}$$

Like ^{18}F, most products of cyclotron reactions tend to be proton-rich. Therefore, they tend to decay via positron emission or electron capture. Atomic number (Z) almost always changes in the reactions that occur in cyclotrons, making it easier to isolate carrier-free product because the target and the product will have differing chemical properties. Cyclotron reactions are even less efficient than neutron activation. This makes sense because we are now trying to get two positively charged particles to hit each other. Coulomb repulsion forces must be overcome for this to happen. As a result, cyclotrons produce nuclides in pretty low yields, and they tend to be a little pricey.

Table 9.2 shows some of the more common nuclides produced by cyclotrons for medical applications. The first three nuclides listed, ^{11}C, ^{13}N, and ^{15}O, all have short half-lives ($\leq$20 minutes), so they must be produced on-site quickly, incorporated into a drug, and then delivered to the patient. Carbon, nitrogen, and oxygen are important elements in biological molecules—if we could place one of these nuclides in a biomolecule, then it would have all the same chemical and biological properties, but the radioactive label would allow us to see where that molecule goes in the body. This is an exciting area of current research in nuclear medicine, with the potential to develop a number of highly effective radiopharmaceuticals.

Another type of circular accelerator that can accelerate particles up to incredibly high energies (TeV = 10^{12} ev) is the **synchrotron**. A schematic drawing of a

TABLE 9.2
Some Cyclotron-Produced Radionuclides Used in Nuclear Medicine

Product Nuclide	Decay Mode(s)	Common Production Reaction(s)	Natural Abundance of Target Nuclide (%)
^{11}C	β^+, EC	$^{14}N(p,\alpha)^{11}C$	99.63
		$^{10}B(d,n)^{11}C$	19.9
		$^{11}B(p,n)^{11}C$	80.1
^{13}N	β^+	$^{16}O(p,\alpha)^{13}N$	99.76
		$^{12}C(d,n)^{13}N$	98.93
^{15}O	β^+	$^{14}N(d,n)^{15}O$	99.63
		$^{15}N(p,n)^{15}O$	0.368
^{18}F	β^+, EC	$^{18}O(p,n)^{18}F$	0.205
		$^{20}Ne(d,\alpha)^{18}F$	90.48
^{22}Na	β^+, EC	$^{23}Na(p,2n)^{22}Na$	100
^{43}K	β^-	$^{40}Ar(\alpha,p)^{43}K$	99.60
^{67}Ga	EC	$^{68}Zn(p,2n)^{67}Ga$	18.75
^{111}In	EC	$^{109}Ag(\alpha,2n)^{111}In$	48.16
		$^{111}Cd(p,n)^{111}In$	12.49
^{123}I	EC	$^{122}Te(d,n)^{123}I$	2.55
		$^{124}Te(p,2n)^{123}I$	4.74
^{201}Tl	EC	$^{201}Hg(d,2n)^{201}Tl$	13.18

synchrotron is shown in Figure 9.7c. Synchrotrons use oscillating magnetic and electrical fields to accelerate particles on a fixed radius. As the particle energy increases, so does the strength of the magnetic field, keeping the particles moving through the same circle. Synchrotrons are enormous. The Large Hadron Collider (LHC) near Geneva, Switzerland, is over 27 km in circumference, and it accelerates protons to 7 TeV. One of the cool aspects of the LHC is that protons can be accelerated to 7 TeV in opposite directions in the same circle before being forced to collide with each other.

9.5 COSMOGENIC NUCLIDES

Just about all background radiation detected on the surface of the planet originates from cosmic radiation. As depicted in Figure 9.8, cosmic radiation originates from our sun, other stars in our galaxy, and other galaxies. In terms of ionizing radiation, these stars emit mostly electrons, protons, other light nuclei (collectively, these particle emissions make up what is called the solar wind), high-energy photons, and neutrinos. About 90% of the particles bombarding our planet are *very* high-energy (up to 10^{10} GeV) hydrogen and helium nuclei. When these screaming nuclear projectiles slam into the atmosphere, they annihilate themselves, each one producing dozens of pions (π). Pions are subatomic particles that are commonly produced in high-energy (>400 MeV) nuclear reactions. In the average nucleus, pions are exchanged between

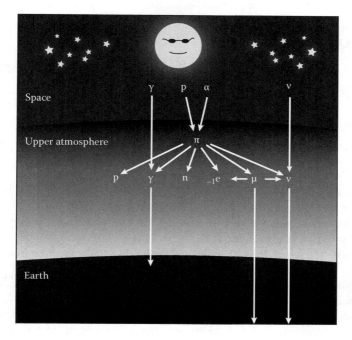

FIGURE 9.8   Cosmic radiation.

nucleons and thereby hold the nucleus together—somewhat like the way electrons hold atoms together in molecules.

Pions don't last long on their own. They decay to neutrinos and muons (μ) with a half-life of only 10^{-6} s. Muons are yet another subatomic particle and are similar to electrons. Somewhere between 50% and 80% of the muons formed make it to the Earth's surface. Muons don't last long either ($t_{1/2} = 2 \times 10^{-6}$ s), decaying to electrons and neutrinos. Even so, much of the cosmic radiation that reaches the surface of our planet is in the form of muons. High-energy photons (γ) and neutrinos (ν) make up most of the rest of the cosmic radiation flux at the planet's surface.

Thanks mostly to the Earth's magnetic field, and to all the matter in our atmosphere, we are shielded from all but about 5% of the cosmic radiation that bombards our planet. The significantly higher radiation flux in space is an important concern for astronauts spending considerable amounts of time there, and it is a major hurdle to long space missions such as travel to other planets both within and outside of our solar system.

Pions also collide with matter in the atmosphere producing electrons, neutrons, protons, and photons. Some of these particles can initiate nuclear reactions forming ^{3}H, ^{10}B, and ^{14}C, providing a constant source of these (and other) radionuclides. As mentioned in Chapter 2, ^{14}C has a half-life of 5715 years, is a soft β^- emitting nuclide ($E_{max} = 157$ keV), and is the nuclide of interest when performing carbon dating. ^{14}C is formed by the ^{14}N(n,p)^{14}C reaction when a thermal neutron is involved. If the neutron is more energetic (fast), formation of ^{12}C is preferred via ^{14}N(n,t)^{12}C.

QUESTIONS

9.1 Complete the following and write balanced equations (A + x → B + y).

$$^{18}O(n,\beta^-) \quad ^{14}N(\alpha,p) \quad ^{141}Pr(\gamma,2n)$$
$$^{238}U(\alpha,n) \quad ^{59}Fe(\alpha,\beta^-) \quad ^{10}B(d,n)$$

9.2 Balance the following nuclear reactions and provide the shorthand for each.

$$^{20}Ne + \underline{\quad} \rightarrow {}^{18}F + {}^4He \quad \underline{\quad} + n \rightarrow {}^{32}P + {}^1H$$
$$^{14}N + {}^1H \rightarrow \underline{\quad} + {}^4He \quad ^{124}Te + {}^1H \rightarrow {}^{123}I + \underline{\quad}$$

9.3 Write out the neutron capture reaction for ^{191}Ir. How much energy would be produced by this reaction?

9.4 Identify the target, the projectile, and the product in the reaction from Question 9.3.

9.5 Define the following: compound nucleus, endoergic, cross section, and resonance integral.

9.6 Write out the alpha,p reaction for ^{12}C. What is the compound nucleus formed in this reaction? If the alpha particle has an energy of 7.69 MeV (from the decay of ^{214}Po), what is the excitation energy of the compound nucleus?

9.7 What is the minimum amount of energy required for the reaction in Question 9.6 to proceed in good yield? What is the excitation plus kinetic energy of the product and emitron (B + y)?

9.8 Calculate the velocity (m/s) and wavelength (nm) of a thermal neutron.

9.9 The reaction that follows occurs in a small cyclotron. Calculate the effective Coulomb barrier for this reaction, the amount of energy produced or required by this reaction (Q), and the kinetic energy of the tritium ion necessary to make this reaction proceed in good yield.

$$^3_1H + {}^{124}_{50}Sn \rightarrow {}^{126}_{51}Sb + {}^1n$$

9.10 ^{111}In is a nuclide used in clinical settings for diagnostic scans. It can be made using an alpha,2n reaction. How much energy is required for this reaction to proceed in good yield? Could this reaction be performed using a cyclotron? Briefly explain.

9.11 A neutron treatment for cancer produces its neutrons by the bombardment of beryllium with 66 MeV protons. Assuming this is a p,n reaction, write out a balanced nuclear reaction for this process. What is the minimum energy required for this reaction to proceed in good yield? If the product of the nuclear reaction is unstable, write out a balanced decay equation.

9.12 What do threshold energy and effective Coulomb barrier have in common? What distinguishes them?

9.13 Suggest a reasonable, low-energy reaction for the production of ^{18}F from $^6Li_2CO_3$ using reactor neutrons (2 MeV). What is the excitation energy of the compound nucleus formed in this reaction?

9.14 Define the following: flux, saturation, dee, and synchrotron.

9.15 A thin sample of Mg (0.100 g) is irradiated for 24.0 h in a nuclear reactor with a neutron flux of 5.82×10^6 n/cm²·s. Calculate the number of ^{25}Mg atoms formed.

9.16 A thin sample of Mn (549 mg) was placed in a neutron howitzer (a laboratory neutron source) with a flux of 1.50×10^4 n/cm²·s for 5.00 h. Calculate the activity of the sample when it was removed from the howitzer. If a product is formed, what is it, and how much (mg) is present when the sample was removed from the howitzer?

9.17 A 0.486 g sample of gold was placed in a cyclotron beam of helium ions (flux $= 1.03 \times 10^4$ ions/cm² · s) for 67 hours, resulting in an activity of 97 dps. Assuming that the only particle emitted from the compound nucleus is ^3He, calculate the cross section (b) for the reaction and the total number of atoms produced (throughout irradiation). Is there a simpler way to perform this reaction? Explain.

9.18 Give at least three nuclear reactions that could produce ^{15}O (used for PET). Based on threshold energy and effective Coulomb barrier, which would be the best one to use?

9.19 A neutron howitzer employs 25 gallons of water for shielding and has a flux of 4.9×10^4 n/cm²·s at the neutron source. If the source is a cylinder that is 4 inches tall and 1 inch in diameter, what percentage (maximum) of the water will be DOH (D = deuterium) after one year? Would you expect a significant amount of TOH (T = tritium) or D_2O to form in this time? Briefly explain.

9.20 A 1 g sample of ^{14}N is placed in a neutron howitzer with a flux of 1.50×10^4 n/cm²·s for five days. Write out the most probable nuclear reaction. Assuming this is the only reaction, what percentage (atom!) of the sample was transmuted?

9.21 5.43 g of lead is irradiated in a nuclear reactor with a flux of 4.74×10^{13} n/cm²·s. How long (days) will it take to obtain a ^{205}Pb activity of 120 dpm?

9.22 A thin 17.3 mg sample of nickel is pummeled by 8.43 MeV deuterons in a cyclotron. If the emitron is a neutron, and the desired product is ^{61}Cu, what is the reaction? After 95 minutes in the cyclotron beam, the sample has a count rate of 1653 cpm under the annihilation peak on a high-purity Ge detector. Assume that decay of ^{61}Cu is the only contributor to this peak. If the detector has an efficiency of 10.8% at 511 keV, and the cross section for this reaction is 78 b, what is the deuteron flux in this cyclotron?

9.23 Briefly describe how a cyclotron works.

9.24 Cold fusion enthusiasts recently reported that the $^{133}Cs(d,\gamma)^{135}Ba$ reaction takes place close to room temperature. As the science editor for a prestigious journal, you are asked to review their work. How would you respond?

9.25 A student places a sample containing an unknown element in a neutron howitzer for 1.5 hours. Upon removal, the student observes that a nuclide was formed with a half-life of about 3.8 h, but the count rates were low. How long should the student keep the sample in the howitzer to achieve maximum activity?

9.26 It has been reported that the reaction of ^{48}Ca with ^{186}W produces ^{40}Mg. If only one other nuclide is formed by this reaction, write out a balanced nuclear reaction. What might be the driving force for the formation of these products?

9.27 How is tritium formed from cosmic radiation? Calculate the minimum projectile energy necessary for this reaction to proceed in good yield.

10 Fission and Fusion

Nuclear fission and fusion are special types of nuclear reactions. Because of their prominence in modern science and society, they will both be discussed in some detail in this chapter. Before we do, we need to cover one more form of decay, spontaneous fission.

10.1 SPONTANEOUS FISSION

Spontaneous fission is like alpha and cluster decay in that a nuclide splits in two. The difference is that with spontaneous fission, the nuclide splits roughly in half. It is only observed for the heavier nuclides and even then is only a primary decay mode for a handful of nuclides. Spontaneous fission can be generically represented by:

$$_{Z}^{A}X \rightarrow _{Z_1}^{A_1}L + _{Z_2}^{A_2}M + \nu n$$

where L and M are **fission products** (or fragments), and νn indicates the release of a small whole number (typically 0–4) neutrons. A wide variety of fission products are formed by the decay of a parent, but products with magic numbers (such as 50 and 82) of neutrons and/or protons have increased odds of formation.

The only naturally occurring nuclide that undergoes spontaneous fission is ^{238}U; however, its branch ratio for this form of decay is quite low (0.00005%). ^{238}U has 92 protons and 146 neutrons and is, therefore, an even-even (*ee*) nuclide. Even-even nuclides are somewhat more likely to undergo fission than nuclides with odd numbers of protons and/or neutrons.

The neutrons released in fission reactions usually escape their immediate surroundings and can interact with material that is a significant distance away. The fission products formed often do not travel far. These products have a great deal of kinetic energy, which is mostly dissipated as heat. The kinetic energies of the fission products are high because of the strong Coulomb repulsion of the two fragments right after they separate. The total kinetic energy (K_T) in MeV can be estimated using Equation 10.1:

$$K_T = 0.80 \times \frac{Z_1 Z_2}{\sqrt[3]{A_1} + \sqrt[3]{A_2}} \tag{10.1}$$

where Z_1 and Z_2 are the atomic numbers of the two fission products, and A_1 and A_2 are their mass numbers. Notice that this formula is similar to the one we used to calculate the Coulomb barrier (Equation 9.2). The only difference is the constant value— it was 1.11 for the Coulomb barrier, and it is 0.80 here. The reason for the difference lies in the distance between the centers of the nuclei as they are separating or joining.

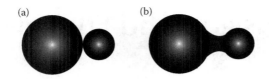

FIGURE 10.1   The differences in nuclear separation for (a) collision and (b) fission.

When calculating Coulomb barrier, we assume the nuclei are spheres that are just barely touching each other (Figure 10.1a). This is when the Coulomb repulsion forces are strongest, and the strong force (between the two nuclei) is just beginning to kick in. At this point, the Coulomb barrier is at a maximum. When determining kinetic energy following fission, it is better to assume some separation between the spheres, as this gives more realistic results (Figure 10.1b). With greater separation, energy (K_T) should be lower, so we use 0.80 to modify the Z/A term (Equation 10.1) rather than 1.11 (Equation 9.2).

The kinetic energies of the individual fragments can then be estimated using the same conservation of momentum formulas used to approximate recoil and alpha particle energies in alpha decay (Section 5.1):

$$K_1 = K_T\left(\frac{A_2}{A_1 + A_2}\right) \tag{10.2}$$

$$K_2 = K_T\left(\frac{A_1}{A_1 + A_2}\right) \tag{10.3}$$

Some of the variables have changed, but these are the same as Equations 5.1 and 5.2.

Example 10.1

Calculate the total energy released, and estimate the kinetic energies of the fission products in the spontaneous fission reaction shown here.

$$^{238}_{92}U \rightarrow\ ^{134}_{52}Te + {}^{102}_{40}Zr + 2{}^{1}_{0}n$$

Note that these fission products are just two of a great many possible spontaneous fission products for ^{238}U. Notice also that ^{134}Te has a magic number (82) of neutrons and, therefore, can be expected to be formed in somewhat higher yield. To find the total energy released by this decay, we need to determine how much mass is lost.

Mass of parent	^{238}U	238.050788 u
Mass of products	^{134}Te	−133.91154 u
	^{102}Zr	−101.92298 u
	2n = 2(1.008665 u) =	−2.017330 u
Mass lost		0.19894 u

Total energy of decay:

$$E = \frac{931.5\,\text{MeV}}{\text{u}} \times 0.19894\,\text{u} = 185.3\,\text{MeV}$$

Now, let's estimate the kinetic energy of the two fragments:

$$K_T = 0.80 \times \frac{Z_{Te}Z_{Zr}}{\sqrt[3]{A_{Te}} + \sqrt[3]{A_{Zr}}} = 0.80 \times \frac{52 \times 40}{\sqrt[3]{134} + \sqrt[3]{102}} = 170\,\text{MeV}$$

$$K_{Te} = K_T \left(\frac{A_{Zr}}{A_{Te} + A_{Zr}} \right) = 170\,\text{MeV} \times \frac{102}{134 + 102} = 73\,\text{MeV}$$

$$K_{Zr} = K_T \left(\frac{A_{Te}}{A_{Te} + A_{Zr}} \right) = 170\,\text{MeV} \times \frac{134}{134 + 102} = 97\,\text{MeV}$$

These formulas provide an estimate only, so it's a good idea to keep the number of significant figures low in the answers. The heavier fragment has less energy than the light one—just like alpha decay, only now the difference in energy between the two fragments is a lot less dramatic. Note that the sum of the two fragment kinetic energies equals the total kinetic energy:

$$K_1 + K_2 = K_T$$

$$73\,\text{MeV} + 97\,\text{MeV} = 170\,\text{MeV}$$

Notice also that the total energy of decay is significantly larger than the kinetic energy of the fragments. The remaining 15 MeV (185 − 170 MeV) is divided among the neutrons and gamma rays emitted as part of the decay and the excitation energy of the products. The gamma rays emitted as part of the decay are referred to as **prompt** gamma photons and are not typically shown when writing out the decay equation.

Because of the low branch ratio, relatively few ^{238}U atoms undergo spontaneous fission. Generally speaking, the odds of spontaneous fission increase with mass number (A). For example, ^{256}Fm decays via spontaneous fission in 92% of all decays. This makes intuitive sense; the bigger the nucleus, the more likely it is to split into two fragments.

10.2 NEUTRON-INDUCED FISSION

If heavy *ee* nuclides are more likely to undergo spontaneous fission, can we get a heavy even-odd (*eo*) nuclide to split if we fire neutrons at it? You bet! In fact, this is a pretty handy way to get fission to occur, rather than sitting around waiting for a nuclide to undergo spontaneous fission. The advantage of adding a neutron to an *eo* nuclide is that we obtain an *ee* nuclide with some excitation energy. Neutron-induced fission reactions (typically referred to simply as "fission reactions") can be generically represented as:

$$A + n \rightarrow [C]^* \rightarrow B + D + \nu n$$

where A is the heavy nuclide, also known as the **fissionable material** (*vide infra*). The compound nucleus is [C]*, and B and D are the fission products (fragments). Neutrons are typically released by fission reactions, which is pretty handy if you've got a lot of A and want more of them to undergo fission. Such a process is known as a **fission chain reaction** and was introduced back in Chapter 4, and it will be discussed some more in Chapter 11. Just as in spontaneous fission, the symbol ν represents a small whole number, typically 0–4.

A fissionable material is any nuclide that will undergo fission when you throw a neutron (thermal *or* fast) at it. As you no doubt remember from Chapter 4, fissile material undergoes fission after being hit by a thermal neutron (our main examples are ^{233}U, ^{235}U, and ^{239}Pu). As such, both ^{235}U and ^{238}U are fissionable (^{238}U can undergo fission with fast neutrons, see Section 4.2), but, of the two, only ^{235}U is fissile. Further complicating fission semantics is the term **fertile**. As defined in Section 4.1.3, a fertile nuclide is one that is not itself fissile but can be converted to a fissile nuclide in a reactor. Examples are ^{232}Th and ^{238}U, which make ^{233}U and ^{239}Pu, respectively.

Fission reactions tend to be pretty messy—hundreds of different nuclides have been identified as fission products for ^{235}U. These products tend to be rather neutron-rich since they originate from a nuclide (^{235}U) with an N/Z ratio that is relatively high when compared to the stable isobars of the fission products. One example of a fission reaction for ^{235}U forms ^{94}Rb, ^{140}Cs, and two neutrons:

$$^{235}_{92}U_{143} + {}^{1}_{0}n \rightarrow {}^{94}_{37}Rb_{57} + {}^{140}_{55}Cs_{85} + 2{}^{1}_{0}n$$

$$N/Z \rightarrow \quad 1.55 \qquad\qquad 1.54 \qquad 1.54$$

The N/Z ratio for each nuclide is given directly below it. Despite the emission of two neutrons in this particular fission reaction, the N/Z ratios of the products are pretty much the same as the target nuclide. However, we know (Section 1.4) that nuclides with lower mass numbers will only be stable with lower N/Z ratios. For example, the only stable nuclides for $A = 94$ are $^{94}_{42}Mo_{52}$ ($N/Z = 1.24$) and $^{94}_{40}Zr_{54}$ ($N/Z = 1.35$). Since fission products tend to be neutron-rich, they will generally undergo β^- decay. In fact, each fission product can form a short (isobaric) decay series on the way to stability. The series formed by ^{94}Rb consists of ^{94}Sr and ^{94}Y before forming stable ^{94}Zr:

$$^{94}_{37}Rb \xrightarrow{\beta^-} {}^{94}_{38}Sr \xrightarrow{\beta^-} {}^{94}_{39}Y \xrightarrow{\beta^-} {}^{94}_{40}Zr$$

The relative amount that each nuclide is formed is known as the **fission yield**, and is expressed as a percentage. It is sometimes difficult to determine the yield of a particular fission product (**independent yield**) because it may have been formed through beta decay of an isobar. Some of these products formed through beta decay series have very short half-lives, making it difficult to run a fission reaction for an amount of time, then analyze the products and get reliable independent yields. Since the products generally retain the same mass number, yields can be determined for all isobars (**cumulative yield**).

The cumulative yields for thermal neutron-induced fission of ^{235}U are collectively graphed in Figure 10.2. Fission products for this reaction distribute themselves into two large peaks. Products with mass numbers of 90–100 u and 130–145 u are formed in the highest yields and are, therefore, preferred. Generally speaking, the mass numbers from these two peaks add up to the mass number of ^{235}U, meaning that nuclides with differing mass numbers are usually formed. Thermal neutron fission of ^{235}U is, therefore, generally **asymmetric**.

Symmetric fission would be the division of the ^{236}U compound nucleus (formed by the interaction of a neutron with ^{235}U) into identical fragments. This clearly does take place but at a much lower yield than asymmetric fission. This is indicated in the middle of Figure 10.2 by the formation of nuclides with mass numbers of approximately 118.

Why is asymmetric fission so strongly preferred? Look closely at the small spike in fission yield at $A = 132$. Something is definitely special about formation of nuclides with this mass number. In particular, $^{132}_{50}Sn_{82}$ is formed, which has magic numbers of both neutrons and protons and therefore has greater stability. Fission reactions prefer to form at least one product with a magic number of protons and/or neutrons because of the stability of those nuclides. Many of the other nuclides formed under the broad peak on the right of Figure 10.2 also have either 50 protons (tin) or 82 neutrons. Some of the nuclides under the left peak are preferentially formed because they have 50 neutrons. The rest of the graph is a mirror to these preferences. For example, the

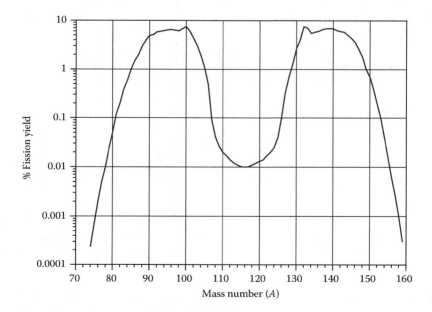

FIGURE 10.2 Fission product yields for thermal neutron fission of ^{235}U.

small spike at $A = 100$ reflects the spike at $A = 132$, and corresponds to a reaction like the one shown here:

$$_{92}^{235}\text{U} + {}_{0}^{1}\text{n} \rightarrow {}_{50}^{132}\text{Sn} + {}_{42}^{100}\text{Mo} + 4{}_{0}^{1}\text{n}$$

^{100}Mo is simply what's left over after the very favorable ^{132}Sn is formed and four neutrons are emitted.

The energy of the incident neutron will have an effect on the distribution of the fission products. As shown in Figure 10.3, higher energy (more than thermal) neutrons will increase the probability of symmetric fission relative to the probability of asymmetric fission for ^{235}U. The valley between the two peaks fills in as neutron energy increases. This can be understood in terms of the excitation energy of the compound nucleus ($[^{236}\text{U}]^*$). The more energy the neutron has, the greater the excitation energy of the compound nucleus. The greater the excitation energy, the less fussy the compound nucleus will be about forming products with magic numbers of nucleons. Formation of nuclides with magic numbers generate more energy for the reaction; however, if the reaction already has some extra energy, it isn't as important to generate more.

The fission product distribution also depends on the nuclide undergoing fission (the target). Figure 10.4 shows how the distribution varies between ^{233}U and ^{239}Pu. Notice that the distribution for the lighter of these two nuclides (^{233}U—solid line) is

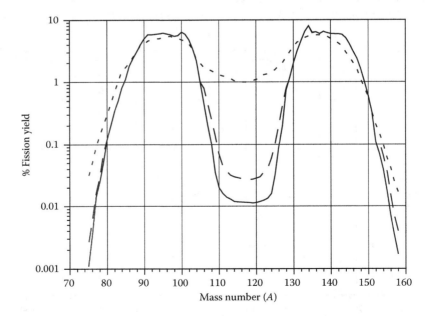

FIGURE 10.3   Fission product distributions for ^{235}U with thermal neutrons (solid line), ~2 MeV neutrons (larger dashed line), and 14 MeV Neutrons (smaller dashes). (Data from Flynn, K. F. and Glendenin, L. E., Rep. ANL-7749 Argonne National Laboratory, Argonne, IL 1970.)

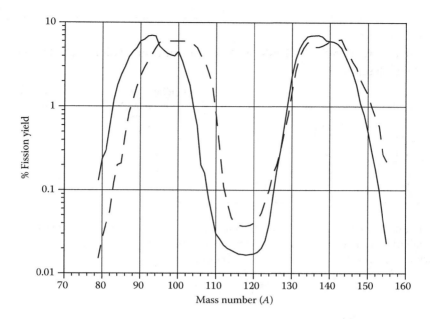

FIGURE 10.4   Thermal neutron fission product distributions for ^{233}U (solid) and ^{239}Pu (dashed).

shifted to lower masses, while it is shifted to higher masses for ^{239}Pu (dashed line), when compared to Figure 10.2. Notice also that the right peak shifts much less than the left peak. This reflects the fact that nuclides formed under the right peak have two choices for magic numbers (50 protons *or* 82 neutrons), while the left peak only has one (50 neutrons).

Not all heavy *eo* nuclides undergo thermal neutron fission, and not all heavy *ee* nuclides decay via spontaneous fission. Why not? There is an energy barrier to fission, which we can think of as the same as the Coulomb barrier discussed in Section 9.1. Just as there is an energetic barrier to bring two positively charged nuclei together, there is also a barrier to splitting them up. It is called the **fission barrier**, and it can be thought of as disruption of the strong force which holds all the nucleons together in the nucleus. Considering this barrier, it is surprising that spontaneous fission takes place at all. The only way it can take place is by avoiding the barrier. It is generally believed that spontaneous fission takes place via tunneling—the nuclide finds a way through the barrier rather than over it. As you will recall, tunneling is a path less taken; therefore, spontaneous fission is a relatively rare form of decay, but it can sometimes occur for some of the heaviest known nuclides.

The binding of an additional neutron to a target nucleus can provide the energy to overcome the fission barrier. Let's look at some relevant information for some common nuclides of the actinide series (Table 10.1). As discussed in Section 9.2, σ_γ is the cross section (probability) for an n,gamma reaction, and σ_f is the cross section for a neutron-induced fission reaction. Both cross-sectional values are given in barns to allow easy comparison. For example, ^{232}Th is *much* more likely to form ^{233}Th when hit by a neutron, while ^{233}U is more likely to undergo fission.

TABLE 10.1

Relative Tendencies to Undergo Neutron-Induced Fission for Selected Heavy Nuclides

Nuclide	σ_f (b)	σ_γ (b)	Binding E of a Thermal ^{1}n (MeV)	Fission Barrier (MeV)
^{232}Th	0.000003	7.34	4.8	7.5
^{233}U	531	46	6.8	6.0
^{235}U	585	98	6.5	5.7
^{238}U	0.000005	2.7	4.8	5.8
^{239}Pu	750	271	6.5	5.0

For each nuclide in Table 10.1, the energy produced by binding a thermal neutron is calculated. This is the same as the excitation energy of the compound nucleus formed by neutron capture. Let's use ^{235}U as an example:

$$^{235}U + {}^1n \rightarrow [{}^{236}U]^*$$

$$Q = \frac{931.5\,\text{MeV}}{u} \times (235.043930\,u + 1.008665\,u - 236.045568\,u)$$
$$= 6.546\,\text{MeV}$$

You may recall that calculation of the excitation energy of a compound nucleus also needs to consider the kinetic energy of the compound nucleus (E_{KC}), which is conveniently calculated with Equation 9.4:

$$E_{KC} \approx E_{KX} \times \left(\frac{A_X}{A_C} \right)$$

The kinetic energy of a thermal neutron is rather low ($E_{KX} = 0.0253$ eV); therefore, the kinetic energy of the compound nucleus formed after thermal neutron capture will be very small relative to the excitation energy. The bottom line is that excitation energy of the compound nucleus is essentially equal to Q for thermal neutron capture.

Notice that the excitation energy is lower ($\sim$4.8 MeV) for $^{232}_{90}$Th$_{142}$ and $^{238}_{92}$U$_{146}$, while it is higher (6.5–6.8 MeV) for $^{233}_{92}$U$_{141}$, $^{235}_{92}$U$_{143}$, and $^{239}_{94}$Pu$_{145}$. This difference arises from the number of neutrons in each nuclide. ^{232}Th and ^{238}U have an even number of neutrons, so addition of another neutron would result in the formation of a compound nucleus with an odd number of neutrons. ^{233}U, ^{235}U, and ^{239}Pu all start with an odd number of neutrons, and adding another neutron forms a compound nucleus with an even N. Remember that even numbers of nucleons mean greater stability, and odd numbers mean less stability (Section 1.4). An appropriate analogy is a rock rolling down a hill into a ravine. Generally speaking, the taller the hill and the deeper the ravine, the more energy the rock will have when it hits bottom. Going from even N to odd N (^{232}Th and ^{238}U) means the hill is lower and the ravine shallower; therefore, less excitation energy results.

The fission barrier for each nuclide is given in the final column of Table 10.1. Those nuclides that generate enough excitation energy from thermal neutron capture to overcome the fission barrier are those that tend to undergo fission rather than an n,γ reaction. Notice that those nuclides with enough energy to overcome the fission barrier also have relatively high σ_f values. Our example nuclide, ^{235}U, looks pretty good for fission, since its excitation energy is 6.5 MeV, but its fission barrier is only 5.7 MeV. As we might then expect, its cross section for a thermal neutron induced fission reaction is high (585 b) relative to its cross section toward an n,γ reaction (98 b). For ^{238}U, the opposite is true. Its excitation energy is too low (4.8 MeV) to overcome its fission barrier (5.8 MeV), and it has a very low σ_f (0.000005 b) relative to its σ_γ (2.7 b). We would therefore expect ^{238}U to predominately form ^{239}U when exposed to thermal neutrons (although not in great yield).

As we learned in Section 9.2, cross sections change with the energy of the neutron. Fast neutrons dramatically increase σ_f while decreasing σ_γ, causing more ^{238}U to fission rather than just form ^{239}U. Again, this can be understood in terms of excitation energy of the compound nucleus. If the neutron has more than about 1 MeV of energy, the excitation energy of the compound nucleus will be greater than the fission barrier, allowing more ^{238}U fission reactions to take place. Therefore, fission is not restricted to excited *ee* nuclides but to any large nucleus with sufficient excitation energy to overcome its fission barrier.

The energetics of a neutron-induced fission reaction can be handled the same way as for spontaneous fission. Don't forget there's now a neutron on the left side of the arrow. You can save a step in the math by canceling it with one of the product neutrons, as shown in the following example.

Example 10.2

Calculate the total energy released by the following fission reaction. Estimate the kinetic energy of the fission products.

$$^{235}_{92}U + {}^{1}_{0}n \rightarrow {}^{141}_{56}Ba + {}^{92}_{36}Kr + 3{}^{1}_{0}n$$

Remember, this is just one of hundreds of possible reactions for ^{235}U fission. Following Example 10.1:

Mass of target	^{235}U	235.043930 u
Mass of products	^{141}Ba	−140.914406 u
	^{92}Kr	−91.92615 u
	net 2n = 2(1.008665 u) =	−2.017330 u
Mass lost		0.18604 u

Energy produced by the reaction:

$$Q = \frac{931.5\,\text{MeV}}{u} \times 0.18604\,u = 173.3\,\text{MeV}$$

Total kinetic energy and kinetic energies of the fission products:

$$K_T = 0.80 \times \frac{Z_{Ba}Z_{Kr}}{\sqrt[3]{A_{Ba}} + \sqrt[3]{A_{Kr}}} = 0.80 \times \frac{56 \times 36}{\sqrt[3]{141} + \sqrt[3]{92}} = 166 \text{ MeV}$$

$$K_{Ba} = K_T \left(\frac{A_{Kr}}{A_{Ba} + A_{Kr}} \right) = 166 \text{ MeV} \times \frac{92}{141 + 92} = 66 \text{ MeV}$$

$$K_{Kr} = K_T \left(\frac{A_{Ba}}{A_{Ba} + A_{Kr}} \right) = 166 \text{ MeV} \times \frac{141}{141 + 92} = 100 \text{ MeV}$$

Just as we saw with spontaneous fission, the total energy is carved up into kinetic and excitation energy of the fragments, neutron energy (velocity), and prompt gamma photons. In this example, only about 7 MeV separate the total energy produced (Q) and total kinetic energy (K_T). What about the neutrons, prompt gamma rays, and excitation energy? Keep in mind: (1) The total kinetic energy is estimated—it could be a little bit off, and (2) the energy values for the emitted neutrons, prompt gamma rays, and excitation vary somewhat from one fission reaction to another. The math doesn't always work perfectly, but it does give us an idea of where all the energy goes.

The fission products in Example 10.2 will both undergo decay. The energy released as part of these decays is called **delayed energy**, as opposed to the **prompt energy** associated with the act of splitting the nucleus. The average values for both are detailed in Table 10.2.

The total kinetic energy for ^{235}U fission products averages about 167 MeV. ^{235}U emits an average of 2.4 neutrons per fission, each carrying about 2 MeV of energy, giving approximately 5 MeV of energy associated with the prompt neutrons. Finally, an average of six gamma rays are emitted per fission of ^{235}U, each packing about 1 MeV of energy.

Since delayed energy is all about the beta decay of the fission products, it is composed of the energy of the beta and antineutrino particles and the gamma photons emitted during decay(s) to a stable nuclide. The 6 MeV of energy listed for the delayed gamma rays consist of gamma emissions that are part of the beta decays and those that are emitted during de-excitation of the fission fragments. On average, the thermal neutron fission

TABLE 10.2
Approximate Average Energy Values (MeV) for the Thermal Neutron Fission of ^{235}U

Prompt energy	Fission products	167
	Neutrons	5
	Prompt gamma rays	6
Total prompt energy		178
Delayed energy	Beta particles	8
	Antineutrino particles	12
	Gamma rays	6
Total delayed energy		26
Total fission energy		204

products of ^{235}U emit about 26 MeV as delayed energy. This brings the total average energy per fission to about 200 MeV. Let's see how our example compares.

Example 10.3

Estimate the delayed energy emitted by the fission reaction given in Example 10.2. All we need to do is look at the two fission products and add up the total energy of their isobaric decays to stable nuclides. Data below are rounded to the nearest tenth MeV because of the qualitative nature of this calculation.

$$^{141}_{56}Ba \xrightarrow[\text{3.2 MeV}]{\beta^-} \, ^{141}_{57}La \xrightarrow[\text{2.5 MeV}]{\beta^-} \, ^{141}_{58}Ce \xrightarrow[\text{0.6 MeV}]{\beta^-} \, ^{141}_{59}Pr$$

$$^{92}_{36}Kr \xrightarrow[\text{6.0 MeV}]{\beta^-} \, ^{92}_{37}Rb \xrightarrow[\text{8.1 MeV}]{\beta^-} \, ^{92}_{38}Sr \xrightarrow[\text{2.0 MeV}]{\beta^-} \, ^{92}_{39}Y \xrightarrow[\text{3.6 MeV}]{\beta^-} \, ^{92}_{39}Zr$$

$$^{141}Ba: \quad 3.2 \text{ MeV} + 2.5 \text{ MeV} + 0.6 \text{ MeV} = 6.3 \text{ MeV}$$

$$^{92}Kr: \quad 6.0 \text{ MeV} + 8.1 \text{ MeV} + 2.0 \text{ MeV} + 3.6 \text{ MeV} = 19.7 \text{ MeV}$$

$$\text{Total delayed energy:} \quad 6.3 \text{ MeV} + 19.7 \text{ MeV} = 26.0 \text{ MeV}$$

Spot on!

Some of the fission products are so neutron-rich they sometimes emit delayed neutrons. Don't confuse delayed neutrons with delayed energy. As mentioned in Section 5.8.4, particles are sometimes emitted from an excited daughter and are called delayed because they are emitted *after* the original decay has taken place. For example, 0.03% of all ^{92}Kr beta decays are followed by the emission of a delayed neutron from the excited ^{92}Rb daughter. Because a short time interval (seconds to minutes) occurs between the beta particle emission and the neutron emission, it is called a delayed neutron. These delayed neutrons are useful in controlling nuclear reactions in power plants, as discussed further in the next chapter.

Other projectiles can induce fission reactions. This makes some sense, since the reasonable occurrence of fission requires only a big, excited nucleus. How that nucleus is formed is somewhat irrelevant. Protons, alpha particles, deuterons, and other small nuclei have caused fission in a variety of heavy targets. Since all these projectiles carry a positive charge, they need to be accelerated before they can strike the heavy target (to overcome the Coulomb barrier). High-energy photons can also induce fission (**photo fission**) simply by transferring their energy to the nucleus. Neutron-induced fission is the most heavily studied because of its applications in nuclear reactors and weapons (Chapters 4 and 11) and because neutrons do not need to overcome the Coulomb barrier to react with a nucleus.

10.3   FUSION

Fission reactions release energy because a large nuclide splits into two that have greater stability in terms of binding energy per nucleon (or lower average mass per nucleon, see Section 3.1). Likewise, the combination of two light nuclides forming a product with greater stability will also release energy. The combination of two light nuclides to form a heavier one is called **fusion**. Stars, such as our Sun, are huge fusion reactors, and the energy they emit comes from fusion reactions. Some of the important fusion reactions taking place inside stars and the total energy produced by each are given here:

$$_1^1H + _1^1H \rightarrow _1^2H + _{+1}e + \nu \quad 0.42\,\text{MeV}$$
$$_1^1H + _1^2H \rightarrow _2^3He + \gamma \quad 5.49\,\text{MeV}$$
$$_2^3He + _2^3He \rightarrow _2^4He + 2_1^1H \quad 12.86\,\text{MeV}$$

It is so scorching hot inside the Sun that matter does not exist as atoms—nuclei and electrons are separated, and, therefore, it is plasma. Therefore, it is more appropriate to think of these reactions as taking place as nuclei rather than atoms. The first reaction fuses two ¹H nuclei (protons) into deuterium. Once formed, deuterium can combine with another proton to form ³He (second reaction). If the first two reactions run twice, we'll have two ³He nuclei to fuse together to form ⁴He and two protons (third reaction). Because nuclei rather than atoms are reacting, it is sometimes best to use nuclear masses when performing energy calculations for fusion reactions. As an example, let's calculate the energy released by the first reaction, the fusion of two protons:

Mass of reactants	2(proton mass) = 2(1.007276 u) =	2.014552 u
Mass of products	²H nucleus = 2.014102 u − 0.000549 u =	−2.013553 u
	Positron mass =	−0.000549 u
Mass lost		0.000450 u

Energy produced

$$0.000450\,\text{u} \times \frac{931.5\,\text{MeV}}{\text{u}} = 0.419\,\text{MeV}$$

Note that the neutrino mass is not necessary in the calculation because it is so ridiculously small. Note also that if this calculation were done with atomic masses, an incorrect value would be obtained. Using atomic masses for fusion reaction energy calculations only work when the atoms on both sides of the reaction arrow have the same number of electrons. Do you understand why?

The positrons produced in the proton-proton reaction will annihilate (there are plenty of electrons in the Sun!), producing even more energy:

$$_{+1}e + _{-1}e \rightarrow 2\gamma \quad 1.022\,\text{MeV}$$

Taken together, the four reactions shown so far in this section make up what is called the **proton cycle**. If we add them together, we will see the *net* reaction and energy for the proton cycle. Remember that the first two reactions need to happen twice, and, therefore, the positron annihilation also happens twice per cycle:

$$2(_1^1H + _1^1H \rightarrow _1^2H + _{+1}e + \nu) \quad 2(0.42\,\text{MeV})$$
$$2(_1^1H + _1^2H \rightarrow _2^3He + \gamma) \quad 2(5.49\,\text{MeV})$$
$$_2^3He + _2^3He \rightarrow _2^4He + 2_1^1H \quad 12.86\,\text{MeV}$$
$$\underline{2(_{+1}e + _{-1}e \rightarrow 2\gamma) \quad 2(1.022\,\text{MeV})}$$
$$2_{-1}^0e + 4_1^1H \rightarrow _2^4He + 2\nu + 6\gamma \quad 26.7\,\text{MeV}$$

The proton cycle accounts for roughly 90% of the Sun's energy output. Fortunately, our Sun is about 70% hydrogen (by mass), so it looks like there's still plenty of fuel up there. Notice that the Sun is also emitting a lot of gamma photons and neutrinos (see Figure 9.8). Are you scared?

We should also be thankful that the proton–proton reaction has a low cross section (probability of success). This slows down the entire proton cycle. If the cross section were higher, the Sun would burn hotter and run out of hydrogen fuel faster—not an attractive prospect for life on Earth.

The fusion reactions shown in this section fit the general characteristics of the nuclear reactions studied in Chapter 9, but which is the projectile, and which is the target? These reactions happen inside the Sun because the reacting nuclei are all moving very quickly, that is, they have a lot of energy. Their energy comes from the extraordinarily high temperatures inside the Sun, which allow these nuclei to overcome the Coulomb barrier and react. Because of this, fusion reactions are often called **thermonuclear reactions**. As such, the projectile/target labels seem inadequate here; it's more like two projectiles colliding with each other.

The proton cycle is generally thought to be impractical for fusion power reactors here on Earth, primarily due to its relatively high **ignition** temperature—the temperature needed to start a fusion reaction. To get an idea of how hot it needs to be to get the proton-proton reaction going, let's calculate its effective Coulomb barrier:

$$_1^1 H + _1^1 H \rightarrow _1^2 H + _{+1} e$$

$$E_{ecb} \approx 1.11 \times \left(\frac{1+1}{1} \right) \times \left(\frac{1 \times 1}{\sqrt[3]{1} + \sqrt[3]{1}} \right) = 1.11 \, \text{MeV}$$

This value can be converted to temperature using Boltzmann's constant:

$$1.11 \, \text{MeV} \times \frac{10^6 \, \text{eV}}{\text{MeV}} \times \frac{K}{8.63 \times 10^{-5} \, \text{eV}} = 1.29 \times 10^{10} \, \text{K}$$

According to these calculations, the Sun would have to be nearly 13 billion Kelvin to get the proton cycle to proceed in good yield. The surface of the Sun is about 6000 K, and the core is believed to only reach ~15 million K. Fortunately, plasma, like other forms of matter, contains individual particles that are moving both faster and slower than they should be at a particular temperature. Temperature is a reflection of the average kinetic energy of all particles, and there is always a distribution of energies about the mean value.[1] Also remember that both particles on the left side of the arrow are moving, so, if they hit head-on, they only need half as much energy each. Therefore, the Sun will have some particles moving fast enough to overcome the Coulomb barrier and fuse. Finally, remember that nuclear reactions can take place even if the particles lack the minimum energy—through tunneling. This has

---

[1] It is an *asymmetric* distribution. It tails off more gradually at the high-energy end, ensuring at least some will have sufficient energy, even at much lower temperatures.

positive implications for human-built fusion reactors here on Earth. These reactors do not need to get quite as hot as calculated using Boltzmann's constant because some reactions will take place at lower temperatures due to tunneling.

Tunneling may have led two chemists to conclude they had observed fusion in 1989. They observed heat being generated during an electrochemical experiment involving a palladium electrode in deuterated water (D_2O). They dubbed the phenomenon "cold fusion," then, circumventing the normal scientific peer review process, announced their results at a press conference. A great deal of popular media attention was given to this result, which seemed to promise an easy fix for our increasing energy needs. Several other scientists tried to duplicate these remarkable results. At best, these efforts could be described as inconsistent, and at worst, they were a clear repudiation of the original results. The scientists were discredited, and the entire scientific community suffered a loss of public confidence. Some research has continued in this area, but proof of low-energy nuclear fusion remains elusive. It is unfortunate that the term "cold fusion" was applied to these poorly interpreted experiments, as it is a *bona fide* area of nuclear science (see Section 10.5).

10.4   STELLAR NUCLEOSYNTHESIS

How were the elements we find here on the Earth made? In the stars! Fusion reactions can make nuclides up to ^{56}Fe. Remember, this is the pinnacle of nuclear stability (binding energy per nucleon—Figure 3.1, or mass per nucleon—Figure 3.2), so fusion up to the peak is exoergic and is endoergic after.

After ^{56}Fe, stars rely on neutron capture to make most of the heavier nuclides. The neutrons come from nuclear reactions taking place in the Sun like this one:

$$^{21}Ne + {}^4He \rightarrow {}^{24}Mg + {}^1n$$

^{56}Fe will capture neutrons until it forms ^{59}Fe, which decays to ^{59}Co. ^{59}Co will capture a neutron and form ^{60}Co, which decays to ^{60}Ni. Through this combination of n,gamma reactions and beta decay, almost all the naturally occurring nuclides from $A = 56$ through $A = 210$ can be prepared. This series of reactions is called the **s-process**, where "s" stands for slow. It's slow because the neutron flux inside of stars is not all that great, and many of the cross sections are small. It takes a long time for significant amounts of nuclides beyond ^{56}Fe to build up.

The s-process also fails to explain the existence of a few stable, but somewhat proton-rich, nuclides such as ^{124}Xe. ^{124}Xe cannot be formed through n,gamma reactions followed by beta decay because it is "shielded" by stable ^{124}Te. Any neutron-rich, $A = 124$ nuclide will undergo beta decay until it forms ^{124}Te, which cannot then form ^{124}Xe through decay. Nuclides such as ^{124}Xe are formed through the **p-process**. Two types of reactions are postulated to be part of the p-process, gamma,n or p,x. Either one can lead to the formation of more proton-rich naturally occurring nuclides. Some of the other nuclides in the same boat are ^{78}Kr, ^{112}Sn, ^{120}Te, ^{144}Sm, and ^{184}Os. These nuclides are all found in lower percent abundances than the more neutron-rich, stable isobars because the p-process is rather unlikely. This is generally true for pairs of naturally occurring isobars between $A = 70$ and $A = 204$—the isobar with more neutrons will often have a higher percent abundance because it can be formed through the more likely s-process.

The s-process cannot make nuclides beyond $A = 210$. It terminates in the following cycle of reactions:

$$^{209}Bi + {}^1n \rightarrow {}^{210}Bi + \gamma$$
$$^{210}Bi \rightarrow {}^{210}Po + {}_{-1}e$$
$$^{210}Po \rightarrow {}^{206}Pb + \alpha$$
$$^{206}Pb + {}^1n \rightarrow {}^{207}Pb + \gamma$$
$$^{207}Pb + {}^1n \rightarrow {}^{208}Pb + \gamma$$
$$^{208}Pb + {}^1n \rightarrow {}^{209}Pb + \gamma$$
$$^{209}Pb \rightarrow {}^{209}Bi + {}_{-1}e$$

How are elements like thorium and uranium made? The rapid process (**r-process**) makes them. As its name implies, it is a quick series of neutron absorptions, leading to an extremely neutron-rich nuclide, which then undergoes a number of beta decays until the N/Z ratio optimizes. An exceptionally intense neutron flux is needed to accomplish this, as the neutrons need to be continually added, faster than the newly formed nuclides can decay. Such a flux is not possible under normal conditions in a star. It is believed that these fluxes are possible in supernovae. Our solar system was formed from the remnants of a supernova, which explains why we have these heavy elements in some abundance. Therefore, we live in a second (or possible third) generation solar system. Makes you wonder if there are alien civilizations out there in older solar systems …

10.5   SYNTHESIS OF UNKNOWN ELEMENTS

Humans have done a pretty good job preparing isotopes of elements not found on the Earth. These elements had been unknown because all their isotopes have half-lives that are short when compared to the age of the solar system ($\sim$4.6 billion years), and they are not part of one of the naturally occurring decay series (Section 1.7). Therefore, if they were formed in the sun that preceded ours, they have since decayed away. The first "artificial" element in the periodic table is technetium (Tc, $Z = 43$), and it was first prepared in 1945 by accelerating a deuteron in a cyclotron and slamming it into a molybdenum target. Its name comes from the Greek word *technetos*, which means artificial. The preparation of technetium was a validation of the configuration of the periodic table as it filled a void in the middle of the transition metals. Technetium has since been made in large quantities as a fission product in nuclear reactors. A significant amount of its chemistry has been investigated, and it generally parallels the chemistry of its heavier congener rhenium. It is an obscure irony that technetium is now more plentiful on Earth than rhenium.

Promethium (Pm, $Z = 61$) is the only other element before lead ($Z < 82$) that was missing from our current periodic table because it did not exist on the Earth. It was first observed among the fission products of ^{235}U in 1945. It is named for the mythological Greek titan who stole fire from the gods and gave it to humans.

Both Tc and Pm have an odd number of protons, which means that any of their isotopes with an odd number of neutrons (oo) will be unstable, but why are there no stable oe isotopes of these elements? A simple way to look at Tc is to state that $^{98}_{43}\text{Tc}_{55}$ ($N/Z = 1.28$) has the best N/Z ratio and the longest half-life of all the known Tc isotopes, but it is oo and, therefore, cannot be stable. What about $^{97}_{43}\text{Tc}_{54}$ ($N/Z = 1.26$) or $^{99}_{43}\text{Tc}_{56}$ ($N/Z = 1.30$)? They have even numbers of neutrons and pretty good N/Z ratios. Examination of the $A = 97$ and $A = 99$ isobars shows that $^{97}_{42}\text{Mo}_{55}$ and $^{99}_{44}\text{Ru}_{55}$ are the only stable nuclides for those sets of isobars. As we saw in Section 5.6 (The Valley of Beta Stability), when $A = $ odd, there can only be one stable isobar. Additionally, a stable ^{97}Tc or ^{99}Tc would violate Mattauch's Rule of no stable, neighboring isobars.

What about the anthropogenic syntheses of the elements beyond uranium? A partial answer was given in Example 9.2 where a plutonium target was hit with an alpha particle to produce a curium isotope:

$$^{240}_{94}\text{Pu} + {}^{4}_{2}\text{He} \rightarrow {}^{243}_{96}\text{Cm} + {}^{1}\text{n}$$

So long as the target nuclide can be isolated and has a reasonably long half-life, we can perform lots of reactions like this one to make nuclides of the elements beyond uranium. We can even mimic the s-process, making new nuclides through n,γ reactions and beta decay, such as the preparation of ^{239}Pu from ^{238}U:

$$^{238}_{92}\text{U} + {}^{1}\text{n} \rightarrow {}^{239}_{92}\text{U} + \gamma$$

$$^{239}_{92}\text{U} \rightarrow {}^{239}_{93}\text{Np} + {}_{-1}\text{e}$$

$$^{239}_{93}\text{Np} \rightarrow {}^{239}_{94}\text{Pu} + {}_{-1}\text{e}$$

The r-process is mimicked in the explosions of thermonuclear weapons. Tremendous neutron fluxes are briefly created in these explosions that can add several neutrons to ^{238}U, which is often used as a casing material for these weapons. The first detonation of a thermonuclear device occurred in 1952 on Eniwetok Island in the South Pacific. Two elements that were first observed in the debris following this explosion were einsteinium (Es) and fermium (Fm). As many as 17 neutrons were absorbed by ^{238}U atoms in the bomb's casing during the explosion, followed by a series of beta decays to form the observed nuclides.

We can only get so far with the preceding reactions. Generally speaking, the half-lives of nuclides get shorter as atomic number increases beyond uranium. This makes intuitive sense—the nuclides are already too big, and making them bigger should push them further from stability.[2] Larger nuclides mean the strong force is having an increasingly difficult time keeping the nucleus together under the burgeoning Coulomb repulsion of increasing numbers of protons. At a certain point,

---

[2] This isn't always true; an ee that is heavier than a nearby oo nuclide might have a longer half-life. We could also make new nuclides (heavier than those known today) with magic numbers of nucleons that have anomalously long half-lives.

the half-lives of potential targets will get too short, so the projectile will have to be bigger if we want to make super heavy nuclides. For example, ^{254}No can be prepared by hitting ^{246}Cm with a ^{12}C projectile:

$$^{246}_{96}\text{Cm} + {}^{12}_{6}\text{C} \rightarrow {}^{254}_{102}\text{No} + 4\,^{1}\text{n}$$

Increasing the atomic number (Z) of the projectile increases the Coulomb barrier and, therefore, the minimum energy for the reaction to proceed in good yield. This also means that the excitation energy of the compound nucleus will also increase. We'll soon reach a point where the compound nucleus will have too much excitation energy and will fall apart in ways that make it difficult to determine if it was actually made. A subtler approach is required.

If the projectile has less than the minimum energy required for the product to be formed in good yield, the product can still be formed; there just won't be very much of it. Thankfully, nuclear decay can be detected, even if only a handful of decays occur—so we only need to make a few product nuclides. This approach has traditionally been referred to as cold fusion, while the reaction of ^{246}Cm with ^{12}C would be called hot fusion. As mentioned at the end of Section 10.3, cold fusion has unfortunately become more widely applied to alleged, low-energy nuclear reactions involving low Z nuclides. The first atoms of element 111 (roentgenium, Rg) were made with the cold fusion method, bombarding a ^{209}Bi target with ^{64}Ni:

$$^{209}_{83}\text{Bi} + {}^{64}_{28}\text{Ni} \rightarrow {}^{272}_{111}\text{Rg} + {}^{1}\text{n}$$

QUESTIONS

10.1 Calculate the partial half-lives for the two modes of decay (alpha emission and spontaneous fission) for ^{238}U. How many alpha and spontaneous fission decays take place every minute inside a 100 g block of ^{238}U?

10.2 Which is more likely to undergo spontaneous fission, ^{256}Fm or ^{254}No? Briefly explain.

10.3 Which is more likely to undergo spontaneous fission, ^{256}Fm or ^{258}Md? Briefly explain.

10.4 $^{256}_{100}$Fm can decay by spontaneous fission yielding two atoms of $^{128}_{50}$Sn. Calculate the total energy released by this decay, and estimate the kinetic energy of the tin fragments. Why is symmetric fission of ^{256}Fm a reasonable possibility?

10.5 $^{240}_{94}$Pu can decay by spontaneous fission yielding ^{103}Mo and ^{134}Te. Write a balanced equation, and estimate the kinetic energy of the fragments. Would you expect the fission yield for these products to be high? Why?

10.6 Why are nuclides with N or Z = 50 or 82 more likely to be formed in fission reactions?

10.7 Is ^{238}U fissile, fissionable, and/or fertile? Briefly explain.

10.8 Just like fission, there's a barrier to alpha decay. Explain the nature of this barrier and why alpha decay spontaneously occurs despite this barrier.

10.9 Write out the decay series that is generated by the formation of ^{140}Cs as a fission product. Calculate the N/Z ratio for each isobar formed.

10.10 Briefly explain why not all heavy *eo* nuclides undergo thermal neutron fission.

10.11 Calculate the excitation energy of the compound nucleus formed by the capture of a thermal neutron by ^{239}Pu and ^{240}Pu.

10.12 If the fission barrier for ^{240}Pu is 5.5 MeV, decide which (if either) nuclide in the previous problem will fission. If either nuclide will not fission, how much energy would need to be brought in by the neutron to overcome the fission barrier?

10.13 Calculate the minimum neutron energy needed for fission of ^{238}U to proceed in good yield. Don't forget to account for the kinetic energy of the compound nucleus.

10.14 ^{233}U has a critical mass that is much lower than ^{235}U. What can you conclude about the number of neutrons produced per fission of ^{233}U?

10.15 One of the possible reactions for thermal neutron fission of ^{239}Pu produces ^{134}Te and ^{103}Mo. Write out a balanced equation, calculate the total energy released, and estimate the kinetic energies of the products.

10.16 Using the reaction in the previous problem, calculate the nuclear binding energy for ^{239}Pu and the products. Are the differences between reactants and products consistent with the results above? Explain.

10.17 Ten years after some pure ^{235}U undergoes fission, what are the major radioactive nuclides present?

10.18 The three fission reactions given in Section 8.4.1 are reproduced here. Would you expect these reactions to have high fission yields? Briefly explain.

$$^{235}\text{U} + {}^{1}\text{n} \rightarrow {}^{135}\text{Sn} + {}^{99}\text{Mo} + 2{}^{1}\text{n}$$

$$^{235}\text{U} + {}^{1}\text{n} \rightarrow {}^{131}\text{I} + {}^{102}\text{Y} + 3{}^{1}\text{n}$$

$$^{235}\text{U} + {}^{1}\text{n} \rightarrow {}^{133}\text{Xe} + {}^{100}\text{Sr} + 3{}^{1}\text{n}$$

10.19 Define the following: fission products, fission yield, thermonuclear reaction, and cold fusion.

10.20 Why are fission of heavy nuclides and fusion of light nuclides generally exoergic?

10.21 Calculate the amount of energy produced in the following fusion (net) reaction based on the mass differences. Also, determine how many MeV/u are produced by this reaction. How does this value compare to a typical fission reaction?

$$4{}^{1}_{1}\text{H} \rightarrow {}^{4}_{2}\text{He} + 2{}_{+1}\text{e} + 2\nu$$

10.22 Home fusion enthusiasts claim they can perform the reaction below in their garage by accelerating deuterons through a 45 kV electric field. Estimate the effective Coulomb barrier for this reaction, and comment on the validity of their claims. If this reaction is taking place in someone's garage, what safety concerns might result? Also calculate the energy produced by this reaction.

$$2{}^{2}_{1}\text{H} \rightarrow {}^{3}_{2}\text{He} + {}^{1}\text{n}$$

10.23 Why can't atomic masses be used in energy calculations for certain fusion reactions?

10.24 In 2002, a scientist at Oak Ridge National Laboratory claimed he was observing the fusion reaction below by exposing deuterated acetone ($(CD_3)_2CO$) to 14 MeV neutrons and ultrasound. Suggest another reaction that could've produced the tritium.

$$2\,{}^{2}_{1}H \rightarrow {}^{3}_{1}H + {}^{1}_{1}H$$

10.25 Starting with ^{131}Xe, give the likely nuclear reactions that might produce ^{132}Xe and ^{132}Ba in stars. Look up the percent abundances of ^{132}Xe and ^{132}Ba. Can you explain the difference between them?

10.26 Rationalize why no stable isotopes of Pm exist.

10.27 Compared to other elements near it on the periodic table, why is gold relatively rare in our solar system? Why are iron and lead so common?

10.28 ^{244}Pu was detected for the first time following the thermonuclear detonation on Eniwetok Island. Show the probable nuclear reactions and decays that likely led to the formation of this nuclide from ^{238}U. Also show how ^{235}U might be formed in a supernova from ^{209}Bi—this time, simplify the process by writing it as a single nuclear reaction.

10.29 ^{266}Mt can be prepared by the reaction of ^{209}Bi with ^{58}Fe. Write out the complete nuclear reaction for this synthesis. What projectile energy would be required to prepare ^{266}Mt in good yield using this reaction? What would the excitation energy of the compound nucleus be in this case? Do you think this reaction is an example of hot or cold fusion? Briefly explain. If the excitation energy of the compound nucleus is limited to 13 MeV, what would the projectile energy be?

11 Applications of Nuclear Science III
More about Nuclear Reactors

Chapter 4 provides an introduction to nuclear power and this chapter seeks to explore it a little deeper. This chapter follows those on nuclear reactions (Chapter 9) and nuclear fission (Chapter 10) because it applies some of that stuff to nuclear power reactors. We'll also take a more in depth look at the different types of nuclear reactors, their safety systems, and the causes of some of the recent nuclear plant accidents. We'll also look at what makes modern nuclear power plants better than those built in the 1960s and 1970s. Finally, we'll examine ideas for future nuclear power reactors including fusion power.

11.1 REACTIONS IN REACTORS

It may seem obvious that "reactions" would take place in "reactors." Even so, a little redundancy is usually a good thing when it comes to nuclear power. The most important reaction that takes place inside a nuclear reactor is the neutron induced fission of ^{235}U. Not only does it produce a lot of energy by splitting the ^{235}U nucleus into two fission fragments, but it also frees up (on average) a couple of neutrons. These newly freed neutrons can then go on to cause more fission reactions to take place, creating a nuclear chain reaction as illustrated in Figure 11.1 (yes, it's the same as Figure 4.1, but it *is* quite pretty!).

Whether a neutron captured by ^{235}U leads to fission or to an n,gamma reaction is governed by their respective cross sections. Table 11.1 (partly a reproduction of Table 10.1) gives the fission and n,gamma cross sections for various fissile and fertile nuclides. Notice that both cross sections are large for ^{235}U (the most common nuclear fuel today), but fission is clearly the more probable outcome of a thermal neutron running into a ^{235}U nucleus, which is why it is fissile. Table 11.1 also lists the ratio of fission cross section to n,gamma cross section (σ_f/σ_γ) to provide an easy comparison of the relative probabilities of these two reactions. Notice that, of the fissile nuclides, ^{233}U has the highest ratio, making it an attractive nuclear fuel. More on this in Section 11.2.5.

Remember also (Section 4.1.3) that nuclear fuel nowadays is only ~3%–4% ^{235}U, and the rest is ^{238}U. What will ^{238}U do with all the neutrons flying around inside of a nuclear reactor? A quick look at Table 11.1 tells us ^{238}U will most likely undergo an n,gamma reaction, eventually forming ^{239}Pu (Section 4.1.3). Even so, the cross section for this reaction is not very high, so only a relatively small percentage of the ^{238}U reacts.

Successful management of a nuclear power reactor is all about what happens to the neutrons, that is, how many induce a fission reaction versus how many end

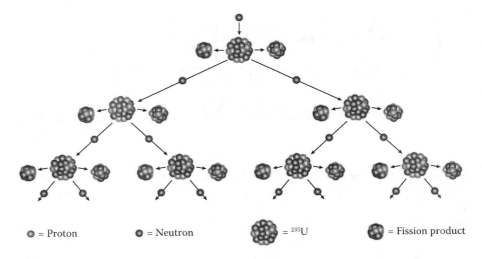

= Proton   = Neutron   = ^{235}U   = Fission product

FIGURE 11.1 Nuclear chain reaction.

up doing something else. To properly maintain the chain reaction, at least one of the neutrons produced by a fission reaction should cause another fission reaction to occur. If too many neutrons are lost to other reactions, the chain reaction will die out. If too many cause new fission reactions, the chain reaction can get out of control, and the reactor will overheat. Therefore, it's probably a good idea to know the average number of neutrons produced per fission reaction. These values ($\bar{\nu}$) are listed in the final column of Table 11.1 but only for the fissile nuclides.

We now have another way to think of **critical mass**. Remember that it is the minimum amount of material necessary to obtain a self-sustaining fission chain reaction. So long as we're talking about the same overall shape of pure material, this value will depend on the average number of neutrons produced per fission and the probability of fission (σ_f). For example, ^{239}Pu produces more neutrons per fission than ^{235}U, has a higher fission cross section (Table 10.1), and, therefore, has a smaller critical mass than ^{235}U.

For a reactor to run for any reasonable amount of time, much more than a critical mass of fuel needs to be placed into the reactor. Placing a supercritical mass all

TABLE 11.1
Select Data for Some Fissile and Fertile Nuclides

Nuclide	σ_f (b)	σ_γ (b)	σ_f/σ_γ	$\bar{\nu}$
^{232}Th	0.000003	7.34	4×10^{-7}	
^{233}U	531	46	12	3.1
^{235}U	585	98	6.0	2.4
^{238}U	0.000005	2.7	2×10^{-6}	
^{239}Pu	750	271	2.8	2.9

in one place would result in a rapidly accelerating fission chain reaction, a lot of neutrons and heat, and would make a bit of a mess. To prevent this, the fuel rods are spread out in the reactor and, initially, are separated by control rods. As the reactor starts up, the rods are partially removed, just enough to obtain criticality, and they are gradually withdrawn over the following year or so as the ^{235}U is used up. Since control rods need to absorb neutrons, they are typically made of elements that have a high cross section for various n,x reactions, such as cadmium ($\sigma_a = 2520$ b), boron ($\sigma_a = 760$ b), and gadolinium ($\sigma_a = 49,000$ b). These elements (B and Gd in particular) can also be dissolved in the moderator and/or coolant and their concentration adjusted to maintain the proper neutron flux within the reactor.

Remember that the purpose of the moderator in a nuclear reactor is to slow the neutrons down (Section 4.1.2). This is because slower neutrons are generally more likely to initiate a reaction (Section 9.2). Unfortunately, the average energy of a neutron produced in the fission of ^{235}U is 2 MeV (Section 10.2). If more fission reactions are desirable, it would be best to slow them down. This is another good reason to separate the fuel rods. Aside from being able to place the control rods between the fuel rods, it would also be a good idea to put some stuff in there that will slow down the neutrons. An ideal moderator has a relatively low neutron absorption cross section (σ_a) yet has a high neutron scattering cross section (σ_s) and is relatively inexpensive. Water (H_2O) is commonly used as a moderating material as its σ_a is 0.66 b, its σ_s is 49 b, and it is very cost effective. Other moderator materials are heavy water, (D_2O, where D = ^{2}H, $\sigma_a = 0.0013$ b, $\sigma_s = 10.6$ b) and graphite (a form of pure carbon, $\sigma_a = 0.0035$ b, $\sigma_s = 4.7$ b). Based solely on their very low neutron absorption numbers, both D_2O and graphite would appear to be superior moderators than regular, or **light water**, but H_2O is less expensive, and it can also be used as a coolant for the reactor.

The coolant carries the thermal energy away from the fuel to be used in power generation (Section 4.1.2). Water (H_2O) is by far the most common coolant used because of its great thermal properties. Water is somewhat unusual in that it takes a great deal of thermal energy (heat) to raise its temperature by a degree. In other words, it has a rather high specific heat capacity. To transfer heat, water's a great choice because it soaks up heat like a sponge! Water is also easy to come by.

That's wonderful, but what about neutron reactions? Well, water has two elements, and either can react with neutrons. Naturally occurring hydrogen consists of two stable isotopes (99.99% ^{1}H and 0.01% ^{2}H) and a ridiculously small amount of tritium (^{3}H, $t_{1/2} = 12$ a). Oxygen occurs as three stable isotopes (99.76% ^{16}O, 0.04% ^{17}O, and 0.20% ^{18}O). ^{1}H, ^{16}O, and ^{17}O can all undergo n,gamma reactions to form another stable isotope, and, given the high percent abundances of ^{1}H and ^{16}O and reasonable cross sections, these are the most likely neutron-induced reactions involving the coolant. The three most likely reactions yielding *radioactive* products are listed here:

$$^2_1H + {}^1n \rightarrow {}^3_1H + \gamma$$

$$^{17}_8O + {}^1n \rightarrow {}^{14}_6C + {}^4_2He$$

$$^{18}_8O + {}^1n \rightarrow {}^{19}_8O + \gamma$$

Deuterium can capture a neutron, forming tritium. ^{17}O undergoes an n,alpha reaction, forming ^{14}C ($t_{1/2} = 5715$ a). Finally, ^{18}O reacts with neutrons to form ^{19}O ($t_{1/2} = 27$ s). All three radioactive products decay by beta emission to form stable daughters. The first two (^3H and ^{14}C) are also naturally occurring. ^{19}O has a relatively short half-life and, therefore, does not represent a long-term radioactive hazard in the coolant. Another source of radioactive contamination could be small amounts of fission products leaked into the coolant from faulty fuel rods.

Neutrons can also interact with structural components of the reactor. Most significant is the neutron activation of cobalt in steel:

$$\,_{27}^{59}Co + \,^1n \rightarrow \,_{27}^{60}Co + \gamma$$

^{60}Co is particularly nasty because it has a half-life of 5.3 years and emits two high-energy gamma photons with every beta decay.

Finally, we should consider the fact that neutrons need to pass through the zirconium alloy (commonly called zircaloy) that makes up the fuel rods. Zirconium typically makes up at least 95% of the alloy because Zr is nearly transparent to neutrons ($\sigma_a = 0.18$ b) and maintains its structural integrity even under high temperatures and in high-radiation environments. While Zr may be unreactive toward neutrons, it can react *chemically* with water to form zirconium oxide and hydrogen:

$$Zr(s) + 2H_2O(g) \rightarrow ZrO_2(s) + 2H_2(g)$$

While this reaction is spontaneous at room temperature, it is inhibited by the formation of a thin layer of zirconium oxide on the surface of the metal. It only becomes a problem in a nuclear reactor at relatively high temperatures (above ~1200°C), that is, when the fuel rods are no longer underwater but are in contact with really hot steam. This reaction is rather exothermic, which exacerbates situations where heat removal from the reactor core is impaired. We're talking nuclear meltdown.

As mentioned in Section 10.2, not all neutrons produced in a nuclear reactor originate in the fission reaction. Some of the fission products emit delayed neutrons sometime after the fission reaction occurs. These neutrons can also cause more fuel to undergo fission and are important in controlling the rate of fission reactions in the reactor. If fission were the only source of neutrons (**prompt neutrons**), the reactor operators would only have a fraction of a second to respond if the number of fission reactions suddenly started to increase. Even computer-operated control rods would not have enough time to respond, and meltdowns would be common. The fact that a small fraction of all the neutrons flying around the reactor are produced some time after fission occurs allows operators and automated safety systems a minute or two to respond to an escalation in the chain reaction. The delayed neutrons provide a nice buffer to sudden changes in the number of fission reactions per unit time.

11.2   OTHER REACTOR TYPES

The attentive reader will have noticed that pressurized water reactors (PWRs) and boiling water reactor (BWRs) only make up roughly 80% of the operable nuclear power plants in the world as of mid-2017. The remaining 20% are mostly pressurized heavy water reactors (PHWRs, ~10% of all reactors), gas-cooled reactors (GCR, ~3% of all reactors), and light water graphite reactors (RBMKs,[1] ~3% of all reactors). The basic operating principles of these reactors will be discussed in this section along with ideas for reactors that are currently being researched and developed. Along the way, we'll also learn a little bit more about PWRs and BWRs.

11.2.1   Pressurized Heavy Water Reactors (PHWR)

The only difference in nomenclature from pressurized water reactors (PWR) is insertion of the word "heavy." PHWRs use heavy water (D_2O, where D = ^{2}H) rather than regular (or "light") water as a moderator and, typically, as a coolant. The advantage to using deuterium rather than hydrogen in the moderator is that it has a much lower neutron absorption cross section (σ_a). This means that far fewer neutrons will be absorbed by the moderator/coolant and are therefore still available to cause fission. This is often referred to as an improved "neutron economy."

There are a few significant ramifications to having a good neutron economy. Most important is that these reactors can run on natural uranium, that is, they don't require enriched uranium. With more neutrons flying around, the concentration of fissile ^{235}U doesn't need to be as high, as enough will still get hit by a neutron and split. Enrichment is expensive, so PHWRs are cheaper to fuel. Greater numbers of neutrons also mean more ^{238}U atoms will get hit, form ^{239}U, then decay to form fissile ^{239}Pu. In fact, this happens roughly twice as often in a heavy water reactor than in a light water reactor. This means more power could be generated from the same amount uranium.

Sounds great! So why aren't there more PHWRs? The heavy water doesn't come easily. Even though deuterium is a naturally occurring hydrogen isotope, separating it from the rest of hydrogen is costly and a little messy. Since several hundred tons are required for moderator and coolant, this significantly adds to the cost of constructing a PHWR. These reactors also tend to be more complex than other types, which also increases the upfront costs. These additional initial expenses are offset by saving on fuel costs, so PHWRs end up costing about the same to build and operate as other reactors. Another drawback is that with so many ^{2}H atoms being bombarded with neutrons, some tritium (^{3}H) is formed. As a result, PHWRs need to routinely remove the ^{3}H as it builds up. Tritium is also formed in light water reactors but at a much lower rate than in heavy water reactors.

The most common commercial PHWR is the Canadian deuterium uranium (CANDU) reactor, which is represented schematically in Figure 11.2. The CANDU reactor is very similar to other pressurized (light) water reactors (PWRs) in that the coolant is kept under pressure, exchanges its heat with another water loop, which boils and drives a steam turbine to generate electricity. The differences are all in the reactor. The main difference is that instead of enclosing all the fuel in one pressure vessel, the fuel is divided up among hundreds of pressure tubes (only three are shown

---

[1] RBMK is the acronym commonly used to characterize reactors cooled with light water and moderated by graphite because almost all reactors of this type were designed in the old Soviet Union. RBMK stands for reaktor bolshoy moshchnosty kanalny, which translates to English as high output channel reactor.

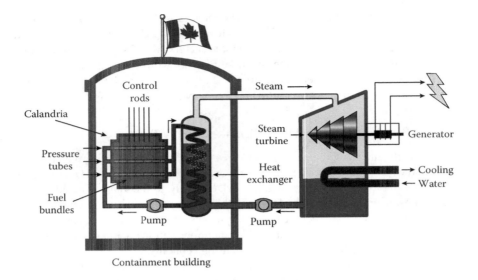

FIGURE 11.2 Schematic of a CANDU reactor.

in Figure 11.2). These tubes are placed horizontally (instead of vertically, as in other reactors) inside a large tank filled with D_2O. This tank is called a **calandria**. The calandria is thermally insulated from the pressure tubes by a larger tube surrounding each pressure tube filled with CO_2. The heavy water in the calandria does not heat up much; it simply acts as the moderator for neutrons traveling through it.

Each pressure tube is loaded with several **fuel bundles**. These are made up of several short fuel rods, bundled together in the shape of a log. The coolant (also D_2O, but could be H_2O or some other stuff) flows over the fuel bundles in the pressure tubes. With this design, moderator and coolant are both water (D_2O) but are physically separated.

The multiplicity of pressure tubes in a CANDU reactor has some advantages. Other reactor designs only have a single reactor pressure vessel and must shut down every year or two to refuel. This takes some time and may not always be convenient. Having hundreds of pressure tubes allows CANDU reactors to refuel while still running. Just shut down one pressure tube and swap out its fuel bundles. CANDU operators also move fuel around within the reactor to maintain more consistent power production throughout the reactor.

With all the shuffling of fuel bundles, the total amount of fissile material in a CANDU reactor stays relatively constant. This is not the case in other reactor designs that are loaded with enriched uranium at startup, which gradually gets "burned up" until the reactor is shut down for refueling a year or two later. To maintain consistent power production in a PWR or a BWR during this time, control rods are (very) gradually removed during that time. Because the CANDU has roughly the same amount of fissile material all the time, there's not much need for control rods. They are only needed to shut the reactor down. Again, this improves the CANDU's neutron economy, that is, fewer neutrons are absorbed by the control rods during operation of the reactor and are, therefore, available to cause fission or create more fissile material ($^{238}U \longrightarrow \longrightarrow \, ^{239}Pu$).

11.2.2   GAS-COOLED REACTORS (GCR)

As their name implies, the coolant is now a gas rather than a liquid. In the reactors we've considered so far, the coolant and the moderator have been the same, either H_2O or D_2O. Because of its low density, a gas is not a very efficient moderator, so something else must be used as a moderator, and graphite is the material of choice. Both CO_2 and He have been used as coolants in GCRs, but He is preferred because of its extreme lack of chemical reactivity. Helium also has a low neutron absorption cross section ($\sigma_a = 7$ mb for thermal neutrons). Other than the gas coolant and the graphite moderator, these reactors are very similar to PWRs.

The main advantage of a gas-cooled reactor is that a gas can be heated much higher than water. This makes some of the thermal transfer from fission to electricity more efficient (less heat lost elsewhere). Heating to about 700°C is common. The super-hot gas can also be used to run turbines directly to achieve even higher thermal efficiencies. Finally, the heat could be used industrially (to heat chemical reactions, that is, splitting water into H_2 and O_2), or to simply provide heat for a group of buildings.

Like PHWRs, gas-cooled reactors have a good neutron economy. Because of this GCRs have used an unusual mix of fissile ^{235}U and fertile ^{232}Th. Fuel pellets can be coated with ceramic materials such as pyrolytic carbon or silicon carbide. These materials are stable to much higher temperatures than the zircaloy used in water-cooled reactors. Alternatively, $^{235}U/^{232}Th$ carbide microspheres can be incorporated into a $\sim$6-cm sphere (about the size of a tennis ball) of graphite, then coated with silica or silicon carbide. This latter form is called a "pebble," even though it is more the size of a rock. These pebbles can be slowly loaded and, when spent, removed from a reactor on a conveyor system. Such a reactor is called a pebble bed reactor. Another advantage of combining fuel and moderator in a relatively stable ceramic form is that it can then be disposed of directly, since the fuel is already in a chemically inert form.

11.2.3   LIGHT WATER GRAPHITE REACTORS (RBMK)

Almost all reactors of this type were built in what once was the Soviet Union. These reactors are very similar to the PHWRs except that the moderator is graphite, and the coolant is regular (light) water (Figure 11.3). Instead of a huge tank filled with heavy water (the calandria), RBMK reactors surround their pressure tubes with tons of graphite. RBMK reactors have hundreds of pressure tubes ($\sim$1500 in a 1000-MW reactor) containing a couple fuel bundles each. As with the PHWRs, this allows for refueling while the reactor is still running and for use of only slightly enriched uranium. Unlike the PHWRs, the fuel bundles are vertical in RBMK reactors, and the coolant water is allowed to boil in the pressure tubes, making these a boiling water-type reactor. It's not shown in Figure 4.3, but all boiling water reactors have a steam separator or dryer. Its function is to separate the gaseous water (steam) from the liquid, so only steam goes on to the turbine. The BWRs such as those pictured in Figure 4.3 have their steam dryer in the top of the reactor vessel.

Most nuclear reactors have what is known as a negative **void coefficient**. This means that the rate of fission reactions decreases as the density of matter (usually

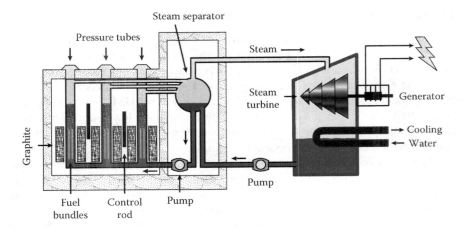

FIGURE 11.3 Schematic diagram for a RBMK reactor.

water) around the fuel decreases. This makes some intuitive sense, as the amount of matter (per unit volume) decreases, there's less for neutrons to interact with, lowering the probability of slowing them down. Faster neutrons mean fewer fission reactions taking place (lower σ_f). This is most significant for boiling water reactors. As water boils, bubbles of steam form, lowering the density of water in the volume that contains the bubble. As such, BWRs tend to be a little sluggish in responding to an increase in power. As the water heats and more boiling takes place, more bubbles form, lowering density and lowering neutron moderation—power then drops.

RBMK reactors are unusual in that they have a large *positive* void coefficient when operating at low (less than 20%) power levels. This stems from the fact that graphite moderates the neutrons. After they are slowed down by the graphite, these neutrons need to travel through the coolant water to get to the fuel, and some are absorbed. When the water is in the gas phase (steam), not as many neutrons are absorbed, and more fission reactions take place. This allowed the Chernobyl 4 reactor to run its power up from below 20% to hundreds of times normal in a matter of seconds, leading to two explosions and a core meltdown. Direct exposure of the red-hot core launched 1.4×10^{19} Bq of radioactive nuclides into the atmosphere. It has been widely assumed that the spread of radioactive material was facilitated by burning graphite, but the nuclear power industry disputes this saying that it is nearly impossible to burn reactor-grade graphite. Regardless, reactors with graphite moderators have become a lot less popular since the Chernobyl accident.

11.2.4 SMALL MODULAR REACTORS (SMR)

This isn't really a new idea, as hundreds of small reactors have been built for the propulsion of naval vessels and as research reactors are located at many universities. What's new is thinking of small nuclear reactors for power and/or steam generation in isolated areas, for a smaller community or an industrial facility. It also represents a rethinking of the economics of nuclear power.

Nuclear power plants are large, complex, and highly regulated. As a result, they are rather expensive to build. Compared with other major power generating plants (coal and natural gas), nuclear plants are far costlier. Where they tend to make this up is in their lower operating and fuel costs. Coal and natural gas plants require a lot more fuel to operate and, therefore, cost more than nuclear to run. Fossil fuel plants also save money by dumping most of their wastes directly into the environment. Because of the high upfront, or capital, costs for nuclear, larger and larger plants were built thorough the 1970s, trying to realize some economy of scale. This was also a time of high inflation, which significantly added to the costs (when you borrow tons of money at a high interest rate, it takes a lot more to pay it back).

Building a large, complex plant is also prone to delays and, therefore, cost overruns. Let's say that you're building a nuclear plant, and you put the wrong kind of rebar into the concrete footings for the plant. You did it because it met the current construction standards for large buildings, but unfortunately, it doesn't match the specifications for the nuclear plant that were submitted to the regulator 10 years ago. It doesn't matter if the new standards are better. You still need to tear it up, then rebuild. Everything else gets delayed. Really big parts must be stored ($$) and construction crews deferred ($$). Now let's say you've also got some changes to the plant design because the folks building an identical plant in China figured out that the changes would make it safer. You can't make the changes until approved by the regulator, which again idles crews and postpones shipments. It's a huge puzzle with a billion pieces, and the time frames keep changing with every misstep or tweak. Quite a headache, eh?

This is where small nuclear reactors suddenly look attractive. They can be built in a factory and shipped by rail. They can be modular so the whole reactor can be replaced instead of refueled (refueling would take place back at the factory), and multiple units can be located at sites with greater demand. Just dig a hole in the ground, drop in the reactor, and hook it up to the local grid. Economy of scale is provided by the quantities produced. While a great deal of research and development is taking place, a few of these sorts of reactors are already operating or are under construction in Russia, India, and China. The Russian reactors have been providing steam and electricity to a remote corner of Siberia since 1976.

11.2.5 THORIUM IN REACTORS

As we saw back in Section 4.1.3, thorium exists entirely as ^{232}Th ($t_{1/2} = 1.4 \times 10^{10}$ a) in nature and is fertile in nuclear reactors undergoing conversion to ^{233}U:

$$^{232}_{90}\text{Th} + {}^{1}\text{n} \rightarrow {}^{233}_{90}\text{Th} + \gamma \quad \sigma_\gamma = 7.34\,\text{b}$$

$$^{233}_{90}\text{Th} \rightarrow {}^{233}_{91}\text{Pa} + {}_{-1}\text{e} \quad t_{1/2} = 22\,\text{min}$$

$$^{233}_{91}\text{Pa} \rightarrow {}^{233}_{92}\text{U} + {}_{-1}\text{e} \quad t_{1/2} = 27\,\text{d}$$

^{232}Th is much more likely to undergo an n,gamma reaction than split when it absorbs a neutron. The ^{233}Th product can then decay via beta to ^{233}Pa, which can also spit out a beta, forming fissile ^{233}U. ^{232}Th could therefore could be used to some extent in just about any reactor. So, what's the big deal? Thorium gets its own subsection because of recent (zealous) advocacy for designing and building reactors that primarily use thorium as a fuel source.

^{232}Th is roughly three times more abundant on Earth than uranium, with the most significant reserves in India, Brazil, Australia, and the United States. At the time of this writing, the United States, India, Canada, and China have shown the most interest in using thorium in nuclear reactors. The United States used thorium blended with HEU or ^{233}U in four different reactors in the 1970s and 1980s. India currently includes thorium in the fuel mix of some of its heavy water reactors (PHWRs).[2] A company in Norway[3] has recently developed fuel pellets that are a mix of ^{232}Th and ^{239}Pu for use in current reactors.

Not only is there a lot of thorium lying around, but ^{232}Th is a more superior fertile material than ^{238}U. The cross section for ^{232}Th(n,γ)^{233}Th with a thermal neutron is $\sim$7 b, but it increases to $\sim$26 b (due to some resonance) for reactor neutrons. These values beat ^{238}U, as shown in Table 11.1. The ^{233}U formed from ^{232}Th is also superior to the ^{239}Pu formed from ^{238}U. ^{233}U cranks out an average of 3.1 neutrons per fission ($\bar{\nu}$, in Table 11.1), while ^{239}Pu generates only 2.9. ^{233}U also has better odds of undergoing fission than an n,gamma reaction. This is represented in Table 11.1 as the ratio of the fission cross section to the n,gamma cross section (σ_f/σ_γ).

Another advantage to thorium is that it is all burned in the reactor, generating more energy per kilogram of fuel. This also means the waste will almost entirely be made up of fission products, which *generally* have shorter half-lives.

Sounds great! There's plenty of thorium, it produces a better fission fuel more easily than from ^{238}U, we get more energy, and waste that might be more socially acceptable. What are the drawbacks? First, uranium is pretty cheap, making the cost of fuel at a nuclear plant a very small percentage of its total costs. Also, we need to remember that we're asking our neutrons to do double duty. Not only do we need to have enough neutrons around to cause fission reactions, but they also must start thorium on its way to becoming fissile by making the ^{232}Th(n,γ)^{233}Th reaction happen. In light water reactors, there aren't enough neutrons to convert all the thorium—too many are soaked up by the water and the control rods. The only currently operating reactors with enough neutrons to spare are the heavy water and gas reactors (PHWRs and GCRs). In light water reactors, this problem can be circumvented by blending in more ^{239}Pu. Can you explain why?

Another problem using thorium as fuel is the formation of so many fission products. Some of these have rather high neutron cross sections. If accumulated in sufficient quantities in the reactor core, they will absorb enough neutrons to significantly affect the rate of fission and conversion reactions (**poison** the reactor). To maximize thorium utilization, the fuel will need to be reprocessed to remove the fission products. Reprocessing is also desirable because of the 27-day half-life of ^{233}Pa. ^{233}Pa is formed along the way of converting ^{232}Th to ^{233}U (see previous equations). Because of its long half-life and high cross section toward an n,gamma reaction (the thermal

---

[2] India uses ^{239}Pu and ^{233}U produced in other reactors to mix in with ^{232}Th. The reactors "burning" thorium will get ~75% of their power from thorium.

[3] Ironic, since there are no nuclear power plants in Norway.

neutron cross section is 1500 b), some of it will likely absorb a neutron and form ^{234}Pa. ^{234}Pa then decays to ^{234}U, which absorbs a neutron and forms fissile ^{235}U:

$$^{233}_{90}\text{Th} + {}^{1}\text{n} \rightarrow {}^{234}_{90}\text{Th} + \gamma \quad \sigma_\gamma = 1500\,\text{b}$$

$$^{234}_{90}\text{Th} \rightarrow {}^{234}_{91}\text{Pa} + {}_{-1}\text{e} \quad t_{1/2} = 24\,\text{d}$$

$$^{234}_{91}\text{Pa} \rightarrow {}^{234}_{92}\text{U} + {}_{-1}\text{e} \quad t_{1/2} = 27\,\text{d}$$

$$^{234}_{90}\text{U} + {}^{1}\text{n} \rightarrow {}^{235}_{90}\text{U} + \gamma \quad \sigma_\gamma = 100\,\text{b}$$

In the end, we still have a fissile nuclide, but we need to use up two more neutrons in the process. We'll also get fewer neutrons from fission of ^{235}U than from fission of ^{233}U (Table 11.1).

In an attempt to address these issues, some have proposed using a liquid fluoride thorium reactor (LFTR) like one developed at Oak Ridge National Laboratory in the 1960s. In this reactor, the ^{232}Th and ^{233}U form compounds that are soluble in a molten salt. Yes, even salts will form liquids if you give them enough heat. In this case, the salt is made up of lithium, beryllium, and fluoride ions. This soup circulates through the reactor in several small tubes separated by the graphite moderator. It heats up, then transfers the heat, which eventually produces electricity. When returning to the reactor, some of the circulating molten salt can temporarily be removed from the loop to separate out fission products and ^{233}Pa. The ^{233}Pa can then be held until it decays to ^{233}U, then reinserted into the molten salt loop. This way waste is continuously removed, and fresh fuel can continuously be introduced. Somewhat sophisticated separations systems will be needed to accomplish all of this under relatively high radiation conditions. Alternatively, solid particles of the fuel wrapped in graphite can be suspended in the liquid salt, forming an emulsion. However, this would not address the issue of ^{235}U formation.

There are also corrosion issues when using molten salts—a leak in this system would not be pretty. Considerable shielding will also be necessary on the molten salt loop, since it will also be circulating the fuel.

As discussed in Section 4.1.7, significant nuclear weapons proliferation risks exist around enrichment and reprocessing activities. An important downside to use of thorium in power reactors is that one or both of these activities are necessarily involved. However, some ^{232}U is also formed when using ^{232}Th in nuclear fuels. ^{232}U is pretty darned toasty making it difficult to work with the fissile ^{233}U, and it also makes detection of these operations easier.

11.2.6 GENERATION IV REACTORS

The liquid fluoride thorium reactor is one of several new commercial reactors that are being proposed for development. Collectively, these new reactor types are known as fourth-generation nuclear reactors. What were the first three generations? The first generation of commercial nuclear power reactors were the early prototypes, built in the 1950s and early 1960s. The second generation comprise the bulk of the ~450 operable reactors today[4]—PWRs, BWRs, CANDU, and RBMK reactors that were built from the late 1960s through the mid-1990s. The third generation are reactors that have recently been constructed or are currently under construction or planned for construction in the next few years. These reactors are basically the same as the second generation, but they have improved safety, efficiency, and reliability. Most significantly, some have passive cooling features that would operate without power or operator intervention.

Generation III reactors represent the summit of our ability to engineer nuclear reactors that use water as a coolant. Generation IV reactors aim beyond water-cooled reactors, look to use more fertile fuel, and to generate less long-lived radioactive waste. Molten salt-cooled reactors (MSR), like the LFTR described in Section 11.2.5, are examples of fourth-generation nuclear power reactors. The following paragraphs briefly describe the other reactor types under active research and development.

The **very high-temperature reactor** (VHTR) is essentially the same as the gas-cooled reactors (GCRs) described in Section 11.2.2, except that these are designed to run hotter (900–1000°C) for greater thermal efficiency and to allow the possibility of splitting water into H_2 and O_2. They are helium-cooled and graphite-moderated and use enriched uranium as the fuel. Since several GCRs have been built and operated around the world (albeit at lower temperatures), this technology appears relatively accessible.

The **gas-cooled fast reactor** (GFR) is very similar except it lacks a moderator, and the neutrons are therefore "fast." This allows fission of various actinides that comprise much of the long-lived radioactive waste generated in more conventional nuclear reactors. Various isotopes of Np, Pu, and Am are produced from neutron capture reactions initiating at ^{238}U (see Questions 11.1 and 11.14 at the end of this chapter). These nuclides make up most of the long-term radioactive components of spent fuel, so burning them with fast neutrons would help make the nuclear waste problem more tractable. Fast reactors can also burn ^{238}U and ^{232}Th. These reactors would use He as a coolant and operate either at ~850°C or at 600–650°C. They could be fueled by ^{239}Pu (15%–20%) mixed with depleted uranium or reprocessed spent fuel from water-cooled nuclear plants. Spent fuel from GFRs would be reprocessed, and any fissile or fertile stuff recovered will be fed back into the reactor, limiting waste to almost entirely shorter-lived fission products. While this is not a new idea, no prototype had been built as of 2017.

Metal-cooled fast reactors, like GFRs, also eschew moderators and can burn nuclear "trash" with their zippy neutrons, but they are cooled by sodium (SFR) or lead (LFR—actually either pure lead or a lead/bismuth 45:55 mix). In both cases, the coolant heats up to about 550°C. Experimental and demonstration models of both reactor types have been built and run, so these types of reactors are closest to commercialization. Small modular reactors of these types are being considered as well as large (1400–1500 MW) plants.

An **accelerator-driven system** (ADS) slams really fast protons into high Z material (W, Pb, etc.), knocking some of their neutrons loose in the process. The high Z stuff is surrounded by fissionable material that splits and produces heat, which is

---

[4] July 2017.

transferred to a coolant, probably lead or a lead-bismuth mix. Like GFRs and metal-cooled fast reactors, ADSs do not moderate the neutrons, so they can burn ^{232}Th, ^{238}U, other actinides in nuclear wastes, as well as the usual fissile nuclides. In fact, an accelerator-driven system (notice "reactor" isn't being used here) could be set up to run on spent fuel from generation I-III reactors. Since the accelerator is providing some of the neutrons, subcritical amounts of fuel can be used. This is a relatively new concept for nuclear energy, and we are not likely to see commercial application before 2050.

The **supercritical water-cooled reactor** (SCWR) is like most of the generation I-III reactors. The distinctive difference is that the water is kept under higher pressure and temperature than conventional modern reactors. The water is pushed past the critical point and is, therefore, supercritical. This means it is no longer a liquid, but it is not a gas either—it is something in between.[5] Cranking up the pressure and the temperature in this way provides greater thermal efficiencies, roughly one-third higher than today's nuclear plants. Like a boiling water reactor, the supercritical water heated in the reactor core directly drives the turbine, generating electricity. Because of the lower density of the coolant, this reactor could be built using thermal or fast neutrons. It can also be built with a single pressure vessel or hundreds (like the CANDU).

As a nuclear nerd, it is easy to get excited about nifty new nuclear plant designs. Keep in mind that the very short descriptions given are limited to how they work and what they could do. When generation I and II nuclear power plants were being built, we were still learning how to build and operate them efficiently and safely. Generation IV plants present a whole new set of problems to solve before we can decide that they will be economical, safe, and reliable.

11.3   REACTOR SAFETY SYSTEMS

This section will focus on the safety systems in the most common reactor types: pressurized water (PWR), boiling water (BWR), and pressurized heavy water (PHWR, which are generally CANDU) reactors. The basic operating principles for PWRs and BWRs are presented in Section 4.1.2 and all other types in Section 11.2. Nuclear power plant safety systems can broadly be divided into **active** and **passive**. While the lines between these two are sometimes a little fuzzy, active systems can be defined as those that require human intervention and/or electrical power to function. Passive systems are, therefore, those that would function without humans or electricity—they are inherent in the design of the reactor. Generally speaking, the more passive systems deployed in a reactor, the safer and more reliable it will be.

The most important safety systems in nuclear reactors are those that shut down (scram) the fission chain reaction. This can be accomplished by reducing the neutron flux within the reactor. As explained in Section 11.2.3, when the water in most light water reactors (LWRs) heats up, its density decreases. This decreases neutron moderation, leading to fewer fission reactions, eventually shutting down the fission chain reaction. This is not a particularly good way to shut down the chain reaction

---

[5] This is not new in power generation. Hundreds of coal-fired power plants heat water past its critical point.

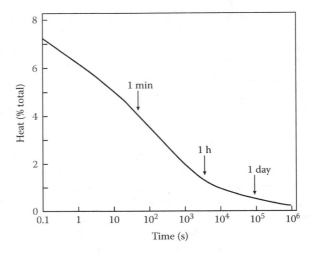

FIGURE 11.4 Residual heat in a nuclear reactor after shutdown.

because it requires running the reactor hotter than normal, and the neutron flux will likely return to normal after the water begins to cool and higher density is restored. Even if shutdown were successful, there is still plenty of residual heat from the decay of the fission products in the fuel rods. This heat is typically equal to roughly 7% of the reactor's power on shutdown (Figure 11.4). Still plenty toasty to melt the fuel assemblies. Figure 11.4 shows how the heat from fission product decay tapers off with time. It decreases rather rapidly after shutdown, cooling to 4% of full power after only a minute—this is still pretty toasty—it would be 60 MW for a reactor that was running at 1500 MW. After an hour, it's down to ~1.5%, and it is about 0.4% after a day. Note that the rate of cooling slows considerably with time. This is why spent fuel must be stored for few years under water.[6] Once again, water's amazing ability to absorb heat comes in handy!

A better way to scram a nuclear plant is by inserting the control rods all the way into the reactor. Remember that control rods are made from elements with high neutron absorption cross section such as Cd, B, and Gd (Section 11.1). These elements could also be injected into a reactor core in some water-soluble form (e.g., sodium borate or gadolinium nitrate). Note that inserting control rods into PWRs and BWRs will meet with resistance as these reactor vessels are both pressurized. BWR control rods are also typically located at the bottom of the reactor vessel and, therefore, also need to fight gravity (active!). CANDU (and other PHWR) reactors have a distinct advantage here in that their control rods drop into the nonpressurized calandria and, therefore, could be rigged to drop automatically upon loss of power (passive).

Figure 11.4 tells us that cooling of the fuel rods is still necessary after the fission chain reaction is shut down. This cooling is typically accomplished by pumping

---

[6] The U.S. Nuclear Regulatory Commission has allowed removal in as little as three years but prefers at least five years and states that the industry norm is 10 years.

water through the reactor and dumping that heat into a convenient heat sink such as a nearby body of water or into the air through cooling towers. Extra water (as a liquid or a solid) is typically stored in one or more places in the reactor building so there is enough to cover modest losses due to a leak. Many reactors store these in tanks located higher than the reactor so gravity can feed it into the reactor, if necessary. CANDU reactors cleverly build a large water tank into the top of containment building. This feature has been duplicated in some of the Generation III reactors such as the Westinghouse AP1000 and the AREVA EPR. The AP1000 also uses natural air convection to passively cool the inside of the reactor containment structure, indirectly exchanging heat with air outside of the containment building. Additionally, water can be sprayed onto the outside of the reactor vessel to facilitate cooling. The steam that is generated can then flow out of the building.

In addition to water stored above the reactor vessel, boiling water reactors (BWRs) use a large body of water in the bottom of the reactor building called a suppression pool (Figure 11.5). Excess heat in the form of steam inside of the primary containment is vented into the suppression pool where it condenses. This pool can serve as a passive, temporary heat sink until more active cooling (pumping water through the core) can be started. The suppression pool also serves as a trap for radioactive materials potentially being released by the core.

Nuclear reactors protect workers, those living nearby, and the environment from ionizing radiation by putting several barriers between the radioactive material and the outside. First, the uranium fuel is in a ceramic oxide form. This is similar to how uranium is found naturally, and it is relatively insoluble and unreactive. Second,

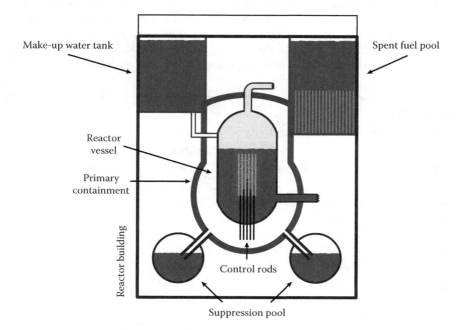

FIGURE 11.5   Some safety systems in a BWR.

the fuel pellets are sealed inside of zirconium tubes (fuel rods). The idea here is to contain the fission products within the fuel rods. Some still leak out, even under normal conditions. Those that get out tend to be isotopes of elements that are gases, namely Xe and Kr. Under more extreme conditions (fuel rod distortion or melting), isotopes of volatile[7] elements (I and Cs) are also released. The fuel rods are grouped into assemblies, which are surrounded by water, and all of this is sealed inside of the reactor vessel. The reactor vessel (and heat exchangers for PWRs) are surrounded by a containment structure. This structure is usually constructed from thick concrete and steel, and it is sturdy enough to withstand a plane crash or a core meltdown (preventing the molten core from getting into the ground). Boiling water reactors also have secondary containment (and additional shielding) from the reactor building (Figure 11.5).

11.4 NUCLEAR POWER PLANT ACCIDENTS

This section will focus on how the safety systems described in the previous section worked or failed in the accidents at Three Mile Island (TMI), Chernobyl, and Fukushima. These accidents are generally described in Section 4.1.4, so the focus here will be on reactor safety systems and operator responses to the accidents. Other nuclear reactor accidents and incidents have occurred, but these three are the most commonly known and resulted in some of the most significant radiation releases. To put the amount of radiation emitted from these accidents in context, these and other significant releases of radioactive material that have occurred are listed in Table 11.2.

By far, the most significant releases have been the detonation of nuclear weapons above ground, first over Hiroshima and Nagasaki, then primarily in the United States and the Soviet Union. Remember that nuclear weapons are inefficient (Section 4.2.1), so a good deal of the fissile material in the weapon is spread around the blast zone and launched into the atmosphere. There's also an intense neutron flux

TABLE 11.2
Significant Releases of Radioactivity

Source	Country	Year	Radiation (Bq)	Important Nuclides
Hiroshima/Nagasaki	Japan	1945	4×10^{16}	Fission products and actinides
Above ground tests	USA/USSR	1945–1963	2×10^{20}	Fission products and actinides
Windscale	UK	1957	1×10^{15}	Noble gases, ^{131}I
Chelyabinsk	USSR	1957	8×10^{16}	^{90}Sr, ^{137}Cs
TMI	USA	1979	1×10^{12}	Noble gases
Chernobyl	USSR	1986	1.4×10^{19}	^{131}I, ^{137}Cs
Fukushima	Japan	2011	9.4×10^{17}	^{131}I, ^{137}Cs

[7] In this context, volatile means simply "easily evaporated." Don't start thinking these elements are somehow explosive.

during detonation, which can cause rapid, multiple neutron absorptions (Section 10.5). When these neutron-rich products undergo beta decay, a variety of actinide and other nuclides are formed.

Looking through Table 11.2, it is clear that 1957 was a bad year for radiation releases. Not only were nuclear weapons being detonated above ground, but there were two other large emissions. The Windscale accident was caused by a sudden burst of energy in a graphite-moderated, air-cooled reactor in the UK. This caused the fuel to overheat and started a fire, releasing gaseous and volatile fission products into the atmosphere. Some few fission products were also released in 1957 at a high-level waste storage facility near Chelyabinsk (southern Russia, near the Kazakh border) after what was believed to have been a chemical explosion caused by mixing nitric acid with organic chemical wastes. The contamination was spread over approximately 1600 km^2. A similar accident may also have taken place in 1967 at the same site.[8]

11.4.1 THREE MILE ISLAND

The meltdown at Three Mile Island (TMI) started with the failure of some relatively minor pumps on the secondary coolant loop (the loop connecting the heat exchanger and turbine—see Figure 4.2). These pumps run the water through a loop (not shown in Figure 4.2) that removes impurities common to hot water running through pipes. One of these pumps was clogged and being repaired when all the others went down. Normally a valve that would bypass this cleaning loop would open at this point, but it didn't. This prevented circulation through the secondary coolant loop and caused the turbines to automatically shut down. This also prevented heat from being properly transferred in the heat exchanger, causing temperature and pressure to rise in the reactor. This caused the reactor to scam by automatically inserting the control rods, shutting down the fission chain reaction. However, the reactor still needed to deal with decay heat from the fission products.

The critical point in this crisis occurred when a pressure relief valve automatically opened to relieve some of the pressure in the reactor coolant loop but failed to close properly when the pressure was sufficiently low. As a result, coolant drained out of the reactor for a couple of hours. A similar incident happened two years earlier at a reactor near Toledo, Ohio. This reactor was the same design as the one at TMI, but it did not meltdown because one of the operators figured out what had happened before too much coolant had been lost. Unfortunately, the Toledo incident was not well communicated to other plants of the same design. The confusion at both plants is understandable because the control room indicator for this valve showed it was closed. Additionally, they had no gauge to tell them water levels in the reactor vessel were lowering. It had to be inferred from other readings, which gave apparently contradictory information.

---

[8] Yet another similar incident took place in the Waste Isolation Pilot Plant (WIPP—a deep geological repository in the United States) in 2014. Organic (instead of inorganic) kitty litter was mistakenly used to pack a drum containing a mixture of radioactive nitrate salt waste. While only trace amounts of radioactive materials were released to the surface, WIPP was shut down for three years at a cost of $500 million.

As water drained from the reactor, alarm systems began to go off, but none directly told the operators that the problem was a stuck valve. The operators became focused on the fact that water levels were too high in one part of the loop (near the stuck valve) and began shutting off pumps circulating water through the reactor. At about this time, one of the operators determined that the pressure relief valve was stuck open, and they closed another valve below it to stop the leaking. Unfortunately, they did not then consider that a great deal of water may have drained out of the reactor (about one-third of the total). Shortly thereafter, the reactor got hot enough to vaporize enough of the remaining water to expose the fuel rods and begin a meltdown. The zirconium cladding around the fuel pellets reacted with the steam to form hydrogen gas, and the fuel rod assemblies began to slump, break apart, and collapse into the bottom of the reactor.

Once the cladding around the fuel is breached, the more volatile fission products are released into the coolant. Radiation alarms started going off all over the place, which generally added to the confusion. It wasn't until ~21 hours after the initial pump failures that one of the reactor coolant pumps was successfully restarted and progress toward stability began. Unfortunately, a fair amount of pressure had built up inside the buildings from the release of fission products from the meltdown. This pressure was vented, resulting in the release of radioactive isotopes of Xe and Kr, as well as a very small amount of ^{131}I. As can be seen in Table 11.2, the amount of radioactive material released was low (compared to others on the table), and because it was almost entirely noble gases, it was of negligible health concern. Because they are chemically unreactive, they do not accumulate in living things. Unfortunately, the releases were not always well communicated, resulting in significant public concern and sometimes gave the appearance that those managing the crisis were incompetent or even malevolent.

11.4.2 CHERNOBYL

Like TMI, the Chernobyl disaster was the result of inadequacies in the reactor design and the inability of its operators to fully understand the consequences of their actions. The reactor was scheduled to be shut down for routine maintenance, and the operators thought it would offer an opportunity to test how long the turbine would provide power to the pumps after a sudden shutdown. In other words, when the steam is cut off from the turbine, it continues to spin for a while, generating electricity until it coasts to a stop. The Chernobyl operators wanted to know if it would give them enough time to start up the emergency backup (diesel) generators and thereby have no interruption of power to the pumps. They had done this experiment before, but the turbine spun down too quickly. They had since changed some of the electrical equipment, thinking this would give them more time.

The shutdown of the reactor ended up being delayed a few hours because the power it was generating was still needed by the grid. When approval for shutdown came, a different shift was operating the reactor instead of the one that had planned to perform the experiment. The first step of the experiment was to slowly lower power to about 25%–30% of maximum. Due to an operator error, it instead dropped to about 1%. Because it was a light water graphite moderated reactor (RBMK), whose main flaw is a large positive void coefficient at low power levels (<20%, see

Section 11.2.3), the operators should have terminated the experiment at this point and completed reactor shutdown.

The operators instead removed control rods to try to bring the power back up. This type of reactor responds a bit sluggishly under these conditions, so the operators bypassed safety protocols and pulled out more control rods than they were normally allowed. Even so, they were only able to bring the reactor power up to ~6% of capacity. Again, safety protocols dictated they should've shut down at this point as the reactor was in a dangerous state (low power, large positive void coefficient—if this were a movie, you'd be hearing the ominous music playing right now!). Instability in the amount and pressure of the steam developed and should've triggered an automatic shutdown but didn't because the operators had also bypassed this safety feature, so it wouldn't mess up their experiment.

They started their experiment by shutting off the steam to the turbine. This reactor was a boiling water reactor, so there's only one coolant loop. Shutting off the steam to the turbine lowered the pump output to the core, resulting in a rise in temperature, creating steam bubbles (voids), increasing the rate of fission reactions. Reactor power and temperature increased rapidly,[9] and the operators tried to insert the control rods, but they were too slow. It appears that most only made it in about halfway before the reactor started to blow itself apart, and a few jammed due to a faulty design. The core overheated, vaporizing the coolant which exploded with considerable force. A few seconds later, a second explosion occurred. It's not entirely clear what caused the second explosion, but it is widely assumed it involved hydrogen produced from the reaction of the super-heated steam with the zirconium cladding. However, the second blast could have been another steam explosion.

The explosions destroyed the reactor core and the upper part of the reactor building. The Soviets did not build containment buildings around their RBMK reactors, so this exposed the red-hot reactor materials to the atmosphere, and the second explosion launched some of the core onto nearby buildings, causing several fires. The fires were contained within a few hours, but only after many of the firefighters received very large doses. Over the next several days, helicopters dropped boron carbide (absorbs neutrons), dolomite (to absorb heat), lead (radiation shielding), and sand and clay (to inhibit release of radioactive materials to the atmosphere). Additionally, cold nitrogen was pumped in to cool and blanket the core materials (to inhibit combustion), and a tunnel was dug under the reactor to install a massive concrete slab with a cooling system. Nine days after the accident, all these measures finally significantly reduced the temperature of the core materials and sharply curtailed the release of radioactive materials.

A massive concrete structure with a cooling system (called the "sarcophagus"), was built over the trashed reactor. It took seven months to complete and was designed to shield and contain. Unfortunately, it was not designed to last. It started allowing rain water into the destroyed reactor, and it was weakening to the point that it is threatening to collapse. A larger, more permanent cover (the New Safe Confinement, or NSC) was built next to the reactor and slid into place in 2016. This structure also has remote control equipment to dismantle the sarcophagus, the old reactor, and remove the radioactive material from the destroyed reactor.

---

[9] Reactor power was estimated to be some hundreds times normal at the time of the explosions.

Like TMI, the Chernobyl disaster could've been averted had the operators taken a safer path at any number of points leading up to the explosions. There were two prior, similar incidents in the Soviet Union with RBMK reactors. They occurred on a much smaller scale, with very little radiation released, but these incidents were kept secret instead of shared with other RBMK reactor operators. Unlike TMI and the previous Soviet incidents, Chernobyl released a considerable amount of radioactive material. It is estimated that all the xenon, half of the iodine and cesium, and roughly 5% of the rest of the radioactive material in the core at the time was released into the environment. Much of it was deposited locally, but measurable quantities were spread over much of the Northern Hemisphere. As described in Section 4.1.4, Chernobyl is the only nuclear power plant accident known to have caused fatalities due to exposure to ionizing radiation. Explosively exposing the reactor core makes a big difference—TMI was pretty mild in comparison to Chernobyl.

About 200,000 people were eventually relocated away from contaminated areas around Chernobyl. Many of the evacuated areas have since reopened for settlement. Roughly 600,000 liquidators were brought in during the years after the accident to decontaminate the area near the reactor and build the various structures, including the sarcophagus. Epidemiological studies of these groups of people in the 31 years since the accident have revealed very few radiation-related deaths.

11.4.3 FUKUSHIMA

Unlike TMI and Chernobyl, it is difficult to point to specific deficiencies in reactor design as major contributors to the radiation releases. As we'll see, the main factors that led to radioactive material being released at Fukushima came largely from poor site design, and (like TMI and Chernobyl) inadequate operator training for such an incident. The reactor design is implicated in some reports, usually simply by stating that they were old with little or no further explanation. It is true that they were built in the late 1960s and early 1970s but did not contain substantially different safety systems than most other BWRs. The only real argument to be made is that they lacked sufficient passive safety systems that could have brought them under control without electrical power. However, such a statement is true of almost all of the hundreds of Generation II reactors still in operation today.

The main reason the Fukushima reactors ended up releasing radioactive materials was the site design (Figure 11.6). There were 15 BWRs along the Japanese coast that were affected by the tsunami, but only three reactors, all at the Fukushima Daiichi[10] site, suffered meltdowns and released radioactive material. In fact, there were six reactors at the Fukushima Daiichi site—the other three did not experience meltdowns. Four of the six reactors (reactors 1–4) at Fukushima Daiichi were located near the ocean. When the tsunami waves hit, they flooded the lower levels of the reactor and turbine buildings disabling the emergency diesel generators and the backup batteries (Figure 11.6) because they were located in the basements of these buildings. The designers of this site had not thought to place these critical backup

---

[10] There are two nuclear power plant sites in the Fukushima prefecture: Daiichi and Daini. Daiichi means "first" and Daini "second." To state the obvious, the Daiichi site was developed first.

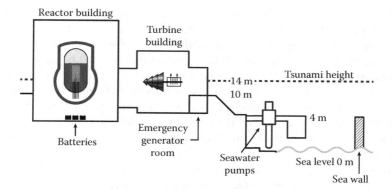

FIGURE 11.6   Schematic view of part of the Fukushima Daiichi site; not drawn to scale.

systems on higher ground or to waterproof their locations. Only one of the 13 diesel generators at the Fukushima Daiichi site survived the tsunami, but because electrical switching equipment was also knocked out, it could only supply power to reactors 5 and 6. Because they had power, these two reactors were able to safely achieve cold shutdown nine days after the tsunami.

This was generally true of the other Japanese reactors that survived the catastrophic events. They all had some backup power that enabled them to shut down their reactors. South of Fukushima, the Tokai Daini plant had just finished waterproofing one of its diesel generators, and it was the only one to survive—but it was enough to safely shut down the reactor. Further north, the Onagawa site was originally planned to be located on lower ground but shifted to higher ground after a local engineer pointed out the tsunami hazard.

It should also be noted that the sea wall at Fukushima was only 4 meters tall, while the highest tsunami wave to hit the site in 2011 was about 14-m high (Figure 11.6). There was historical evidence of tsunami waves of this height hitting this part of Japan. Japanese nuclear authorities were also twice warned by the International Atomic Energy Agency about the potential dangers of flooding. The first came after some flooding at a nuclear plant in France in 1999 illustrated vulnerabilities to floods, and the second after a tsunami caused an Indian plant to shut down in 2004. One of the most significant outcomes from the Fukushima disaster is the importance of a strong, independent nuclear regulatory agency. A clear flaw prior to the 2011 earthquake and tsunami was the cozy relationships between and insular confidence of the Japanese government, nuclear regulatory agency, and the nuclear industry.

The earthquake and tsunami knocked out all power to reactors 1–4 at Fukushima Daiichi (all BWRs). Reactor 4 had no fuel in its reactor vessel at the time as it was undergoing routine maintenance. As described in Section 11.3 and illustrated in Figure 11.5, these reactors could transfer excess heat from the reactor to a suppression pool without the need for electrical power. Reactors 2–4 had an additional safety system that used the steam produced from the decay heat to inject water into the core. Both safety systems operated in reactors 2 and 3 for about 1.5 and 3 days respectively—which is significantly longer than they are designed to run.

The valves connecting the containment structure with the suppression pool in reactor 1 were closed when the tsunami hit, which meant the decay heat had no place to go. Heat and pressure built up within the reactor vessel, and within a few hours, the reactor vessel had lost enough coolant that the fuel began to melt, and the zirconium cladding started reacting with steam to form hydrogen. The pressure (from the heat and the release of gases) soon overwhelmed the primary containment structure, sending volatile fission products and hydrogen into the reactor building. Twenty-five hours after the tsunami, a hydrogen explosion blew the top and much of the sides of the reactor building off, releasing a fair amount of radioactive Xe, I, and Cs.

Somewhat similar sequences of events occurred in reactors 2 and 3 when the suppression pools and steam-driven water injection failed. These systems can only operate for a limited amount of time because there's only so much water in the make-up tank, and the suppression pool eventually gets too hot to absorb more reactor heat. Normally the excess heat would've been transferred to the ocean, but the seawater pumps were knocked out by the tsunami. Like reactor 1, the top of part of the reactor 3 building blew out with a hydrogen explosion (three days after the tsunami). You'd think that any additional hydrogen gas produced by reactor 3 would've vented directly to the atmosphere at this point. It was instead vented into the reactor 4 building through some shared ductwork. The top of the reactor 4 building exploded (Figure 4.5) a day after the reactor 3 explosion. Reactor 2 managed to avoid a hydrogen explosion because a large panel in the exterior wall near the top of the building was removed, allowing the hydrogen (and volatile fission products) to escape to the atmosphere. It is believed that most of the radioactive releases came from building 2. Ironic, since it was the only one that did not explode.

The Fukushima reactors had other emergency core cooling systems designed to remove decay heat and add water as necessary, but these systems required electrical power, which was not available for several days following the earthquake/tsunami. The operators also had an option to pump water into the reactors from fire trucks, which were on the scene shortly after the tsunami. This required depressurizing of the reactor vessel, which the operators were unable to do because the valves required electricity or compressed air to operate. In some instances, there were valves that could be opened manually, but the workers were prevented by high levels of radiation or debris from the tsunami. By the time they could start depressurizing, it was too late, and containment was already lost. Had they been able to act more quickly to depressurize the reactor vessels and pump water in from the trucks, the reactors may have been saved with little to no release of radioactive material.

It wasn't until five to six months after the tsunami that adequate cooling was established at all four damaged reactors. It was a full nine months before radioactive emissions from the plant were minimized, and "cold shutdown" was declared. The top floors of each of the damaged reactors (also known as the service floor) have since been rebuilt to provide containment, protection from the elements, and allow the removal of the stored spent fuel (in the spent fuel pools—see Figure 11.5). All spent fuel that was in the reactor 4 pool has been removed, and the reactor 1 and 3 pools should be empty by the time this edition is published. Removal of the melted fuel will not begin until 2021, to allow for fission product decay.

11.5   FUSION REACTORS

The proton cycle discussed in Section 10.3 is generally thought to be impractical for fusion power reactors here on Earth, primarily due to its relatively high ignition temperature—the temperature needed to start a fusion reaction. Back in Section 10.3, we calculated the temperature needed for ignition of the proton-proton fusion reaction as nearly 13 billion Kelvin! In reality, fusion reactors do not need to get quite this hot because some reactions will take place at lower temperatures due to the distribution of particle energies at a particular temperature or because tunneling will occur. At a certain point (several million Kelvin), enough reactions are running that they generate enough heat for the reactions to be self-sustaining. Even so, reaction ignition is a major roadblock in the development of practical fusion power reactors. All test reactors built so far require more energy to be put in for ignition than is obtained from the subsequent fusion reactions.

Currently, the best hope for a practical fusion reactor lies in the so-called "D + T reaction." It is the reaction of deuterium (^{2}H) and tritium (^{3}H):

$$^2_1H + ^3_1H \rightarrow ^4_2He + ^1n$$

This reaction produces 17.6 MeV of energy, and its Coulomb barrier is only 0.68 MeV, thereby minimizing the ignition temperature. The D + T reaction has been successfully performed in several research reactors. A problem here is sustaining the fusion reaction. For this to happen, fuel needs to be fed into the reactor while products are removed. One way around this problem is to build a very large reactor, one that can hold enough fuel to provide more energy than was necessary to initiate fusion.

Some fusion enthusiasts like to state that it does not produce radioactive waste. This is not true. Short half-life nuclides with low mass numbers will be formed in any fusion reactor. The fact that they are short-lived means that fusion plant operators just need to wait a little while for them to decay to stable nuclides. The neutrons produced in the D + T reaction are of greater concern. As with fission reactors, these neutrons could interact with some of the metals used to build the reactor, making them radioactive through n,gamma reactions. For example, cobalt (often used in high-performance metal alloys) would form ^{60}Co, a beta and high-energy gamma-emitting nuclide with a half-life of 5.3 years:

$$^{59}Co + ^1n \rightarrow ^{60}Co + \gamma$$
$$^{60}Co \rightarrow ^{60}Ni + _{-1}e + 2\gamma$$

Another big technical challenge for a practical fusion reactor was how to confine the very hot plasma that would be generated. The confinement issue has been solved by using very strong magnetic fields. At these temperatures, the fuel exists as plasma, bare nuclei, and free electrons in a gaseous soup. Since everything is charged, it can be suspended in a magnetic field and will never touch the walls of its container. The optimal shape for this plasma is like a donut (toroid).[11]

---

[11] German scientists are working on a variation of the toroid, called a stellarator. It's a toroid with twists and curves—a cruller? A UK company is also exploring a small reactor that uses spherical confinement—a donut hole?

Because of the enormous challenges in researching and developing a practical fusion reactor, an international consortium is working together to build the International Thermonuclear Experimental Reactor (ITER), a prototype designed to demonstrate the feasibility of fusion as a power source. The ITER is currently under construction near Cadarache in the south of France and plans to be fully operational by 2035.

Finally, you probably noticed that fusion produces a lot less energy per reaction than fission. The D + T reaction is one of the best, but it only generates 17.6 MeV per reaction, while a typical fission reaction produces roughly 200 MeV. This may just be a matter of perspective. If we look at how much energy is released in terms of MeV per mass unit of "target" nuclide, fusion rules! Fusion cranks out ~6 MeV/u, while fission only produces ~0.8 MeV/u. This argument is only important for weapons, where delivering a payload (mass) costs money and energy—placed on the same rocket, more bang can be generated from a fusion weapon than a fission warhead with the same mass.

QUESTIONS

11.1 Starting with ^{239}Pu(n,γ), show the most likely pathway forming ^{241}Am and ^{244}Cm in a nuclear reactor. Include appropriate cross sections and/or half-lives.

11.2 Explain why ^{235}U is better fuel for a nuclear reactor than ^{238}U.

11.3 Of the nuclides listed in Table 11.1, which has the smallest critical mass?

11.4 Beryllium has a low neutron absorption cross section ($\sigma_a = 0.0076$ b) and a high neutron scattering cross section ($\sigma_s = 5.9$ b). Would it be a good material for a fuel rod, control rod, moderator, and/or coolant in a nuclear reactor? Briefly explain your answer.

11.5 Beryllium is used as a neutron "mirror" around the core of reactors that need a higher neutron flux. Using the information from Question 11.4 briefly explain how this works. Is "mirror" a perfect term? Briefly explain.

11.6 In a nuclear reactor, what are delayed neutrons? Why are they helpful?

11.7 Why can't light water be used as a moderator/coolant in CANDU reactors?

11.8 Explain why moving fuel bundles around in a CANDU reactor is necessary to maintain more consistent power production throughout the reactor.

11.9 What is the main advantage in using light water in a reactor? Heavy water?

11.10 Why is tritium produced at lower rates in light water reactors than in heavy water reactors?

11.11 Define the following: calandria, fuel bundle, void coefficient, and reactor poison.

11.12 Why can't RBMK reactors use natural uranium as a fuel?

11.13 Based on their design, would you expect CANDU reactors to have a positive or negative void coefficient? Briefly explain.

11.14 Assuming the initial reaction is n,2n, give the likely series of reactions and decays that results in the formation of ^{232}U in thorium reactors.

11.15 Why does ^{239}Pu make a good "mix-in" with ^{232}Th fuel?

11.16 Why are elements like He, Pb, and Bi selected as coolants in generation IV reactors?

11.17 Why don't ADSs need control rods?

11.18 Define "scram" in the context of nuclear power plants. After a scram, is the reactor then safely shut down? Explain.

11.19 Classify the scram methods described in Section 11.3 as active or passive.

11.20 Assuming all the radioactive material released from TMI was ^{133}Xe, calculate its volume under standard temperature and pressure. If a typical nuclear medicine lung scan uses 12.5 mCi of ^{133}Xe, how many lung scans would the TMI releases represent?

11.21 Assuming all the radioactive material released from Fukushima was ^{131}I, calculate the total mass emitted. Calculate its activity after three months.

11.22 The reaction in Question 11.23 has been proposed for use in reactors here on the Earth. Its proponents say the only drawback is the lack of ^{3}He. What is the percent abundance of ^{3}He on the Earth? Would you expect natural abundance of this nuclide to be low? Explain. The moon is known to have ample supplies of ^{3}He at or near its surface. Why?

11.23 Based on the energy produced by and the effective Coulomb barrier of this reaction, does it look like a viable reaction for fusion power production?

$$^2_1\text{H} + {}^3_2\text{He} \rightarrow {}^4_2\text{He} + {}^1_1\text{H}$$

11.24 The most common alpha particles emitted by ^{235}U have an energy of $\sim$4.4 MeV. Is this enough energy for ^{9}Be to be an effective neutron initiator in a fission weapon?

11.25 Why is a ^{239}Pu fission bomb so much more efficient than one using ^{235}U?

11.26 Calculate the energy produced by the reaction below, which is part of how fusion weapons detonate. Is a similar reaction also likely for the other naturally occurring isotope of lithium? Briefly explain.

$$^6_3\text{Li} + {}^1\text{n} \rightarrow {}^4_2\text{He} + {}^3_1\text{H}$$

12 Radiation Protection

Ionizing radiation is frightening, in part, because the risks of exposure are expressed as probabilities. If a person dies of cancer, it is impossible to point to an exposure to ionizing radiation 20 years in that person's past and state with absolute certainty that the cancer was a result of that exposure. Despite this uncertainty, radiation hazards and risks can be understood and controlled. This chapter will help you understand how ionizing radiation interacts with humans, when to get excited about it, and what to do about it. This chapter builds on what you've learned about the detection of ionizing radiation (Chapter 7) and its interactions with matter (Chapter 6).

12.1   TERMS

It is important to distinguish between exposure and dose. **Exposure** is the amount of ionizing radiation striking an object. **Dose** is the amount of energy absorbed by an object as a result of radiation exposure. As humans, we are exposed to a lot of radiation in the form of high-energy photons, which are part of cosmic radiation. Because they are in the X-ray/gamma ray part of the electromagnetic spectrum, their odds of interacting with us (we are made up of mostly low Z atoms!) are relatively low. Therefore, our *dose* from cosmic radiation is relatively low, even though our *exposure* is high.

As noted back in Section 1.6, the biggest dose to humans living in the United States (and most other places in the world) comes from radon (^{222}Rn). Radon is in the air we breathe and in the water we drink (and it has always been there!), so we ingest a fair bit. ^{222}Rn is an alpha emitting nuclide, with a relatively short (3.8 d) half-life—so there's probably some inside of us right now, decaying away. Those alpha particles will deposit all their energy inside our bodies, since alpha particles have high specific ionization (SI) and linear energy transfer (LET) values and very short ranges (Section 6.2). Even though it is the source of our largest dose, it is not because we are exposed to relatively large quantities of radiation from ^{222}Rn; it is because the energy of the decay is rather efficiently transferred to our bodies. Remember that exposure and dose are different—exposure is all ionizing radiation around something, but dose will depend on the type of radiation and the nature of the matter it interacts with.

Exposure can be more strictly defined as the amount of charge (ion pairs) delivered to a mass (or volume) of air. Its International System of Units (unfortunately, also uses the acronym "SI") are coulombs per kilogram (C/kg). It is sometimes expressed in units of röntgen (R). One röntgen is equal to 2.58×10^{-4} coulombs per kilogram and produces about 2×10^9 ion pairs in 1 cm^3 of air:

$$1\,\text{R} = \frac{2.58 \times 10^{-4}\,\text{C}}{\text{kg}}$$

Likewise, dose can be defined as the amount of energy delivered per unit mass. The SI unit for dose is the gray (Gy), which is equal to a joule of energy dumped into one kilogram of matter:

$$1\frac{J}{kg} = 1\,Gy$$

In the United States, units of *r*adiation *a*bsorbed *d*ose (rad) are commonly used. It is defined as 100 erg/g, which happens to be equal to 1 centigray:

$$1\,rad = \frac{100\,erg}{g} = \frac{10^{-2}\,J}{kg} = 1\,cGy$$

Example 12.1

How much energy (MeV) is deposited in 91 kg by a 1.00 Gy dose?

$$1.00\,Gy = \frac{1.00\,J}{kg}$$

$$\frac{1.00\,J}{kg} \times \frac{MeV}{1.60 \times 10^{-13}\,J} \times 91\,kg = 5.7 \times 10^{14}\,MeV$$

What is this dose in rad?

$$1.00\,Gy \times \frac{rad}{10^{-2}\,Gy} = 100\,rad$$

We know that ionizing radiation interacts differently with different materials. For example, relatively low-energy X-rays are much more likely to interact via the photoelectric effect with high Z materials and via Compton scattering with low Z materials. We also know that alpha particles will dump all their energy in a very short distance, even in gases, while gamma photons *could* pass right through without interacting at all. Dose is nice, but how can we account for the different interactions we see with different types of ionizing radiation with different materials? Specifically, can we know what kind of dose humans get?

Dose equivalent is the answer. We really just need to modify dose so that it depends on the type of radiation. This is done by the **quality factor**, Q, also known as the weighting factor or relative biological effectiveness (RBE). It's a bit like linear energy transfer (LET). Remember that, in terms of LET, alpha particles deposit the most energy per centimeter, followed by beta, while X-rays and gamma rays deposit the least. The quality factor (Q) weights the dose for damage to biological systems. For γ- and X-rays, and most (>30 keV) beta particles, the value of Q is 1. Beta particles with less than 30 keV have a Q of 1.7 because they are more likely to interact with matter. For thermal neutrons (relatively low-energy, only 0.025 eV), it is 5. The Q value varies significantly with the energy of the neutron. The Q value for alpha particles is 20.

We can now define dose equivalent as the amount of biological damage caused by ionizing radiation. Mathematically, it is simply dose (sometimes called "absorbed dose") times Q:

$$H = \text{dose} \times Q \qquad (12.1)$$

The SI units for dose equivalent is the sievert (Sv):

$$\text{Sv} = \text{Gy} \times Q$$

In the United States, units of rem (*R*öntgen *e*quivalent *m*an) are sometimes used. One rem is the same as one centisievert or 10 millisieverts:

$$1 \text{ rem} = 10^{-2} \text{ Sv} = 1 \text{ cSv} = 10 \text{ mSv}$$

Since Q is 1 for γ-rays, X-rays, and most beta particles, dose is numerically equal to dose equivalent for these types of radiation. For this reason, the simpler term "dose" is a common colloquialism for "dose equivalent."

$$1 \text{ Sv} = 1 \text{ Gy} \quad \text{and} \quad 1 \text{ rad} = 1 \text{ rem}$$

We've defined a lot, and it should be enough ... but there's another way to look at dose and dose equivalent: **kerma**. It stands for *k*inetic *e*nergy *r*eleased in *m*edia. As its name implies, kerma is another way of looking at dose, but instead of energy absorbed by the material, we're looking at energy released by the radiation—a subtle distinction at best. It even has the same units as absorbed dose (Gy). Differences between kerma and absorbed dose are only seen when significant amounts of secondary radiation (bremsstrahlung, secondary electrons) escape the material, thereby getting some of the energy out of the material that was originally transferred to it. Under most circumstances, kerma ends up numerically equal to dose and exposure.

To relate different media, we can use a simple ratio to adjust dose, or kerma, to a medium. That ratio is called the *f* **factor** (also known as f_{medium} factor), and is similar to the quality factor, Q:

$$f = \frac{\text{dose in medium}}{\text{dose in air}} \qquad (12.2)$$

Figure 12.1 shows how the f factor varies with photon energy for water, muscle, and bone. Dose depends mostly on the mass attenuation coefficient (μ_m—Section 6.4), which will depend on the atoms (Z) and the density of the matter. If we are only concerned with humans, we know that water and muscle are made up of atoms with

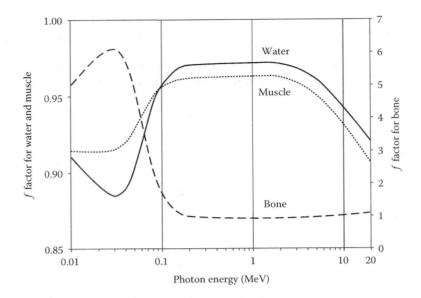

FIGURE 12.1 Variation in the *f* factor for water, muscle, and bone with photon energy. (Data from Hubbell, J. H. and Seltzer, S. M. (2004). *Tables of X-Ray Mass Attenuation Coefficients and Mass Energy-Absorption Coefficients* (Version 1.4). Gaithersburg, MD: National Institute of Standards and Technology. https://www.nist.gov/pml/x-ray-mass-attenuation-coefficients.)

similar atomic numbers (Z). Therefore, we would expect them to have similar *f* factors. Using the *y*-scale on the left side of Figure 12.1, and plotting the ratios of the mass attenuation coefficients versus photon energy, the *f* factor is very close to 1 for most biological materials, regardless of photon energy. This should not be much of a surprise; the atomic numbers of atoms in air molecules are pretty much the same as those in water and our muscles.

The *f* factors for bones are also plotted on Figure 12.1 using the right *y*-scale. Notice that it is also close to 1 for all but the lowest energy photons. Therefore, the *f* factor is close to one for most biological systems under most conditions, just as *Q* is close to one for common forms of ionizing radiation.

As is evident from Figure 12.1, each of the body's organs will respond a little differently to ionizing radiation. While the differences are subtle, they are sometimes distinguished. **Effective dose equivalent** (sometimes just called **effective dose**) is a dose equivalent weighted according to susceptibility of a particular organ to ionizing radiation. Going through detailed weighting schemes for different organs is beyond the scope of this text. The reader should simply know that dose can be further refined to a specific organ.

12.2 REGULATIONS AND RECOMMENDATIONS

The International Commission on Radiological Protection (ICRP) develops recommendations and standards for radiological protection, which can then inform national regulatory and legislative bodies. For example, in the United States, the National Council on Radiation Protection and Measurement (NCRP) uses ICRP standards and recommendations to develop its own recommendations. The primary regulatory authority in the United States is the Nuclear Regulatory Commission (NRC). If all the acronyms are making you a big dizzy, get used to it; such is the way of regulatory bodies.

The NRC *regulates* based on *recommendations* by the ICRP and the NCRP. Table 12.1 shows NCRP recommendations for dose limits. The NRC controls the use and the distribution of radioactive materials in the United States. It has primary regulatory authority and is a federal agency. Many individual states have been delegated some of the NRC's responsibilities—Wisconsin is one. In Wisconsin, the state Department of Natural Resources (DNR) handles much of the regulation of radioactive materials.

Actual regulations vary from state to state and depend on the nature of the facility. For example, the regulations will be much more extensive for a nuclear power plant with tons of toasty fuel than for an academic laboratory that handles only very small amounts of radioactive materials.

All doses in Table 12.1 are over and above those from natural background sources. The first set of limits are for radiological workers—notice that the dose for

TABLE 12.1
Effective Dose Limits Recommended by the NCRP

Occupational Exposures
Effective dose limits

(a) Annual	50 mSv
(b) Cumulative	10 mSv × age
Equivalent dose limits for tissues and organs (annual)	
(a) Lens of eye	150 mSv
(b) Skin, hands, and feet	500 mSv
Guidance for emergency occupational exposure	500 mSv
Public exposures (annual)	
Effective dose limit, continuous or frequent exposure	1 mSv
Effective dose limit, infrequent exposure	5 mSv
Effective dose limits for tissues and organs	
(a) Lens of eye	15 mSv
(b) Skin, hands, and feet	50 mSv
Education and training exposures (annual)	
Effective dose limit	1 mSv
Equivalent dose limit for tissues and organs	
(a) Lens of eye	15 mSv
(b) Skin, hands, and feet	50 mSv
Embryo-fetus exposures (monthly)	
Equivalent dose limit	0.5 mSv

Source: NCRP Report #116, Limitation of Exposure to Ionizing Radiation, 1993.

extremities (500 mSv) is much higher than the annual (whole body) dose (50 Sv). Our hands, feet, and skin can take a much greater dose because the cells in these parts of are body can better repair or replace damaged ones.

The "emergency occupational exposure" is designed for life or death situations. If a life can be saved, or the spread of significant radioactive contamination be contained, then up to a 500 mSv (one-time, whole body) dose can be allowed.

Public exposures refer to nonradiological workers who may be exposed while touring a facility where they might get a little extra ionizing radiation. It is also exposure to those who live nearby due to releases from these facilities. Again, these exposures are over and above those from natural background.

Educational dose limits are designed mostly to allow students to perform experiments with small amounts of radioactive material or radiation generating devices. They are also set low because while the body is still growing it is more susceptible to radiation damage. The human fetus is an extreme example, so dose limits are set quite low for pregnant females—to protect the fetus. Note that the human fetus is surrounded by a fair amount of shielding (also known as Mom) and is, therefore, naturally somewhat protected from external ionizing radiation.

How do we know what kind of dose we're getting? By measuring exposure! Exposure is measured with TLD (Section 7.3.2) or film badges, which are worn between the neck and the waist. This gives a pretty good idea of whole body exposure (and dose if we're dealing with photons and/or beta particles). Exposure to the hands is often measured separately using a TLD ring. The ring should be rotated so that the TLD crystals are closest to the source of ionizing radiation. Pocket dosimeters can also be used, especially if the potential for significant short-term exposure is high. These are typically ionization chambers (Section 7.1.1) for analog measurement or compact Geiger-Müller (G-M) counters (Section 7.1.3) for digital measurement.

In areas where radioactive materials are used, regular swipe tests are performed to test surfaces for radioactive contamination. Typically, a round piece of filter paper is wiped over a random spot on a surface then counted using an organic scintillation counter (Section 7.2.3).

Personnel leaving an area where radioactive materials are handled are required to frisk (monitor) their hands and feet with an appropriate detector (usually a G-M tube). If the potential for contamination of other parts of the body exists, those parts must also be monitored. In these situations, a monitoring portal is typically used. Monitoring portals surround all or most of the body with detectors. It was monitoring portals at Swedish and Finnish nuclear power plants that first alerted the world to the Chernobyl disaster. Workers at these plants were *arriving* at work with enough contamination to set them off.

In areas where radiation-generating devices are used, exposure is determined by performing a radiation survey around the equipment while energized. In some cases, TLD badges may be placed in certain locations to measure exposure over longer periods of time.

Records are meticulously maintained for all measurements of exposure. As is typically required by law, radiological workers must receive regular (usually annual) reports of their exposure. You should always know what you are getting, and it will likely be well under the limits.

At what dose levels do bad things happen? Data are available from the survivors of the Hiroshima and Nagasaki bombings, the Chernobyl disaster, occupational exposures, as well as a number of rather questionable experiments performed by the U.S. government in the 1950s and early 1960s.[1]

The following lists human health effects at various acute dose equivalent levels. Keep in mind that these are dose equivalents for a single exposure event with the entire dose received more or less instantaneously. Exposure over time allows the body to repair damage, so each of the thresholds would need to be higher for chronic exposures.

- **250–500 mSv** (25–50 rem): This is when the first measurable biological effects are observed in humans. A decrease in the white blood cell count is seen in a few hours. Later, a measurable decrease in the red blood cell count and in the number of platelets will be observed. Temporary sterility (both genders) is also commonly observed. Complete recovery can be expected in a matter of weeks or in a couple months. No one dies from this dose.
- **2 Sv** (200 rem): Reversible bone marrow damage occurs. Nausea, vomiting, fatigue, and hair loss also happens. Some will die.
- **4–6 Sv** (400–600 rem): A complete shutdown of the bone marrow function (production of blood cells) happens but is only temporary. Severe gastrointestinal distress is experienced. There is only a 50:50 chance of survival.
- **>7 Sv** (>700 rem): The bone marrow function is irreversibly damaged. Death is almost certain.
- **10 Sv** (1000 rem): You will get really, really sick and die. Death will occur within one to two weeks.
- **20 Sv** (2000 rem): The nervous system is damaged. Loss of consciousness happens within minutes, and death typically follows in a matter of hours (a day or two at the most).

Ew, sounds bad! So, what are we exposed to in daily life? Figure 1.8 gives the average dose equivalent to humans living in the United States from a variety of sources. Note that the units on the y-axis are mSv per year. Radon alone gives the average American a 2.3 mSv dose every year. The total average dose in the United States is about 6.2 mSv per year. Roughly 50% of that dose comes from natural sources and about 48% from medical procedures. Almost all the remaining 2% is due to consumer products containing radioactive materials.[2] Only about 0.05% comes from humans spreading radioactive materials around the planet, such as nuclear power plant accidents, weapons testing, etc.

Not all medical procedures zap you to the same extent. Computed tomography (CT) scans generally result in the highest doses (30–100 mSv, or 3–10 rem)—depending on how much of you gets scanned. A chest X-ray gives you only ~0.1 mSv

---

[1] Some argue that the U.S. government is continuing to experiment. Veterans of the Persian Gulf wars are currently being monitored for long-term health effects due to exposure to depleted uranium used in armor-piercing weapons (Section 4.1.3). So far, there is no epidemiological evidence to suggest these veterans are suffering any adverse health effects due to ionizing radiation.

[2] Mostly from building materials, air travel, cigarettes, mining and agriculture, fossil fuels, glass and ceramics.

(~10 mrem). Nuclear medicine procedures result in a dose ranging from 1 mSv (0.1 rem) up to 10 mSv (1 rem).

We can compare these procedures to other doses you might experience in daily life. The new airport security scanners that peek under your clothes using X-rays[3] deliver a dose less than 0.25 μSv (weak!). During an airline flight, humans receive ~5–10 μSv/h (~0.5–1 mrem/h) because planes fly up where there's not much atmosphere to shield cosmic rays. Fallout from above-ground nuclear weapons testing is responsible for less than 3 μSv/year of our annual dose. The worst doses received because of the Three Mile Island accident in 1979 were about 1 mSv (~100 mrem)— well below the threshold to cause immediate health effects.

Individual doses will vary depending on which medical procedures are performed on them, whether they fly a lot, or where they live. In the United States, those who live in parts of the mountain West see the highest levels of background radiation. This is due to a combination of high uranium concentrations in the ground (higher ^{222}Rn) and higher altitude (higher cosmic radiation levels from less atmospheric shielding). For example, folks living in Denver experience ~2–4 × as much background radiation than people living at sea level in the United States. There are parts of the world where background radiation levels are higher: Finnish citizens average roughly 8 mSv per year, while parts of France, China, Iran, and India see natural doses upwards of 50 mSv each year. It should be noted that people living in these areas of the world do not have higher cancer rates, nor is any adverse health effect observed.

12.3   RISK

Taken collectively, the numbers at the end of the preceding section should provide us with some reassurance concerning the doses received in medical procedures or even living through a nuclear power plant accident of the same magnitude as TMI or Fukushima. Even with this safety in numbers, some apprehensions likely remain. Risk is something that can be quantified and understood, but for most of us, it is our *perception of risk* that affects our behavior. We'll eventually come back to a more quantitative understanding of the risks associated with ionizing radiation, but first let's examine how it is perceived.

Table 12.2 gives the relative risk of various activities and technologies. Starting from a list of 30, a group of 186 Wisconsin college students in four radiation physics classes were asked to quickly rank them from 1 to 30, with 1 being the riskiest and 30 being the safest. Their collective results appear under "Class." Only the top 10 and selected others are shown. These results can be considered "perceived" risk.

Table 12.2 also gives rankings of some of the same activities and technologies by professionals that study risk ("Experts"). These rankings can be considered "actual" risk. By and large, the collective intelligence of the class correctly identified most of the top 10 riskiest things. There are, however, some fascinating anomalies in the overall rankings by the two groups. These anomalies are discussed in the following paragraphs.

---

[3] The other type uses microwaves—see Question 1.7. For more on this, see the paper by Hoppe and Schmidt in the bibliography.

TABLE 12.2

Rankings of Risky Activities and Technologies

Class	Experts
1. Smoking	1. Motor vehicles
2. Motorcycles	2. Smoking
3. Handguns	3. Alcoholic beverages
4. Firefighting	4. Handguns
5. Motor vehicles	5. Surgery
6. Surgery	6. Motorcycles
7. Police work	7. X-rays
8. Alcoholic beverages	8. Pesticides
9. Mountain climbing	9. Electric power (non-nuclear)
10. Large construction	10. Swimming
12. Pesticides	13. Large construction
14. Nuclear power	17. Police work
15. X-rays	18. Firefighting
17. Electric power (non-nuclear)	20. Nuclear power
27. Swimming	29. Mountain climbing

Source: "Expert" rankings are from Slovic, P., Radiat. *Prot. Dosimetry*, *68*, 165, 1996. With permission.

Familiarity breeds contempt. The class ranked swimming as much safer than the experts, probably because it is something most have done safely and likely with some pleasure. The stats say otherwise. Lots of people drown every year while swimming, making it an inherently dangerous activity. The students also ranked X-rays quite a bit lower than the pros. Again, this is because many have experienced X-rays and suffered no ill effects. However, X-rays are a form of ionizing radiation and are potentially dangerous when very large doses are delivered to biological systems like us humans. Similar arguments could be made for the students ranking motor vehicles and alcoholic beverages lower than the experts.

Lack of control creates fear. Most people would rank nuclear power as much riskier (higher on the list) than the experts, and despite being in a radiation physics class, the students followed suit. With the exception of X-rays, anything involving ionizing radiation creates an unusual amount of apprehension in the general population. As a result, even groups with some higher education in radiation physics will list nuclear power as more dangerous than is borne out by experience. Ionizing radiation is feared, in part, because we cannot directly sense it, and we know that it *can* cause adverse health effects, sometimes decades in the future. Our lack of control over exposure to and effects of ionizing radiation causes it to be seen as a higher risk than is warranted. Additionally, nuclear power is feared because of its potential for catastrophe—a nuclear plant burning is potentially more dangerous to the surrounding public than a coal plant catching fire. Nuclear power has proven itself much safer than other major forms of electricity generation—even taking into account the deaths caused by Chernobyl.

Training and experience make certain activities safer than we think. The students ranked police work, firefighting, large construction, motorcycles, and mountain climbing as more dangerous than they really are. While there is real risk associated with these activities, they are less dangerous than we think because of the extensive training, detailed procedures, and considerable regulation associated with each. The same is true of radiation therapy and nuclear medicine. Because both professions involve work with ionizing radiation, and because we (as a society) fear radiation, they are highly regulated. Only well-educated professionals are allowed to perform these procedures, and they do so with a remarkable safety record. These professions are among the safest of all careers! The same can be said of nuclear power. Because it is more extensively regulated than many other forms of power generation, it is inherently safer.

We've discussed risk in general terms and have tried to show that there is often a difference in the perception of risk and the actual risk. We've also looked at the health effects of receiving a single large dose. Those of us working with radioactive materials and radiation generating devices would much rather know: what are the health risks at low doses, especially those received over a long period of time?

As discussed in Section 1.6, there is still some debate to the answer of this question. The uncertainty is illustrated nicely in Figure 12.2 (a reproduction of Figure 1.6). The solid line represents acute doses greater than 250 mSv, where measurable biological effects begin. As we learned in the previous section, the severity of the effects increases with dose, in a (very) roughly linear fashion. It is below the 250 mSv threshold where there is still controversy. Regulatory agencies assume a linear extrapolation of this trend (straight dashed line in Figure 12.2), which is called the linear no-threshold (LNT) hypothesis. While there is no scientific evidence to support this hypothesis, it is used by regulators as a conservative approach to working with ionizing radiation. Unfortunately, it leads to the assumption that any exposure to ionizing radiation can have deleterious effects and, therefore, should be avoided. This generates unnecessary fear and anxiety, even among well-educated professionals.

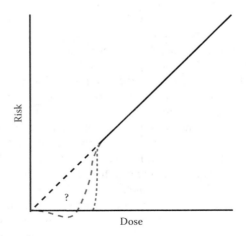

FIGURE 12.2   Risk vs. dose.

Scientific evidence tends to suggest that either there's a threshold below which no harmful effects are observed (dotted gray line in Figure 12.2), or that there's a health benefit (radiation hormesis) to low doses of ionizing radiation (dashed gray line). Evidence comes from the relatively large number of people who live in parts of the world with higher background radiation. Even though ionizing radiation levels are 10 to 30 times higher than where most other people live, the folks getting zapped more do not show any less longevity or higher cancer rates. In fact, the opposite is sometimes observed. The same is true with other groups that received higher doses than normal yet lower than what is known to cause adverse health effects. This includes many of the atomic bomb survivors in Nagasaki and Hiroshima who received low-level doses.

The Chernobyl accident provides some of the best evidence against the LNT model. Hundreds of thousands of people received low-level doses because they lived in an area that became contaminated, or they worked to help clean up after the disaster. Based on the LNT model thousands should've died in the decades since the accident, yet only about a dozen have. These victims generally died from thyroid cancer (or complications resulting from it) brought about from drinking milk heavily contaminated with ^{131}I. This only happened because the Soviet government was in deep denial in the first couple days following the accident and failed to warn its citizens about potential contamination in local milk.

Studies on people who work around ionizing radiation show no difference with standard populations or experience an apparent health benefit. This positive result may be due to higher education and/or greater access to health care in the group studied. What's important is that no additional risk appears to be associated with the sorts of low-level doses experienced by those working with ionizing radiation. This makes some sense. Our bodies have mechanisms to deal with the billion or so radioactive decays that naturally take place inside of us every day, as well as damage done from external (natural) sources of ionizing radiation. Like all living things on the Earth, we've adapted to being radioactive and living in a radioactive environment.

Our cell repair/replacement mechanisms are fairly quick; therefore, a significant dose (more than 250 mSv) received over a period of time is less likely to cause biological damage than if it was received all at once. In fact, the same dose delivered over hours or days, instead of instantaneously, reduces the chance of cancer ~4 times. That same dose delivered over longer periods of time results in a ~10 times lower cancer risk.[4] This is important for those living in, or returning to, areas with some contamination from Chernobyl or Fukushima. We already know that people do fine in parts of the world where external sources of background radiation are 10 to 40 times normal.[5] It is possible that humans can live in areas with external sources 50 or even a few hundred times normal without ill effect. The Chernobyl contaminated areas open for resettlement in the Ukraine and Belarus are well within this limit (<1 mSv/a additional dose). The Japanese government is allowing people to return

---

[4] These factors apply to gamma, X-ray, and beta but not alpha or other "heavy" energetic particles (protons and other nuclei).

[5] "Normal" is about 0.9 mSv/a worldwide for all external sources. Total average dose from all natural sources (internal and external) is 2.4 mSv/a.

to areas with less than 20 mSv/a additional dose. Considering the above, these levels seem rather conservative.[6]

Likewise, an additional occupational dose limit of 50 mSv/a also seems conservative, but this is done to provide an ample margin between what workers receive and the somewhat uncertain threshold below which ill effects might be observed. Because radiation workers have a low annual dose limit, they must be conscious in their work of ways to minimize dose, even if it appears that a higher dose poses no greater risk.

All radiological workers face exposure to ionizing radiation—how do we manage it? As mentioned earlier, we monitor it closely through badges, rings, swipe tests, and radiation surveys. When working with ionizing radiation, we practice **ALARA**. As mentioned in Section 1.6, ALARA stands for "*as low as reasonably achievable*." We want to apply this concept to ourselves, those we work with, and to the environment in terms of contamination. Just because safety limits are set conservatively doesn't mean that we have to run our doses up to them. For example, if there is a procedure than can be done more than one way with equal effectiveness, cost, and benefit, we should pick the procedure that results in the lowest dose and contamination.

Exposure can be lowered by minimizing **time** near the radioactive source, maximizing **distance** to the source, and increasing **shielding** around the source. Exposure and dose will change in a linear fashion with time. If time near a source is cut in half, dose will also be halved.

Example 12.2

If the dose rate near a source is 6.6 mSv/h, what dose will be received over 30 minutes?

$$\frac{6.6\ mSv}{h} \times 0.50\ h = 3.3\ mSv$$

Exposure or dose varies inversely with the square of the distance. If distance is doubled, exposure is decreased by a factor of 4. If distance is tripled, exposure decreases by a factor of 9. This effect is known as the **inverse square law**.

Example 12.3

If the dose rate 1.2 m from a source is 6.6 mSv/h, what will the dose rate be 4.5 m from the source?

$$\frac{6.6\ mSv}{h} \times \frac{1}{\left(\frac{d_2}{d_1}\right)^2} = \frac{6.6\ mSv}{h} \times \frac{1}{\left(\frac{4.5\ m}{1.2\ m}\right)^2} = \frac{0.47\ mSv}{h}$$

---

[6] Some have suggested that we return to the threshold set in the 1920s of 680 mSv/a, that is, anything below that is safe. The safe threshold dose is likely higher, probably between 1 and 10 Sv/a. Applying a 1 Sv/a standard to Fukushima would mean *all* contaminated areas were safe for resettlement within a year of the meltdowns.

TABLE 12.3
Exposure and Dose Rate Constants

Nuclide	Exposure Rate Constant (C·m²/kg·MBq·s)	Dose Rate Constant (mSv·m²/MBq·h)
^{13}N	1.13×10^{-12}	1.53×10^{-4}
^{18}F	1.10×10^{-12}	1.49×10^{-4}
^{60}Co	2.50×10^{-12}	3.38×10^{-4}
^{99m}Tc	1.54×10^{-13}	2.06×10^{-5}
^{125}I	3.38×10^{-13}	4.34×10^{-5}
^{131}I	4.26×10^{-13}	5.65×10^{-5}
^{133}Xe	1.10×10^{-13}	2.78×10^{-5}
^{137}Cs	6.64×10^{-13}	8.90×10^{-5}
^{192}Ir	8.91×10^{-13}	1.19×10^{-4}
^{201}Tl	8.72×10^{-14}	1.16×10^{-5}

Source: From Smith, D. S. and Stabin, M. G. (2012). *Health Phys.*, *102*(3):271–291. With permission.

We can combine time and distance into a single equation:

$$X = \frac{\Gamma \cdot A \cdot t}{d^2} \tag{12.3}$$

where X is exposure in units of coulombs per kilogram (C/kg), Γ is the exposure rate constant which *can* have units of C·m²/kg·MBq·s. With these units of the exposure rate constant, A is activity in megabecquerels (MBq), t is time in seconds (s), and d is distance in meters (m). Any of these units can be changed (e.g., mCi instead of MBq for activity), but should be consistent with the units on the exposure rate constant. Equation 12.3 can be used for dose or dose equivalent, again, so long as the units are changed appropriately. Exposure and dose rate constants for selected nuclides are listed in Table 12.3.

Example 12.4

A human sits 1.5 m from 925 MBq of ^{137}Cs 40 hours each week. What is her exposure each week?

The exposure rate constant is in Table 12.3. Using Equation 12.3:

$$X = \frac{6.64 \times 10^{-13} \dfrac{C \cdot cm^2}{kg \cdot MBq \cdot s} \times 925\ MBq \times 40\ h \times \dfrac{3600\ s}{h}}{(1.50\ m)^2} = 3.9 \times 10^{-5} \frac{C}{kg}$$

Shielding is a little less straightforward. Appropriate shielding should be added when necessary. Low Z materials (such as plastics) should be used to shield pure beta-emitting nuclides. If there's a fair amount of the beta-emitting nuclide, significant bremsstrahlung photons may be produced from the interactions of beta particles with higher Z materials that may be nearby. The use of a low Z material (such as plastics) for shielding will, therefore, minimize the amount of bremsstrahlung. If bremsstrahlung is unavoidable, or the nuclide emits gamma rays as part of its beta decay, a high Z, high-density material such as lead should surround the plastic shielding.

High Z, high-density materials are the best choice for shielding energetic photons such as X-rays and gamma rays. Remember from Section 6.4 that the probability of photon interaction increases with the atomic number (Z) and the amount of material it passes through (density). Also recall that half-value layer (HVL) is the thickness of material that will attenuate half of the photon beam passing through it, and that TVL is the tenth-value layer—the thickness of material that will cut a photon beam to one-tenth of its original intensity. If two TVLs are used as shielding, the photon beam will be cut to 1/100 of its original intensity.

Example 12.5

How much lead is needed to cut the exposure in Example 12.4 to 5.2×10^{-7} C/kg per week? The linear attenuation coefficient is 0.964 cm^{-1} for Pb exposed to gamma rays from ^{137}Cs.

^{137}Cs is a beta- and gamma-emitting nuclide. Let's assume the beta particles are all absorbed and don't create any bremsstrahlung. This allows us to deal with ^{137}Cs as a pure gamma source and to use Equation 6.16:

$$\ln \frac{I_0}{I} = \mu x$$

$$\ln \left(\frac{3.9 \times 10^{-5} \, \frac{C}{kg}}{5.2 \times 10^{-7} \, \frac{C}{kg}} \right) = 0.964 \, \text{cm}^{-1} \, x$$

$$x = 4.5 \, \text{cm}$$

How many HVLs is this?

$$\text{HVL} = \frac{\ln 2}{\mu} = \frac{\ln 2}{0.964 \, \text{cm}^{-1}} = 0.719 \, \text{cm}$$

$$\frac{4.5 \, \text{cm}}{0.719 \, \text{cm}} = 6.2 \, \text{HVLs}$$

How many TVLs is this?

$$\ln \frac{10}{1} = \mu \times \text{TVL}$$

$$\text{TVL} = \frac{\ln 10}{\mu} = \frac{\ln 10}{0.964 \, \text{cm}^{-1}} = 2.39 \, \text{cm}$$

$$\frac{4.5 \text{ cm}}{2.39 \text{ cm}} = 1.9 \text{ TVLs}$$

We can incorporate shielding into the exposure equation (Equation 12.3):

$$X = \frac{\Gamma \cdot A \cdot t}{d^2} B \qquad (12.4)$$

where B is the **transmission factor**—the fraction of radiation making it through the shielding. Transmission factor (B) has no units and is equal to the intensity of photons making it through divided by the initial photon intensity (I/I_0).

Example 12.6

A human works 40 hours per week 1.7 m from a 0.12 mg ^{60}Co source. What thickness of lead would lower his weekly exposure to 2.58×10^{-5} C/kg? The half-value layer for lead is 15.9 mm when exposed to radiation from ^{60}Co.
First, we need to calculate the activity of the source in MBq:

$$A = \lambda N$$

$$= \frac{\ln 2}{5.271 a \times \left(5.259 \times 10^5 \frac{\text{min}}{a}\right)} \times \left(0.12 \text{ mg} \times \frac{\text{g}}{1000 \text{ mg}} \times \frac{\text{mol}}{59.9 \text{ g}}\right)$$

$$\times \frac{6.022 \times 10^{23} \text{ atoms}}{\text{mol}}\right) = 3.0 \times 10^{11} \text{ dpm}$$

$$\frac{3.0 \times 10^{11} \text{ decays}}{m} \times \frac{m}{60 \text{ s}} \times \frac{1 \text{Bq}}{1 \text{dps}} \times \frac{1 \text{MBq}}{10^6 \text{ Bq}} = 5.0 \times 10^3 \text{ MBq}$$

Now we can solve Equation 12.4 for the transmission factor (B):

$$B = \frac{X \cdot d^2}{\Gamma \cdot A \cdot t} = \frac{\dfrac{2.58 \times 10^{-5} \text{ C}}{\text{kg}} \times (1.7 \text{ m})^2}{\left(2.50 \times 10^{-12} \dfrac{\text{C} \cdot \text{m}^2}{\text{kg} \cdot \text{MBq} \cdot \text{s}}\right) \times (5.0 \times 10^3 \text{ MBq}) \times 40 \text{ h} \times \dfrac{3600 \text{ s}}{\text{h}}} = 0.041$$

Before we proceed, we'll need to calculate the linear attenuation coefficient (μ):

$$\mu = \frac{\ln 2}{\text{HVL}} = \frac{\ln 2}{15.9 \text{ mm}} = 0.436 \text{ mm}^{-1}$$

Since $B = I/I_0$:

$$\ln \frac{I}{I_0} = -\mu x \quad \ln 0.041 = -0.436 \text{ mm}^{-1} \times x \quad x = 73 \text{ mm} = 7.3 \text{ cm}$$

Let's look at one more problem, but this time let's look at dose equivalent rate (DR) rather than exposure. Keep in mind that dose equivalent rate is dose equivalent divided by time (t). Γ is now the dose equivalent rate constant:

$$\frac{\text{dose equivalent}}{t} = DR = \frac{\Gamma \cdot A}{d^2} \tag{12.5}$$

Example 12.7

Calculate the dose equivalent rates at 10 cm and 300 cm from a syringe containing 27 mCi of ^{99m}Tc.

First, we'd better convert activity to MBq:

$$A = 27 \text{ mCi} \times \frac{\text{Ci}}{1000 \text{ mCi}} \times \frac{3.7 \times 10^{10} \text{ dps}}{\text{Ci}} \times \frac{1 \text{ Bq}}{1 \text{ dps}} \times \frac{\text{MBq}}{10^6 \text{ Bq}} = 1000 \text{ MBq}$$

At 10 cm = 0.10 m:

$$DR = \frac{2.06 \times 10^{-5} \dfrac{\text{mSv} \cdot \text{m}^2}{\text{MBq} \cdot \text{h}} \times 1000 \text{ MBq}}{(0.10 \text{ m})^2} = 2.1 \frac{\text{mSv}}{\text{h}}$$

At 300 cm = 3.0 m:

$$DR = \frac{2.06 \times 10^{-5} \dfrac{\text{mSv} \cdot \text{m}^2}{\text{MBq} \cdot \text{h}} \times 1000 \text{ MBq}}{(3.0 \text{ m})^2} = 0.0023 \frac{\text{mSv}}{\text{h}} = 2.3 \frac{\mu \text{Sv}}{\text{h}}$$

QUESTIONS

12.1 Define the following and give their SI units: exposure, dose, and dose equivalent.

12.2 Briefly explain why most humans receive their greatest radiation exposure from cosmic rays, yet their greatest dose comes from radon.

12.3 Calculate the number of ion pairs formed in 1.00 cm^3 of air (density = 1.29 g/L) exposed to 1.00 R of X-rays.

12.4 Assuming that all the X-ray energy in Question 12.3 is deposited in the air, and the average ionization energy of air can also be expressed as 33.85 J/C, what dose is deposited? Calculate your answer in both Gy and rad.

12.5 The average American receives a 2.28 mSv dose equivalent from radon each year. Assuming you receive this dose, and it all comes from the 5.49 MeV alpha particle emitted by ^{222}Rn, how much energy is deposited in your body each year from radon? Approximately how many decays does this represent?

12.6 How much energy (MeV) is deposited in a 75 kg human from a 0.055 cGy dose of alpha particles? What is the maximum number of ion pairs that could be formed in air from this same dose?

12.7 Calculate the dose equivalent in humans, in both rem and mSv, for the dose in the previous problem of α radiation. Do the same for 0.055 cGy of X-radiation.

12.8 A scientific paper measured doses received from external sources by people in Ramsar, Iran. The highest external dose is 260 mSv/a in a part of Ramsar noted for its hot springs. The paper noted no positive or negative health effects among people living in this part of town. How many times higher is background in this part Ramsar than the global average? Which dose/risk hypothesis does this work most closely support? What are the ramifications for the values in Table 12.1 or the people living in areas contaminated by Chernobyl and Fukushima?

12.9 During a routine radiation therapy procedure, a therapist is exposed to 8.0×10^{-5} C/kg·hr. If the procedure takes 4 minutes, what is her exposure? These data were obtained while she was standing 0.91 m from the source. If she stands 0.61 m from the source, she can complete the procedure in 1.5 minutes. Would her exposure change? If so, by how much?

12.10 A concrete and steel cask is used to store fuel assemblies after they were used in a nuclear reactor. The dose rate 0.61 m from the cask is 10 µSv/h. If a person lives exactly 0.5 km from this cask, what annual dose are they receiving from it? Assume this person spends exactly half their time in the house; the rest is spent far from the cask. Compare this to something more tangible.

12.11 How does ALARA apply to a patient about to undergo a radiological procedure?

12.12 Using Figure 1.8, estimate the average total dose for someone living in the United States. What percentage of this dose is anthropogenic? Compare to the annual public effective dose limit established by the NCRP. Briefly explain any discrepancy.

12.13 The recommended annual occupational effective dose limit for the whole body is 50 mSv but is 500 mSv for skin, hands, and feet. Briefly explain why these numbers are so different.

12.14 By what factor will dose change if the distance to the source is doubled and the time spent is halved?

12.15 How far would a human need to be from a 925 MBq ^{137}Cs source to keep the exposure rate below 0.31 mR/h?

12.16 A person works 7.32 m from a 1.14 mg (tiny!) sample of ^{60}Co. Assuming a 40-h work week with two weeks off per year and no shielding, what would her annual dose be (mSv)? The TVL in lead for the gamma rays emitted by ^{60}Co is 45.3 mm. Calculate the thickness of lead needed for a 10 mSv annual dose. Briefly state two other ways exposure could be decreased.

12.17 A patient is apprehensive about receiving the 0.15 mSv dose you are about to administer. What should you do?

12.18 A patient receives a 570 kBq of ^{131}I as part of a medical procedure, then goes home to sleep. What maximum dose equivalent could the patient's wife receive overnight if they sleep 48 cm apart for 8 hours and 20 minutes? Assume that the gamma emissions from the ^{131}I travel an average of 15 cm before leaving the patient's body. The linear attenuation coefficient for ^{131}I in humans is 0.11 cm^{-1}.

12.19 The exposure 182 cm from a source is 5.4×10^{-6} C/kg over 5 minutes. What is the exposure 122 cm from the same source for 1.5 minutes?

12.20 The dose 0.91 m from a source is 2.6 mSv over 4.0 hours. What is the dose 0.33 m from the same source for 1.7 hours?

12.21 Calculate the dose equivalent rate 3 m from a 1900 MBq ^{192}Ir source.

13 X-ray Production

X-rays are used extensively in diagnostic and therapeutic medical applications. This chapter starts by looking at how X-rays are generated for diagnostic applications, such as those performed in hospitals and dentist offices, and then moves into how high-energy X-ray beams are produced for radiation therapy. Interestingly enough, knowledge of how electrons interact with matter (Section 6.2) is key to understanding how X-rays come about.

13.1 CONVENTIONAL X-RAY BEAMS

For diagnostic medical applications, such as an X-ray of your foot to see if any bones are broken, we'd like to consistently produce high-energy photons on demand—a flip of the switch to provide the same energy and intensity every time would be great. It is also desirable for our photon generator to be safe and affordable. The only source that satisfies all these requirements is a particle accelerator. A **conventional X-ray tube** is a small particle accelerator (Section 9.4). Let's see how it works.

A schematic of a typical conventional X-ray tube is shown in Figure 13.1. The filament is a tungsten wire, much like those used in (old-school) incandescent light bulbs. If we run a little electric current through the filament at a high enough voltage, we can get electrons to boil off the wire. The filament is, therefore, a cathode.

After they jump off the cathode, the electrons are accelerated toward the positively charged anode (target). The maximum energy they attain will be equal to the potential difference (tube voltage) between the two electrodes. For all this to happen, both electrodes need to be under vacuum, so the whole X-ray tube is enclosed and pumped out. This is important because the electrodes would rapidly corrode if exposed to oxygen and because the electrons could interact with any matter (gases) on their way from the filament to the target.

Electrons won't boil off the cathode unless the filament is pretty hot. The current of electrons flowing through the wire (**filament current**) does this—just like a toaster! Since it will get hot, it's important to make the filament out of a metal with a high melting point. Tungsten melts higher than any other metal (3370°C), so it works quite well. The filament current is usually set at a few amps. The other current flowing in the tube is made up of the (relatively few) electrons moving from the cathode (filament) to the anode (target). This is called the **tube current** and is usually a few tenths of an amp (few hundred milliamps).

The **focal spot** is place on the target where the electrons hit the target. A smaller focal spot creates a smaller beam, which is sometimes desirable, for example, minimizing the dose to surrounding tissue when trying to kill a small tumor. A smaller focal spot can be accomplished by using a smaller filament (makes for a smaller electron beam). Unfortunately, this also creates a weaker (less intense) X-ray beam.

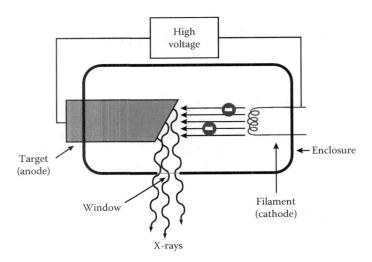

FIGURE 13.1   Schematic of a conventional X-ray tube.

So, how does an X-ray tube make X-rays? When the electrons hit the target, there're two ways they can make X-rays (see Section 6.2 for more details).

1. Bremsstrahlung—the continuous energy spectrum of X-rays produced through inelastic scattering of the electrons in the tube current by the nuclei in the target. This can always happen, regardless of the energy of the incoming electrons. The maximum possible X-ray energy would be equal to the energy of the electron (tube voltage).
2. Characteristic X-rays—if the beam electrons have enough energy to excite or ionize a K shell electron in the target material, an electron vacancy will be created in that shell. When an L or M shell electron drops down to the K shell to fill the vacancy, a characteristic X-ray is emitted. Remember, a characteristic X-ray has a specific energy because the target electron energy levels are quantized.

When we put these two types of interactions together, we see an interesting distribution of photon energies—as shown by the solid line in Figure 13.2. The rather broad part of this spectrum (almost all of it!) is bremsstrahlung radiation. Notice that it maxes out at fairly low-energy then gradually tails off as energy increases until just below the tube voltage. The probability of forming lower-energy photons via bremsstrahlung is higher because it is more likely that an electron will interact with an atom at a longer rather than at a shorter distance. The bremsstrahlung feature tails off sharply at low-energy because the very low-energy X-rays are more efficiently filtered by the tube enclosure. The dashed line in Figure 13.2 shows what the bremsstrahlung spectrum would look like without this filtration.

The sharp peaks on top of the bremsstrahlung spectrum are the characteristic X-rays produced from ionizations and excitations. The energies of these peaks depend on the target material. Tungsten is used as a target for medical applications because it

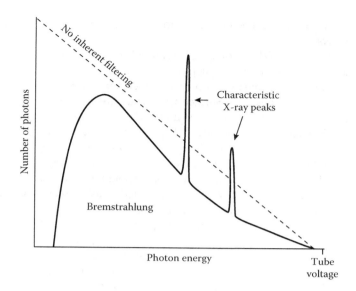

FIGURE 13.2 X-ray spectra with (solid line) and without (dashed line) inherent filtering.

is high melting and because its characteristic X-rays are relatively high in energy. The energy increases because the spacing between electron orbitals generally increases with atomic number. Higher energy means more penetrating—less likely to interact with matter. If only low-energy (soft) X-rays were used, they would be almost entirely absorbed by the patient. A higher atomic number for the target also means that the intensity of the X-rays produced via bremsstrahlung will be higher. Remember that as Z increases, the more likely it is for an electron to interact (Section 6.2).

X-rays can be produced at different depths inside the target—not just at the surface. If one is produced a little way in, it could interact (scattered or absorbed) with the target material on its way out.

We've already seen that the X-ray tube enclosure filters the X-rays to some extent. Because the heat produced in the process of making X-rays needs to be dissipated, X-ray tubes are usually surrounded by oil or water that is circulated. Finally, there is usually an exit window for the X-rays to travel through before leaving the tube. The total attenuation due to the target, enclosure, coolant, and exit window is referred to as the tube's **inherent filtration**. There's nothing we can do about this; it is a result of the tube's design.

The inherent filtration is usually equivalent to placing ~1 mm of aluminum in the X-ray beam. Like all forms of photon filtration, inherent filtration will harden the X-ray beam. Remember that this means the overall intensity is lower, and the average energy is higher after filtration (Section 6.4). Additional hardening can be performed after the beam exits the tube by placing additional filters in the beam's path. This is desirable for most medical applications since the low-energy photons will tend to deposit all their energy in the first centimeter or so of the matter (flesh!).

What happens to the X-ray spectrum when we mess with the current? Basically, it just changes the overall intensity. Intensity increases with current. Increasing filament current increases tube current—more electrons bombard the target, creating more X-rays.

How about voltage? When voltage is increased, the average energy of the photons produced is also increased. Sometimes low-energy X-rays are desirable. If we wanted to treat a surface malady, such as skin melanoma, then we'd want as many low-energy photons as possible. Surface treatments are usually performed with a tube voltage of 5–15 kV. If we want to go a little deeper (~5 mm below the surface), then 50–150 kV would be appropriate. If we need to penetrate further, we need to keep cranking up the voltage. Note that people working with X-rays will typically refer to the photon beam energy using the tube potential, for example, "treatment will be with a 150 kV X-ray beam." While conventional, it is misleading and incorrect. The photons are not 150 keV (notice corrected units); rather they are a broad continuum of energies with a maximum a little below 150 keV.

The shape of the target is also important to consider. In Figure 13.1, the electron beam hits the anode at an angle. This angle serves two purposes. First, electrons with less than 1 MeV of energy (true for all conventional tubes) *tend* to produce X-rays in a direction that is 90° from the electron beam. This process is called **reflectance** because it *looks* like light reflecting on a mirror. You should know this is not true! First, we're starting with electrons, not photons, and second, the photon direction is a tendency; in reality they are produced in all directions.

Second, a focal spot on an angle to the electron beam will produce a narrower, more intense X-ray beam. The angle allows a wider electron beam to hit the target (higher-intensity X-rays produced), while producing an X-ray beam that is smaller in area—as pictured in Figure 13.3. The length A of the focal spot (approximately equal to the width of the electron beam) is larger than the length a (width of the X-ray beam). In fact, the mathematical relationship between the two lengths is:

$$a = A \times \sin\theta \tag{13.1}$$

where the angle θ is the angle of the anode, which is typically 6° to 17° for a diagnostic unit. In Figure 13.3, this angle is drawn at 17°. If A is 0.10 cm, then:

$$a = 0.10 \, \text{cm} \times \sin 17° = 0.029 \, \text{cm}$$

As mentioned previously, reflectance only happens with electron energies below 1 MeV. When electrons have more than 1 MeV, X-rays *tend* to be produced in the

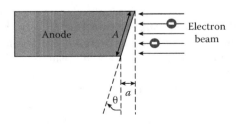

FIGURE 13.3   Geometric reduction of the beam size relative to the focal spot size.

FIGURE 13.4   X-rays produced by a transmission target.

same direction as the electron beam. This process is called **transmission** because it appears that electrons are "transmitted" through the target, magically turning into X-rays (Figure 13.4). As you might guess, the process of converting an electron beam to an X-ray beam is a bit more complex (Section 6.2) than suggested by Figure 13.4!

Keep in mind that X-rays only *tend* to be traveling in the directions shown in Figures 13.3 and 13.4. In reality, X-rays travel out of the target in lots of different directions, regardless of how they are made. We are interested in generating a *beam* of X-rays. That means we just want those photons traveling in a particular direction and soak up all the others. For this reason, X-ray tube housings are usually pretty thick and are made up of high Z metals. To de-select the photons that are traveling close to the correct direction, they are passed through a metal collimator, as shown in Figure 13.5. The length and aperture of the collimator will determine the consistency (how parallel are the photons?) and the size of the beam.

Only part of the projectile electrons' kinetic energy goes into X-ray production. Remember that the maximum electron energy in keV is equal to the tube's electric potential in kV. We can then get a handle on efficiency using:

$$f = 3.5 \times 10^{-4} ZE \qquad\qquad (13.2)$$

where f is the fraction of electron energy that is turned into photon energy, Z is the atomic number of the target material, and E is the maximum electron energy in MeV. For example, with a tungsten anode and a tube potential of 140 kV:

$$f = 3.5 \times 10^{-4} \times 74 \times 0.140\,\text{MeV} = 0.0036 \quad \text{or} \quad 0.36\%$$

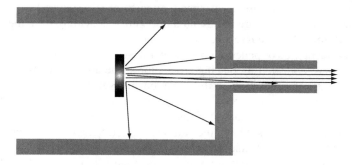

FIGURE 13.5   The use of a collimator to form an X-ray beam.

Only 0.36% of the energy being deposited by the electron beam into the anode ends up as X-ray photons! This is so inefficient it makes SUVs look pretty green. The rest of the electrons' energy ends up as heat. This is the source of all the heat that needs to be dissipated by the coolant mentioned earlier. If we repeat this calculation for a 4 MeV electron beam, we'll get 10.4% efficiency, and 25 MeV gives 65%. As the tube potential increases, so does the efficiency—in rather dramatic fashion. If more photons are desirable, then cranking up the voltage is one way to get 'em.

13.2   HIGH-ENERGY X-RAY BEAMS

As described in Section 8.2, the history of the use of X-rays for radiation therapy shows ever increasing photon energies. Increasing photon energy means more efficient X-ray production (higher intensity) and allows for treatment deeper inside the patient.

Conventional tubes were commonly used in the first half of the twentieth century. Instruments using conventional tubes have several names, usually related to the energy of the X-rays produced. Superficial therapy machines operate in the 50–150 kV range and were only useful in treating the surface or near surface of the patient's body. **Orthovoltage** machines were in common use in the middle of the last century, but they are generally limited to about 400 kV. They were also known as "deep therapy" machines, because of their ability to treat lesions below the skin. They have now been completely replaced by higher-energy linear accelerators, the so-called **megavoltage** machines (or linacs). Why?

Reason #1: Lower-energy photons are more likely to interact with matter. Basically, they are not penetrating enough—too much energy is dumped in the first centimeter (or so) of tissue. If the tumor is deeper than that, we'll end up giving the healthy tissue between the skin and the tumor too much of a dose. To deliver more of a dose deeper into the patient, we need more power!

We already know this is an issue, but how can we imagine this in terms of dose delivered to humans? This is pictured in Figure 13.6. **Percent depth dose** (PDD) is the percentage of the maximum dose remaining at a particular depth in a human. If the maximum dose is delivered 1 cm below the surface, and the dose at 2 cm is 80% of the maximum, then the PDD at 2 cm is 80. Figure 13.6 shows how PDD drops off, as the high-energy photons travel through matter. This is exactly what we expect—the thicker the matter, the greater the beam attenuation (Section 6.4).

Several different curves are represented in Figure 13.6. Going from the orthovoltage curve to the 18 MV curve, the average energy of the photon beam increases, and so does the degree of penetration—greater doses are delivered to greater depth. These data were collected by placing a detector at the appropriate distance underwater while exposing the water to the X-ray beam. Water is an excellent surrogate for humans since we are composed of mostly water, and it generally interacts with high-energy photons such as human tissues (similar Z and density).

The ^{60}Co curve lies just above the orthovoltage curve. ^{60}Co emits photons with energies of 1.2 and 1.3 MeV and, therefore, represents a higher energy beam than an orthovoltage machine. The remaining three curves are from megavoltage machines of increasing energies. Notice that the percent depth dose starts at 100% at the

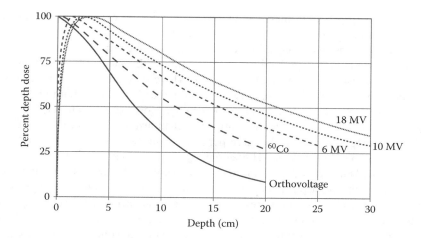

FIGURE 13.6 Variation of percent depth dose (PDD) with depth in water and energy of the X-ray beam.

surface of the water (or the patient's skin) for the orthovoltage and ^{60}Co curves. It is not until we get to the higher-energy machines that the peak moves below the surface, that is, a greater dose is deposited at 1 to 3 cm depth than at the surface. Since most tumors are below the surface, this is desirable.

As you might remember from Section 8.2, frying the patient's skin was used as an indication they had been zapped enough back in the early days of radiation therapy. Delivering the maximum dose below the surface of the skin is known as **skin sparing**. This phenomenon is simply illustrated in Figure 13.7; the darker the shading, the higher the dose delivered at that depth.

If you stop and think about it, this doesn't make sense. Based on what we learned about photons interacting with matter (Section 6.4), photon intensity should decrease as it travels through matter. Therefore, the highest dose should always be delivered at the skin because that's where photon intensity is always greatest. With high-energy photons, it is a little more complex than just attenuation.

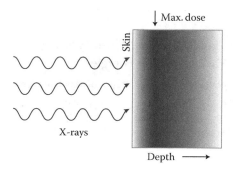

FIGURE 13.7 Skin sparing by higher-energy X-ray beams.

Higher-energy photons interact with low Z matter almost exclusively via Compton scattering (Figure 6.9). When this happens, both the scattered photon and the Compton electron tend to move in the same general direction as the incident beam. Because of their higher probability of interacting with matter, the Compton electrons will deposit their energy fairly close to where they are generated, but it will be *deeper* into the body. Patient dose at depth is therefore due to both photon *and* Compton electron interactions within their body. More photons and electrons are depositing dose at a depth of a few centimeters than just photons at the surface.

Higher-energy beams will be more penetrating and will transfer more energy to the electrons they knock loose. Higher-energy electrons will travel further into the patient's body before they deposit much of their energy (under the Bragg peak, Section 6.2). As a result, the depth of the maximum dose increases with the energy of the incident beam. Table 13.1 illustrates this trend—photon energy increases going down the first column, and the depth of maximum dose increases down the second column.

It turns out that the Compton electrons are key to figuring out where the maximum dose is delivered. The maximum dose is deposited at the point inside the body where the number of electrons knocked loose is equal to the number of electrons being stopped. This is called **electron equilibrium**. Before this point, more electrons are being knocked loose, and after it, fewer electrons are stopping. The tissue between the skin and the maximum dose is appropriately known as the **dose buildup region**.

Reason #2: Penumbra size (or beam edge unsharpness) is larger for lower-energy beams. Penumbra is the fuzzy edge around the beam. This is true for any beam of photons traveling through matter (Figure 13.8a). Even laser light has a fuzzy edge to it. Penumbra is bad for radiation therapy because the tumor needs to be treated with a uniform dose, and penumbra creates nonuniformity at the beam edge.

Some of the penumbra is due to scattering of photons as they travel through matter (usually air); the rest is due to the size of the source. As can be seen in Figure 13.8b, a certain area of the skin will see the entire source and will receive the maximum dose. As we begin to move outside of that area, the collimator blocks parts of the source, so that area will receive a lower dose. Eventually we move too far away, and the source is completely blocked by the collimator. This area sees no dose. Increasing the size of the source (focal spot in X-ray machines) will increase the size of the penumbra (compare Figures 13.8b and 13.8c).

TABLE 13.1
Maximum Dose Depth in Water for Various Photon Energies

Photon Energy	Depth of Max Dose (cm)
Orthovoltage	0.0
^{137}Cs (0.7 MeV)	0.1
^{60}Co (1.2 MeV)	0.5
4 MV	1.0
10 MV	2.5
25 MV	5.0

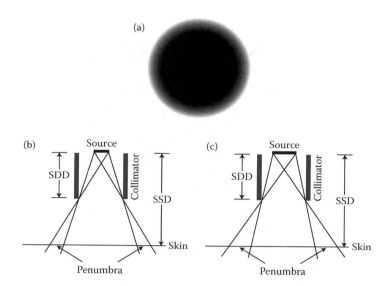

FIGURE 13.8   (a) Penumbra viewed around a cylindrical beam. (b) Geometric origin of penumbra. (c) Effect of on penumbra of increasing source size.

Decreasing the **source to collimator distance** (SDD) also increases penumbra. Finally, increasing **source to surface distance** (SSD) increases penumbra. Decreasing the collimator–patient distance decreases penumbra, but a certain amount of distance (>15 cm) is a good idea. Photons hitting the collimator can kick out electrons—if the collimator is too close to the patient, they can get a pretty good skin dose from these secondary electrons. Having some distance also facilitates movement of the source around the patient.

Reason #3: Can't use **isocentric techniques**. Rotating the source around the patient with the target (what you're aiming your photons at, that is, the tumor) at the center of the circle will maximize the dose to the tumor while minimizing the dose to the healthy tissue surrounding the tumor. The point (or volume) of the beams' convergence is the **isocenter**. Figure 13.9 shows the effect of using four different source positions on the dose received—again, the darker the shading, the greater the dose. As you might imagine, using more source positions will further maximize dose to the area of convergence (the tumor) while minimizing elsewhere (healthy tissue).

Why can't orthovoltage machines use isocentric techniques? Remember that X-rays are very inefficiently produced at orthovoltage energies. Therefore, the intensities of these beams are also low—in order to get a decent dose, orthovoltage machines need to be close to the patient (~50 cm) and cannot be safely rotated around them. Efficiency and intensity increase dramatically at the voltages used by megavoltage machines, allowing them to be far enough (~100 cm) from the patient to be rotated. An added bonus to moving the source further from the patient is that the extra air the beam now travels through will act as a filter and harden it (lowering the surface dose).

At this point, you should be pretty well sold on the fact that X-ray machines operating in the MV range will be better for treating most tumors than X-rays generated

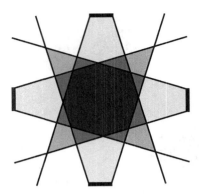

FIGURE 13.9 The effect of using isocentric techniques to maximize dose in the area of convergence, while minimizing elsewhere.

by kV machines. So, how are we going to generate these higher energy photons? Unfortunately, conventional tubes cannot accelerate electrons into the MeV range. We need a more powerful particle accelerator. A quick glance back to Section 9.4 reminds us that we can use linear accelerators to accomplish this.

As we saw in Section 9.4, most modern-day linear accelerators manipulate the phase (peaks and troughs) of microwaves to accelerate electrons. The microwaves can be generated using the same technology as a microwave oven. **Magnetrons** are used in microwave ovens to heat up your leftovers *and* in 4–10 MV accelerators. **Klystrons** are used in higher-voltage machines. Both use electrons to create microwaves via resonance. It works very much the same as blowing air across a pop bottle top or air being blown through a pipe organ. Some of the energy of the moving air is converted into sound through resonance. The same thing happens inside a magnetron or a klystron. Instead of air, it is electrons moving near a cavity, and microwaves are created instead of sound.

Once generated, we want to make sure our waves don't lose any energy. When a broadcast antenna emits radio waves, they spread out in all directions at once. The farther away from the antenna, the lower the intensity of the waves, and the harder it is to tune in on your radio. In a linear accelerator, the microwaves are moved from one point to another, with no loss of intensity, by **waveguides**. These can be as simple as copper tubes but can be more complex (Figure 13.10). The waveguide does just what its name implies; it guides the wave along a specific path rather than allowing it to spread out in all directions at once. Thereby, the same intensity of waves is seen at the end as is put in at the beginning.

Remember that electromagnetic radiation behaves a lot like waves in water. We can think of phase in relation to the peaks and troughs of a wave. A peak represents one phase, and the trough represents the opposite phase. The electron is pushed and

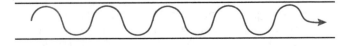

FIGURE 13.10 An electromagnetic wave moving through a waveguide.

FIGURE 13.11   A standing wave accelerator guide.

pulled by these phases, so if we can arrange them just right, they will accelerate the electron. As also mentioned in Section 9.4, there's another way to look at the acceleration of charged particles inside a waveguide. As the EM wave moves through the guide, it will induce temporary positive and negative electrical charges on the inside surface the guide. If these temporary charges move just right through the waveguide, they can be used to accelerate charged particles.

Figure 13.10 shows a wave moving from left to right through a cylindrical waveguide. This can be used simply to transfer the wave's energy from one point to another, or it can be used to accelerate a charged particle, such as an electron. When used to accelerate, it is called a **traveling wave accelerator guide**. We can also use two waves traveling in opposite directions through the same waveguide to set up an interference pattern, as illustrated in Figure 13.11. This is called a **standing wave accelerator guide**. If the interference is done just right, electrons will be accelerated as they move through the guide.

Remember that *constructive interference* happens when two waves are "in phase"—for example, the peak of one wave is at the same point as the peak of the other wave. At this point, the peak height would be the sum of the two peaks. *Destructive interference* happens when the waves are "out of phase"—the peak of one wave coincides with the trough of the other. In this case, the two would cancel out each other. When two waves interfere, as shown in Figure 13.11, a complex pattern results. Our concern is not with how to get this just right to accelerate an electron to specific velocity but rather to know that this is how it is done.

Conventional X-ray tubes can only accelerate an electron to a velocity of approximately one-half that of the speed of light ($0.5c$), and cyclotrons can't do much better. Linear accelerators are needed to get 'em going faster than that. Keep in mind that particles cannot travel at, or faster than, the speed of light, and as we accelerate a particle to speeds approaching that of light, increasing amounts of energy are required. Using the formulas from Section 6.2, we can calculate the relativistic kinetic energy for an electron at various velocities (Table 13.2). We can see that as the velocity approaches the speed of light ($v/c \rightarrow 1$, moving down the first column), the amount of energy (second column) goes through the roof! More and more energy is required for smaller increases in velocity. In the context of generating X-rays, we're not all that interested in the velocities of the electrons as we are their energies. It is the electron energy that determines the energy spectrum of X-rays that are produced.

Where do we get our electrons for linear accelerators? The same way they are produced for conventional tubes—boil them off a hot, high-voltage filament under vacuum. This time we drill a hole in the anode right where the focal spot would be, so the accelerated electron passes right through (Figure 13.12) into the accelerator guide. This is the same technology used in cathode ray tubes (CRT—old style televisions and computer monitors). When placed at the end of a linear accelerator, it is called an **electron gun**. Muy macho.

TABLE 13.2
Relativistic Kinetic Energies for an
Electron at Velocities Approaching
the Speed of Light

v/c	KE (MeV)
0.50	0.079
0.90	0.66
0.990	3.11
0.9990	10.92
0.9999	35.63

We can now put all our components together. Figure 13.13 shows a schematic for a megavoltage X-ray machine that uses a magnetron microwave generator. The microwaves travel through the waveguide to the accelerator guide. There, they accelerate the electrons emerging from the electron gun and slam them into the target. A transmission target is used for all megavoltage machines. Its thickness is a careful balance between the need to stop all the electrons (thicker is better) and the desire to avoid absorbing too many of the photons that are generated (thinner is better). Targets for higher-energy machines will be thicker than those operating at lower energies. Do you know why?

Notice how the accelerator guide fits vertically in the machine's head in Figure 13.13. For higher-energy accelerators, a longer accelerator guide is needed, so it is placed horizontally, as pictured in Figure 13.14. Both types of machines use a thin ionization chamber (Section 7.1.1) as a **monitor** to measure the radiation flux and thereby determine the patient's dose. An ionization chamber is used precisely because it is very inefficient in detecting high-energy photons. We want to get an idea of the flux without significantly attenuating the beam. Both Figures 13.13 and 13.14 have collimators (also known as jaws) to help shape the beam. They are typically made from high Z materials (such as Pb). Finally, there's a treatment platform (slab!) for the patient to lie on.

In higher-energy machines (Figure 13.14), the electron beam needs to be bent by 90° so it will hit the target at an appropriate (right?) angle. Remember from our study of cyclotrons (Section 9.4) that magnetic fields cause charged particles to move in a circular path. If we choose our magnet carefully (field strength), we can get the electron beam to go through a 90° *or* a 270° turn, sending them toward the target.

Unfortunately, the electrons are not all traveling at exactly the same velocity—they have slightly different energies. When they hit the 90° magnetic field, those that

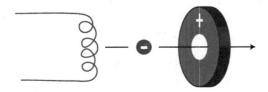

FIGURE 13.12   A cathode ray tube, also known as an electron gun.

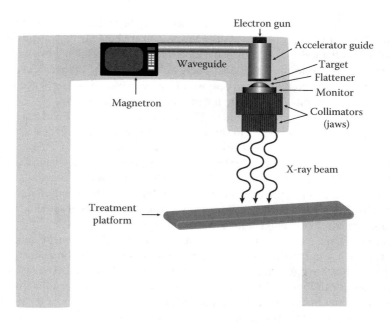

FIGURE 13.13   A MV X-ray machine with a magnetron and a vertical accelerator guide.

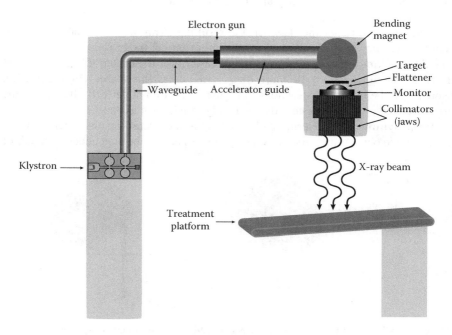

FIGURE 13.14   A MV X-ray machine with a klystron and a horizontal accelerator guide.

are traveling a little faster than the rest ($E+$) are bent through an angle slightly larger than 90°. Those that are traveling a little slower ($E-$) are bent through an angle that is somewhat less than 90°. This causes the electron beam to broaden a bit, with lower-energy electrons at one end and higher-energy electrons at the other (Figure 13.15a). The effect is similar to that of a prism, which spreads out visible light according to energy. For this reason, 90° bending magnets are called **chromatic**.

The same problem is encountered in 270° bending magnets; the electrons are still spread out according to energy. However, running them through 270° causes them to converge at a certain point after exiting the magnet (Figure 13.15b). If left alone, the electrons would again diverge after this point. This point is an excellent place for the target. By doing this, we can obtain a very small focal spot, which results in a narrow X-ray beam and minimal penumbra. Because of this, 270° bending magnets are called **achromatic**. As you may have already figured out, the focal spot and therefore penumbra on chromatic magnet machines are larger.

Achromatic magnets are more commonly used in medical linear accelerators, and since such a narrow beam is rarely needed, it needs to be broadened. This is done by the **flattener** (or flattening filter), which is sandwiched between the target and the monitor in Figures 13.13 and 13.14. Another reason to use a flattening filter is that the photon beam tends to be rather intense in the center, and it drops off rapidly from that point. For therapeutic applications, it is better to have a more uniform beam. Flattening filters accomplish both tasks. Because they need to scatter high-energy photons (which have a low probability of interacting with matter), flatteners are chunks of metal that have different shapes depending on the energy of the photon beam—generally speaking, they are thicker in the center where the photon beam is most intense (Figure 13.16).

After going through the flattener, the resulting X-ray beam should be flat and symmetric. They are both measures of the uniformity of the beam. **Flatness** is measured by looking at the variations in the intensity over the center 80% of the beam. Intensity should not vary by more than ±3% across this region (Figure 13.17a). Notice that the beam cross sections in Figure 13.17 are not perfectly flat across the top. The greater intensities near the beam edge are known as "horns."

Beam **symmetry** is determined by looking at the overall intensity between two halves of the beam—it should not vary by more than 2% (the area under each half of

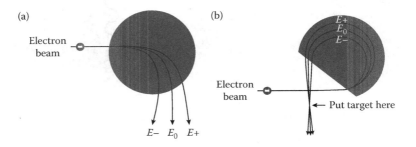

FIGURE 13.15 (a) Energy dispersal effect of a 90° bending magnet. (b) Energy convergence effect of a 270° bending magnet.

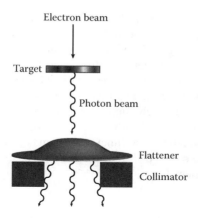

FIGURE 13.16   The dispersal effect of a flattening filter on an X-ray beam.

the beam cross section in Figure 13.17b). If the flattener is not properly aligned to the beam, an asymmetric beam will result.

We should also mention **safety interlocks**—an important component of any radiation-generating device. Safety interlocks are switches that will shut down the X-ray beam or the entire machine if a particular safety mechanism is not in place. For example, there is a switch on the door to the treatment room to ensure it is closed before exposing the patient. This way no one can wander in during therapy and be exposed. There are also switches to detect proper placement of filters and collimators, as well as functioning warning lights. Instrument software will also have safety interlocks checking flatness, symmetry, and patient dose.

Finally, you should feel a little uncomfortable with the energy characterization of the photon beams emerging from megavoltage machines. They are commonly referred to as "X MV photon beams," where X typically ranges from 4 to 25. As mentioned earlier (Section 13.1), this nomenclature comes from the tube potential settings of orthovoltage lower-energy X-ray machines. With megavoltage machines, there isn't a tube potential the electrons are accelerated through; rather this energy value is referring to the energy of the electrons hitting the target and should therefore use units of MeV rather than MV. The reader should also remember that a broad-energy spectrum of photons is produced, with a maximum slightly below the electron energy.

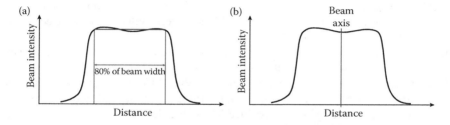

FIGURE 13.17   Measurement of (a) beam flatness and (b) beam symmetry.

Electron beams can be generated by the same machines discussed here—just remove the target so the electron beam hits the patient instead of X-rays. Electron beams will also need to be broadened. As they emerge from the accelerator, these beams are only 1–2 mm in diameter. A thin metal foil will generally do the trick (think back to your freshman chemistry and physics classes—remember Rutherford's gold foil scattering experiment?). These are called **scattering foils**. They are thin because electrons have a relatively high probability of interacting with matter, and thicker chunks of metal would absorb all the electrons. Additionally, an **electron applicator** or **cone** is used to help collimate the beam beyond the jaws. Electrons are more likely to be scattered by the air they travel through, so these cones will need to be fairly close to the patient.

QUESTIONS

13.1 Increasing the current on a conventional tube increases the X-ray intensity. If greater intensity is needed, why can't the current be cranked up as needed?

13.2 Define the following: focal spot, tube current, transmission target.

13.3 Why is bremsstrahlung more likely further from a nucleus?

13.4 On a single plot, overlay the three X-ray spectra produced by the same conventional X-ray tube set at three different voltages.

13.5 Why is tungsten commonly used as a target in medical X-ray generating devices?

13.6 Tungsten, molybdenum, and copper are all are used for X-ray production. For each anode material, calculate the electron energy needed for 100% efficiency of photon production. For the highest value, also determine the electron velocity. Does this seem reasonable for a clinical setting?

13.7 An electron beam measures 3.4×1.0 mm. If this beam strikes a target that is at a $17°$ angle relative to the beam direction, what is the cross-sectional area of the photon beam produced? Assume the longer dimension of the electron beam hits the target at $17°$.

13.8 Typically, when an electron hits a target, only part of its energy goes into forming X-rays. What happens to the rest of the energy?

13.9 What energy X-ray is produced by a conventional X-ray tube operating at 350 kV and 150 mA with a tungsten ($Z = 74$) target?

13.10 Give two reasons why a ^{60}Co irradiator would be preferred for radiation therapy over a 400 kV orthovoltage machine.

13.11 Explain the general trend of greater PDD with X-ray energy observed in Figure 13.6 for the five curves. Why do the highest energy curves peak between 0 and 5 cm depth?

13.12 Give three reasons why megavoltage machines are favored for radiation therapy over orthovoltage units. Briefly describe each reason.

13.13 Carefully make drawings like Figures 13.8b and 13.8c, except change only the SDD. Which gives a larger penumbra? Repeat, but this time move the patient a little closer to the source (reduce SSD). Which of these latter two drawings shows a smaller penumbra? You might want to use a computer-drawing program or graph paper to insure consistency between your drawings.

13.14 Briefly explain the main similarity and the difference between the Bragg peak associated with charged particles interactions with matter and the fact that maximum dose is delivered inside a patient by high-energy photons.

13.15 Briefly explain how radiation therapy patients can be exposed to electron radiation during a photon treatment by a 4 MV linac machine, that is, the target and flattener are in place. Calculate the minimum possible LET (in air) of an electron produced by a 4 MV machine with SSD = 100 cm.

13.16 Give two unrelated reasons why the electrons produced in Question 13.15 are very likely to have a higher LET.

13.17 What is the purpose of a magnetron? A klystron?

13.18 Why are targets for higher-energy X-ray machines thicker?

13.19 Sketch a diagram of a bent-beam accelerator. Label all major components.

13.20 Inorganic scintillator and semiconductor detectors are much more efficient at detecting high-energy photons. Why aren't they used in place of the ionization chamber in megavoltage machines?

13.21 What is an X-ray flattening filter? How does it differ from a scattering foil?

13.22 Chromatic also has meaning in music. Is there an analogy between this definition and the bending magnets described in this chapter? Explain your answer.

14 Dosimetry of Radiation Fields

Radiation therapy is all about zapping tumors with ionizing radiation (Section 8.2). Since tumor cells are rapidly dividing, they are much more susceptible to death by radiation than other cells. X-rays are typically used, but electron and proton beams are also used for certain types of tumors. Since it is difficult or impossible to directly measure dose inside a patient during treatment, radiation therapists need to calculate doses, not only to the tumor, but also to the surrounding (healthy) tissue. When done properly, dose to the tumor is maximized, while dose to healthy tissue is minimized.

Measuring doses is generally done using a **phantom**, which is simply a human surrogate—something that behaves more or less the same way toward ionizing radiation but allows placement of a detector inside. Often it is simply a tub of water. Since humans are mostly water, and our remaining elemental makeup has a similar average atomic number (Z) and density, water is an excellent and inexpensive stand-in for human tissue. It can also be plastic (again, similar average Z to human tissue) where detectors can be placed below various thicknesses.

These measurements are usually expressed as a fraction of the amount of radiation reaching the detector relative to another value. Several different terms are used; we'll look at three different ways at looking at relative dose: percent depth dose (PDD), tissue–air ratio (TAR), and tissue maximum ratio (TMR). Initially, we'll consider only X-rays for human therapy and save electrons and protons for the end of this chapter.

14.1 PERCENT DEPTH DOSE

As we've already seen (Section 13.2) percent depth dose (PDD) is the percentage of the maximum dose remaining at a particular depth inside a human. This is done by making two measurements inside a phantom: the dose at the depth of interest (D_n) and at the depth where dose is maximized (D_{max}). Mathematically it is defined as:

$$\%D_n = \frac{D_n}{D_{max}} \times 100 \tag{14.1}$$

where D_n is dose at depth n, D_{max} is the maximum dose, and $\%D_n$ is the percent depth dose. The fraction of D_n/D_{max} is known as the **depth dose**. As shown in Figure 14.1, dose measurements are made along the beam axis (center of the beam) directly below the radiation source.

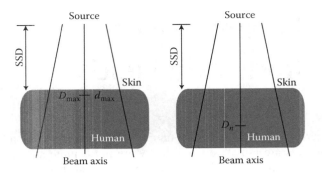

FIGURE 14.1 Percent depth dose measurements.

Measuring PDD is represented schematically in Figure 14.1. The human/phantom is represented as a rounded rectangle, an approximation of a cross section of a human torso. Notice that the maximum dose (D_{max}) is shown as being at a distance of d_{max} below the skin. Remember that this is only true for the megavoltage machines (also known as linear accelerators, or linacs), as shown in Figure 13.6. For lower energy photons, D_{max} would be at the surface ($D_{max} = D_0$). Note also that the radiation source (where the X-rays are produced) is located at a distance of SSD from the patient's skin. As we saw in Chapter 13, SSD stands for **source to surface** (or skin) **distance**. The importance of this distance will be discussed later, but it should be apparent that increasing this distance will decrease dose, according to the inverse square law (Section 12.3).

Example 14.1

A 6 MV linac[1] produces a beam with a PDD of 77.7% at 7 cm depth. If a 0.20 Gy dose is required at 7 cm, what is the maximum dose?

If we rearrange Equation 14.1 by solving for D_{max}, we can plug in the rest and solve:

$$D_{max} = \frac{D_n}{\%D_n} \times 100 = \frac{0.20\,\text{Gy}}{77.7} \times 100 = 0.26\,\text{Gy}$$

Notice that the maximum dose is higher than the dose delivered at 7 cm, as it should be. A 6 MV machine delivers its maximum dose at a depth of 1.5 cm—you can sort of see this on Figure 13.6.

We can calculate percent depth dose, and we have some understanding of how it is measured. What sorts of things have an effect on PDD? First, let's consider beam energy. As the average energy of the photons in the beam increases, the maximum dose will be delivered deeper inside the patient. For example, an X-ray beam from a 20 MV linear accelerator has a d_{max} at 3 cm (compare to the 1.5 cm we saw for 6 MV). We know this from looking at Figure 13.6 and Table 13.1 and (hopefully)

---

[1] Remember that "6 MV" is not referring to the actual energy of the photons generated. It is a broad continuum with a maximum value slightly less than 6 MeV.

from reading some of the text around them. As a reminder, this is because higher-energy photons interact with low Z matter by scattering photons and electrons in the same direction as the beam. When we reach electron equilibrium, the dose from photons + Compton electrons reaches a maximum (see Figure 13.7). How does this affect PDD below d_{max}? Since d_{max} is deeper, PDD values at a specific depth inside the patient will be higher for a higher-energy beam.

Second, let's consider **field size**. Field size is quite simply the cross-sectional area of the beam, typically measured at the end of the collimator (also known as the jaws). The beam widens as it travels down to the patient, but since distances between the source and patient can vary, it's less confusing to state the field size as the beam emerges from the collimator. The bottom line is that a wider beam means a larger field size, and interestingly, this will tend to increase percent depth dose. Remember that PDD is measured along the central beam axis, so a wider field will increase the probability that photons will be scattered to the detector from another part of the beam.

Imagine we first measure a dose from a beam that is narrower than our detector is wide (Figure 14.2a). As this beam travels through the air and the patient, some of its photons will be scattered out of the beam and miss the detector. Now, we increase the field size so that it is much larger than the area of our detector (Figure 14.2b). Now photons that would've missed the detector, had they not been scattered, have a chance of being scattered *into* the detector. Since there are many more photons that start out in a direction that would've missed the detector than those that start out headed for the detector, the odds are that more photons will end up being scattered into the detector than get scattered away from it. The end result is that the percent depth dose increases somewhat with field size.

Photons entering the detector directly from the source (not scattered) are called **primary (1°) radiation**, while those that are scattered into the detector are **secondary (2°) radiation**. The observed dose at any point within a wide beam is due to both primary and secondary radiation. Note that scattering can be due to X-ray interactions with the monitor, collimators, air, or patient.

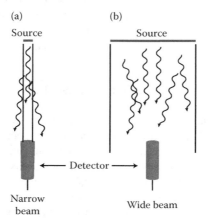

FIGURE 14.2 The effect of field size on PDD: (a) Narrow beam (b) Wide beam.

This increase in PDD becomes less pronounced for very high-energy fields (>15 MV) because higher-energy photons tend to scatter through smaller and smaller angles (scatter forward). At a certain point, these angles get too small for field size to have an effect on PDD.

Example 14.2

Determine the percent contributions of primary and secondary radiation at 16 cm depth for a 10 × 10 cm² 10 MV field with an SSD of 100 cm. The percent depth dose at this depth is 58.8%.

We need one other piece of information to do this problem, the PDD for a "0" field at 16 cm depth for a 10 MV field at SSD = 100 cm. In other words, keeping everything else constant, shrink the field size down to something so narrow that we'll only see primary radiation. For this problem, PDD is 54.4% for a zero field. Notice that this value is smaller than the PDD for the 10 × 10 cm² field. The difference is secondary radiation.

The percent contribution of primary radiation is:

$$\frac{54.4\%}{58.8\%} \times 100\% = 92.5\%$$

The percent contribution of secondary radiation is what's left over.

$$100\% - 92.5\% = 7.5\%$$

Okay, secondary radiation doesn't provide a huge contribution, but it's something.

How can we correct for changes in PDD when the size of the beam changes? Field size corrections (C_{fs}) will do this for us, but, unfortunately, they vary from one machine to another. Therefore, a field size correction specific to a particular instrument must be applied, and it must be determined for each individual machine. It is measured by comparing the dose at a certain depth (typically d_{max}) along the beam axis in any field to the dose in the same place in a **standard field**. A standard field is usually 10 × 10 cm². Field size correction values range from ∼0.90 to ∼1.10 and depend on the collimator size (jaw settings), not SSD or photon energy. This is another reason why field size is measured at the jaws, not at the patient.

$$C_{fs} = \frac{D_n \text{ for field size (fs)}}{D_n \text{ for the standard field}}$$

Percent depth dose is also affected by **field symmetry**, which is simply the shape of the field. Figure 14.3 shows three different shapes superimposed on each other: a circle, a square, and a rectangle. All three have the same area, but note that the square and especially the rectangle have parts of their areas that are further from the center than the circle. For objects of equal area, those with more of their area further from the center will be lower in symmetry. Therefore, the square is lower symmetry than the circle, and the rectangle is lower symmetry than the square.

When symmetry is lowered, PDD generally goes down. Think about a simple change, from square to rectangle. If they are the same area, the rectangular field will

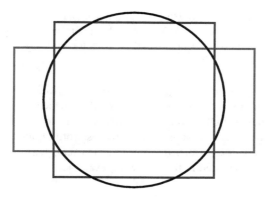

FIGURE 14.3 A circle, a square, and a rectangle, all with equal areas but different symmetry.

scatter fewer photons into the beam axis. In the rectangle, more photons are further away from the beam axis, and the further away, the lower the probability of scattering all the way back to the beam axis. There are ways to convert PDD values from one simple shape to another (like rectangle to what's called an "equivalent square"), but we won't bother with that here. Modern-day radiation therapy often uses complex (and sometimes changing) shapes, and computers do dose calculations. We'll stick with square fields to keep things simple, and use them to illustrate how these calculations are performed, and allow us to focus on what other factors go into the math.

Percent depth dose is also dependent on the source to surface distance (SSD). Unfortunately, this dependence is a bit more complex than beam energy, field size, or symmetry. Ignoring the attenuation of air, increasing SSD will lower the dose rate at any specific point inside the patient according to the inverse square law.

$$\text{Dose rate} \propto \frac{1}{d^2} \quad \text{or} \quad \text{Dose Rate Change} = \frac{1}{\left(\dfrac{d_2}{d_1}\right)^2} \tag{14.2}$$

In these equations, d is distance, d_1 is the original distance, and d_2 is the new distance. We already know from the inverse square law (Section 12.3), that when the distance is doubled, the dose rate will be cut by one quarter. Let's try this out with Equation 14.2:

$$\text{Dose Rate Change} = \frac{1}{\left(\dfrac{2}{1}\right)^2} = \frac{1}{4}$$

Example 14.3

What happens to the surface dose rate if we increase SSD from 100 to 150 cm?

$$\text{Dose Rate Change} = \frac{1}{\left(\dfrac{150\text{ cm}}{100\text{ cm}}\right)^2} = 0.44$$

Meaning that the surface dose will be 0.44 times lower at 150 cm than at 100 cm. We could also say that the dose will be 44% lower at the new distance, and treatment time would need to be increased by a factor of $1/0.44 = 2.3$ in order to achieve the same dose at 100 cm.

Doses at all depths (including d_{max}) will decrease as SSD increases. As a result, we might intuitively expect PDD values below the surface to also decrease. As it turns out, the values of PDD actually *increase* as SSD increases. If we look again at Equation 14.1, we can see that as SSD increases, both D_n and D_{max} will decrease, but if D_{max} *decreases at a faster rate* than D_n, then $\%D_n$ will increase. This idea is mathematically demonstrated in the following example.

Example 14.4

Using the inverse square law, calculate the changes in D_{max} and D_n when SSD is increased from 100 to 150 cm. Assume $d_{max} = 0$ (i.e., D_{max} is at the surface), and D_n is at 10 cm depth. Also assume at least two significant figures in all values.

From Example 14.3, we already know that increasing SSD from 100 to 150 cm will decrease D_{max} (D_0) by a factor of 0.44. We can then calculate the change in D_{10} in the same way. Remember that we are now 10 cm below the surface, which means our original distance to the source is 100 cm + 10 cm = 110 cm, and our final distance is 150 cm + 10 cm = 160 cm.

$$\Delta D_n = \frac{1}{\left(\dfrac{160\,cm}{110\,cm}\right)^2} = 0.47$$

D_{max} decreases to 44% of its original value while D_{10} decreases to 47% of its original value. Both doses decrease, but D_{max} decreases to a greater extent than D_{10}. As you might guess, dose will always decrease at a faster rate for any point that is closer to the source.

Let's see how using a simpler example, shown in Figure 14.4. Initially, worried guy is 1 m from a radioactive source, and happy guy is 2 m away. The source is then moved 2 m further away from the guys, making it 3 m from worried guy and 4 m from happy guy. The distance between worried guy and the source tripled, cutting his dose to 1/9 of the original. However, the distance doubled for happy guy meaning his dose decreased only to 1/4 of original. Worried guy is still getting a higher dose, but his dose is dropping at a faster rate than happy guy's. In general, whatever is closer to a source will see a more dramatic decrease in dose as the source moves away because it also sees a more dramatic change in distance to the source.

Now, let's return to the work we did in Examples 14.3 and 14.4 and calculate percent depth doses at 10 cm depth for both 100 cm SSD and 150 cm SSD. We also need to know that $D_{max} = 100$ cGy and $D_{10} = 65$ cGy when SSD = 100 cm. Note that these values are specific to a particular instrument with certain settings and

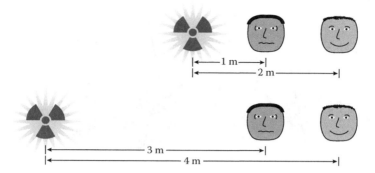

FIGURE 14.4   Effect of distance on dose.

cannot be generalized to all X-ray machines. PDD at 100 cm SSD is just a matter of plugging in values to Equation 14.1:

$$\frac{65\,\text{cGy}}{100\,\text{cGy}} \times 100 = 65\%$$

At SSD = 150 cm, we'll need to first calculate the doses at the surface and at 10 cm depth. We can use the dose rate changes of 0.44 at the surface and 0.47 at 10 cm that were calculated in Examples 14.3 and 14.4:

$$100\,\text{cGy} \times 0.44 = 44\,\text{cGy}$$
$$65\,\text{cGy} \times 0.47 = 31\,\text{cGy}$$

Now we can determine PDD at 150 cm SSD using Equation 14.1:

$$\frac{31\,\text{cGy}}{44\,\text{cGy}} \times 100 = 69\%$$

PDD ($\%D_{10}$) *increases* as SSD increases! If you want to treat a tumor at any depth below d_{max}, you would want to increase SSD (and therefore PDD) as much as possible in order to deliver more dose at the tumor (assuming it is below d_{max}). But wait, remember that PDD is a *relative* dose, that is, the dose at the tumor relative to the dose someplace else. Dose rate (or photon flux) *decreases* with increasing SSD, which would mean longer treatment times to get the same dose delivered to the tumor. Deciding on an optimal SSD is, therefore, a compromise between PDD and dose rate.

We've already looked at how scattering affects PDD due to changes in field size and symmetry. When we did, we only considered scattering in the forward direction (same direction as the beam). What about backscatter? Remember that Compton scattering can occur through any angle up to 180°, and backscatter is scattering in the direction opposite to the beam (Section 6.3.1). Therefore, a photon can be traveling

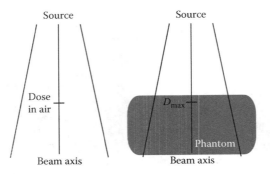

FIGURE 14.5 Determination of backscatter factor.

through a patient, undergo backscattering, travel back the way it came, then interact again, depositing energy at less depth. We can determine the extent that backscatter contributes to dose through an experiment illustrated schematically in Figure 14.5. First, the dose in air is measured at a specific distance from the source. Second, a phantom is inserted into the beam such that D_{max} is observed at the same distance from the source, and dose is measured again. The ratio of these two dose values is the backscatter factor (BSF):

$$\text{BSF} = \frac{D_{max}}{\text{Dose in air}}$$

If there's no backscatter, this ratio is 1.0. If there is some, it'll be greater than 1—in other words, backscattering only adds to dose. Generally speaking, it ranges from 1.0 to 1.5 and is only greater than 1 when the maximum photon energy is below 2 MeV. Above 2 MeV, Compton scattering is almost all in the forward direction (relatively low angle), and backscattering does not contribute significantly to dose. Therefore, this whole concept is pretty useless with most modern-day equipment. Despite its lack of utility with megavoltage machines, it is still applied, although it is called **peak scatter factor** (PSF), and it is almost always 1.0.

Now let's get a little interventionist. Sometimes it's desirable to place stuff in the beam. Why? The main reason would be to harden the beam. Remember (Section 6.4) that this means increasing the average energy by filtering out more of the lower-energy photons. This is easily, and commonly, accomplished by placing a plastic tray in the beam. There are also cases when having an obstruction such as the patient's bed or the treatment platform in all or part of the beam is unavoidable. To account for this attenuation, the **equipment attenuation factor** (C_{attn}) is calculated from the ratio of dose without the obstruction to dose with the obstruction. These values are always less than 1, since the dose will always be lower with less beam intensity:

$$C_{attn} = \frac{\text{Dose with stuff in beam}}{\text{Dose without stuff in beam}}$$

We can now put everything we've learned in this section together and mathematically describe tumor dose based on depth dose and corrections to it:

$$\text{Tumor Dose} = N_{\text{mu}} \times C_{\text{cal}} \times \frac{D_n}{D_{\text{max}}} \times C_{\text{fs}} \times C_{\text{attn}} \tag{14.3}$$

where N_{mu} is the number of **monitor units** (mu)—translated into English, a monitor unit is the amount of time it takes the machine to deliver certain dose. This "certain dose" is C_{cal}, also known as the **calibration factor**. It is usually set at 1.000 cGy/mu delivered at d_{max} at the calibrated field size (usually 10×10 cm²) and the calibrated SSD (typically 80 or 100 cm). Note that multiplying the number of monitor units (N_{mu}) by the calibration factor (C_{cal}) is a measure of the (raw) dose, in cGy at d_{max}.

Normally we know what dose is needed, based on tumor imaging. What we need to calculate is the number of monitor units, that is, the treatment time. Solving Equation 14.3 for N_{mu} gives Equation 14.4:

$$N_{\text{mu}} = \frac{\text{Tumor Dose}}{C_{\text{cal}} \times \dfrac{D_n}{D_{\text{max}}} \times C_{\text{fs}} \times C_{\text{attn}}} \tag{14.4}$$

Example 14.5

Determine the treatment time (in monitor units) to deliver 200 cGy to a tumor 8 cm below the surface of a patient exposed to a 7×7 cm² 6 MV field at 100 cm SSD. This accelerator is calibrated to deliver 0.998 cGy/mu, has a PDD of 74.6%, and a field correction factor of 0.978, under these conditions. There's nothing further attenuating the beam.

From the problem, we can determine that depth dose is 0.746 (just divide PDD by 100), and assume $C_{\text{attn}} = 1.00$ (since there's nothing added to attenuate the beam).

$$N_{\text{mu}} = \frac{200 \text{ cGy}}{\left(0.998\,\dfrac{\text{cGy}}{\text{mu}}\right) \times 0.746 \times 0.978 \times 1.00} = 275 \text{ mu}$$

What is the dose at d_{max} (D_{max}) in Example 14.5? Remember that we delivered 200 cGy at a depth of 8 cm with a PDD of 74.6%. With this wealth of knowledge, all we need to do is rearrange Equation 14.1, solving for D_{max}:

$$D_{\text{max}} = \frac{D_8}{\%D_8} \times 100 = \frac{200 \text{ cGy}}{74.6} \times 100 = 268 \text{ cGy}$$

As expected, the dose at d_{max} is greater than the dose at 8 cm (D_8).

Exit dose is the dose the patient receives where the photon beam leaves the body. With megavoltage machines, the exit dose is often greater than the skin dose, where the beam enters the patient, and is therefore an important consideration in treatment planning. Assuming the patient is 30 cm thick ($\%D_{30} = 22.7\%$) where treatment is taking place in Example 14.5, we can again use a rearrangement of Equation 14.1 to calculate exit dose:

$$\text{Exit dose} = D_{30} = \frac{D_{max} \times \%D_{30}}{100} = \frac{268 \text{ cGy} \times 22.7}{100} = 60.8 \text{ cGy}$$

This is a fairly mild exit dose. It will be higher for skinnier patients (or body parts that aren't as thick) and for higher energy beams.

Example 14.6

What dose will be seen at d_{max} if 250 cGy is delivered at 9.5 cm by a 10×10 cm^2 6 MV field at an SSD of 100 cm? Assume $\%D_9 = 71.3\%$ and $\%D_{10} = 67.7\%$ for this instrument configuration.

This problem provides a great deal of unnecessary information, but it does not give us a percent depth dose at 9.5 cm depth. We can make a pretty good guess by simple linear interpolation:

$$\%D_{9.5} = 67.7\% + \frac{1}{2}(71.3\% - 67.7\%) = 69.5\%$$

Now we can plug into Equation 14.1:

$$D_{max} = \frac{D_{9.5}}{\%D_{9.5}} \times 100 = \frac{250 \text{ cGy}}{69.5} \times 100 = 360 \text{ cGy}$$

You might think the linear interpolation in Example 14.6 is needlessly complex. Why not just average the two PDD values? You'd be right, but the process illustrated here is general for any linear interpolation. For example, what if we needed to know the percent depth dose at 9.2 cm? Note that 9.2 cm is 1/5 of the way from 9.0 to 10.0 cm.

$$\%D_{9.2} = 67.7\% + \frac{1}{5}(71.3\% - 67.7\%) = 68.4\%$$

Before we finish our long look at percent depth dose, let's consider the effect on treatment time by changing source to surface distance (SSD). Notice that SSD is not a part of Equation 14.4; rather it is incorporated into the depth dose (D_n/D_{max}) value. Remember that PDD changes if SSD is changed, so if we've measured all of our PDD values for a particular machine at 100 cm SSD, we can't simply change SSD. We wouldn't know what depth dose value to plug into Equation 14.4.

Example 14.7

At a depth of 10 cm and an SSD of 150 cm, the $\%D_n$ is 71.3% for a 10×10 cm^2 beam from a 6 MV linac. What change in the number of monitor units (treatment time) is required to deliver the same dose as at 100 cm SSD, where the $\%D_n$ is 67.7% for the same field size?

Before we get into this, notice that PDD decreases with decreasing SSD, as expected. Now, as we try to wrap our brains around this problem, realize that there are two, competing effects in this problem. First, PDD is decreasing. A lower percentage of the dose delivered at depth means we'll need a longer treatment time. Second, the decrease in distance to the source means that dose rate will be going up, and treatment time will have to be shortened. Ouchies. Let's take them on one at a time. How much will we have to increase treatment time due to the change in PDD? It is simply the ratio of the two values. The larger value goes on top because treatment time will increase due to the change in PDD.

$$\frac{71.3\%}{67.7\%} = 1.05$$

Treatment time will have to be about 5% longer due to the decrease in PDD. Next, we should use the inverse square law to figure the change in treatment time due to the change in distance between the source and the tumor. Note that the distances are source to tumor, not source to surface. In other words, we'll need SSD plus 10 cm. Applying the inverse square law:

$$\frac{1}{\left(\frac{100\,cm + 10\,cm}{150\,cm + 10\,cm}\right)^2} = 2.12$$

What? Treatment time should be shortened by decreasing SSD, not doubled! Absolutely right. What we've calculated here is the effect on dose—remember *dose* follows the inverse square. If dose is roughly doubled, then treatment time will be approximately cut in half. Let's do it properly and invert 2.12:

$$\frac{1}{2.12} = 0.473$$

That's better. Treatment time will be cut to ~47.3% of its original value by the change in distance. We can combine the two conflicting effects by multiplying them together.

$$1.05 \times 0.473 = 0.498$$

This tells us that decreasing SSD from 150 to 100 cm will decrease treatment time by a factor of 0.498 (roughly half the original time).

The bottom line on depth dose is that it is sometimes an awkward measure of the fraction of radiation delivered to the target volume. The awkwardness can be seen in Example 14.7—dealing with changes in SSD can be more difficult than they need to be. We'll look (more briefly!) at two other ways of doing dose calculations (patient attenuation) that do not depend on SSD.

14.2 TISSUE–AIR RATIO

Percent depth dose is determined by measuring dose at two places inside the patient/ phantom and, thus, is a measurement of how much the patient attenuates the beam. It can also be considered as a *relative* dose, that is, it is the dose in one place inside the patient relative to another. As we've seen, PDD changes (along with dose) with SSD in a manner that requires measuring PDD at every possible SSD. As we saw back in Section 8.2, it is desirable to rotate the source around the patient during treatment. To do so, and maintain the same SSD, the patient would have to be shaped like a cylinder, and the tumor would need to be located in the center of the cylinder.

To avoid the complexity of having to repeat all relative dose measurements at every conceivable SSD, it would be more convenient to measure patient attenuation another way. As shown in Figure 14.6, tissue air ratio (TAR) is determined by first measuring the dose in air a specific distance from the radiation source. A second measurement is taken, at the same distance, but now with the patient in the beam. This ratio then reflects how much the patient (tissue) attenuates the beam relative to air.

If this looks a bit familiar, then you've been paying pretty close attention! If not, you should work on your active reading skills. Tissue–air ratio is determined a lot like backscatter factor (Figure 14.5), except we now measure anywhere along the beam axis inside the phantom/human instead of just at d_{max}. The distance from the source to the point of measurement is the **source to axis distance** (SAD). Not the best acronym, but it could be worse. In radiation therapy, we're most interested in the dose at the tumor, so we could instead define this distance as source–tumor distance (STD).

Mathematically, TAR is defined as the dose to the patient at SAD (D_n) divided by the dose in air at the same distance from the source:

$$\text{TAR} = \frac{\text{dose @ SAD in phantom}}{\text{Dose in air @ SAD}}$$

Because dose changes with distance, so does the tissue–air ratio. Unlike PDD, we don't need to re-measure TAR at the new distance; all we need to do is apply the

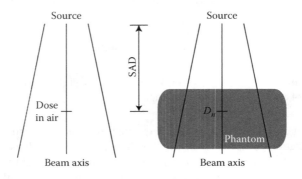

FIGURE 14.6 Determination of tissue–air ratio.

inverse square law. TAR is like PDD in that both increase with beam energy (less likely to interact, more dose delivered deeper) or field size (larger field means greater probability of scattering back to the center).

While it was commonly used in the early days of isocentric therapy, TAR is of relatively little importance today as it was developed when ^{60}Co therapy dominated— as an alternative to PDD. As a result, we won't do any math with it. For reasons we don't need to get into, it just didn't work well with linear accelerators. Another method of determining patient attenuation was necessary.

14.3   TISSUE MAXIMUM RATIO

This is the most commonly used patient attenuation factor today is the **tissue maximum ratio** (TMR). Like the others, it comes from two measurements (Figure 14.7), and like PDD, both are made inside the phantom. First, the dose at d_{max} (D_{max}) is measured at a specific distance from the radiation source (SAD). The second dose measurement is made with the patient/phantom moved up in the beam (closer to the source) so that depth of interest (typically the tumor depth) is now at the same distance from the source (SAD). Like TAR, the dose measurements are both made at the same distance from the source.

Mathematically, TMR looks exactly like depth dose, except that the doses are measured at the same SAD instead of the same SSD. Tissue maximum ratio is the dose at some depth n inside the patient (D_n) divided by the dose at d_{max} (D_{max}):

$$\text{TMR} = \frac{\text{dose @ depth } n}{\text{Dose @ } d_{max}} = \frac{D_n}{D_{max}} \tag{14.5}$$

Like TAR, there is no dependence for TMR on SSD, since we are making measurements at the same distance from the source. TAR and TMR depend only on patient attenuation. Like TAR, if TMR is determined at 1 SAD, but treatment takes place at another, we'll need to use the inverse square law to perform treatment time calculations. For example, the treatment time equation (Equation 14.4) can be adapted

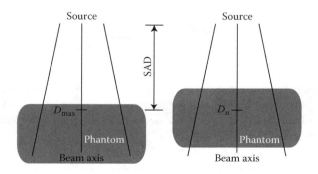

FIGURE 14.7   Determination of tissue maximum ratio.

to tissue maximum ratio by replacing depth dose with TMR and adding the inverse square law as an adjustment to TMR, as shown in Equation 14.6:

$$N_{mu} = \frac{\text{Tumor Dose}}{C_{cal} \times C_{fs} \times C_{attn} \times \text{TMR} \times \dfrac{1}{\left(\dfrac{SAD_{treat}}{SAD_{cal}}\right)^2}} \tag{14.6}$$

SAD_{treat} is the source to tumor distance during treatment, and SAD_{cal} is the source to dosimeter distance during calibration. Equation 14.6 can be rearranged to a more tractable form:

$$N_{mu} = \frac{\text{Tumor Dose}}{C_{cal} \times C_{fs} \times C_{attn} \times \text{TMR}} \times \left(\frac{SAD_{treat}}{SAD_{cal}}\right)^2$$

Example 14.8

Determine the treatment time (in monitor units) required to deliver a 250 cGy dose to a tumor 10 cm below the skin exposed to an 11×11 cm^2 6 MV field at 100 cm SSD. This accelerator is calibrated at SAD = 105 cm to deliver 1.007 cGy/mu under these conditions. A 0.6 cm Plexiglas®[2] tray is also placed in the beam ($C_{attn} = 0.97$). The 15×15 cm^2 field size correction factor for this machine is 1.031. The TMR values are 0.800 (for a 10×10 cm^2 field) and 0.810 (for a 12×12 cm^2 field).

The values of C_{fs} and TMR need to be interpolated. Remember that C_{fs} for a 10×10 cm^2 field is exactly 1.000:

$$C_{fs} = 1.000 + \frac{1}{5}(1.031 - 1.000) = 1.006$$

$$\text{TMR} = 0.800 + \frac{1}{2}(0.810 - 0.800) = 0.805$$

Okay, so the second interpolation could've been done in your head. At any rate, we're now ready to plug into Equation 14.6. Note that $SAD_{treat} = SSD + n = 110$ cm:

$$N_{mu} = \frac{250 \text{ cGy}}{1.007 \dfrac{cGy}{mu} \times 1.006 \times 0.97 \times 0.805} \times \left(\frac{110 \text{ cm}}{105 \text{ cm}}\right)^2 = 347 \text{ mu}$$

Finally, it should be noted that the doses calculated in the examples in this chapter are typically given over several treatments spread over several days to weeks. This is called **dose fractionation**, and it is done to allow the body time to recover from these relatively high doses.

---

[2] Plexiglas is poly(methyl methacrylate) and is a registered trademark of ELF Atochem, a subsidiary of Arkema in the United States. It is a trademark of Evonik Röhm GmbH elsewhere.

14.4 ISODOSE CURVES

Up to this point, we've been looking at dosimetry in only one dimension—along the line defined by the beam axis (the center of the beam). We can easily go to two dimensions by looking at **isodose curves**: two-dimensional dose distribution lines representing equal absorbed dose (Figure 14.8). The numbers in Figure 14.8 indicate the percentage of the maximum dose. They are like the lines you sometimes see on weather maps indicating lines of equal temperature (isotherms) or equal pressure (isobars—not to be confused with nuclides with the same mass number, or drinking establishments that look alike). Basically, you imagine looking at the patient's cross section and draw lines where you see specific dose values.

The isodose curves in Figure 14.8 have good flatness and symmetry (Section 13.2). Notice also that they spread further apart as the dose decreases, that is, 50% is further from 60% than 80% is from 90%. The extent that they are spread depends on the energy of the beam. Isodose curves for lower-energy beams are more bunched together, while those for higher-energy beams are more spread out. This is again a result of higher-energy beams having a lower probability of interacting with matter, depositing more dose deeper within the patient.

For sub-megavoltage machines, these curves will also show a wider penumbra, since nonaccelerator instruments generally have a larger source (focal spot). Penumbra would appear in Figure 14.8 as more angle in the parts of the curve that parallel the beam axis (more width in the curves if the beam axis were vertical). Sub-megavoltage isodose curves also show more bulging, that is, the beam is wider, at greater depths. This is because lower-energy photons have a greater probability of scattering through larger angles.

If you look closely at the higher dose percentage curves in Figure 14.8, you'll notice they have horns (greater dose rates at the beam edge, Section 13.2). This comes from making the flattener a little too thick in the center, and it is quite deliberate. As the X-ray beam travels through the patient, more and more photons are scattered. Because there are no photons outside of the beam (beyond the edge), more photons are scattered into the center of the beam than are scattered into the edge. As a result, the isodose curves flatten out at greater depth. Since flatness is usually more desirable at depth (where treatment takes place), the flatteners are designed to initially overcompensate, attenuating more photons in the center of the beam.

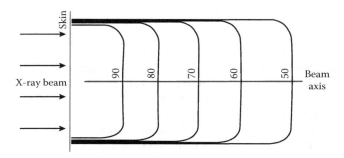

FIGURE 14.8 Isodose curves.

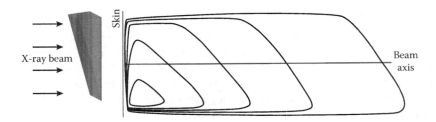

FIGURE 14.9   Shaping dose with a wedge.

How are isodose curves measured? If the phantom is just a box of water, it's pretty easy to move the detector around within it to map out the doses in two (or three) dimensions. You could also use an array of detectors inside a solid (or liquid) phantom and measure everything at once.

We can change the shape of the isodose curves by placing a **wedge** filter in the beam, as shown in Figure 14.9. The filter itself is simply a chunk of metal in the shape of a wedge—it attenuates the beam more dramatically where it is thicker, and less so where it is thin. This is pretty handy when the tumor is thicker in some parts than others. Similar sorts of dose shaping can also be done by adjusting the collimators during treatment, such as is done with intensity-modulated radiation therapy (IMRT, Section 8.2).

In this section, we've made the jump from seeing dose in one dimension into two dimensions, but in reality, doses are delivered in three dimensions. Instead of points or lines, we should really try to visualize isodose surfaces. It would be a bit like a bunch of identical drinking glasses (or plastic food storage boxes) stacked one inside of another. The glasses (or plastic) represent the isodose surfaces. If viewed from the side, their outlines would look like Figure 14.8.

14.5   MOVING FIELDS

Radiation therapy photon sources generally move around a patient, maximizing tumor dose while minimizing dose elsewhere. As already mentioned, isocentric techniques would be much simpler if humans were perfect cylinders and tumors were always located along the central axis. Unfortunately, that isn't normally the case. Tumors locate where they please, and in cross section, humans are a bit on the oval side.

Reality is more like Figure 14.10. The human is represented in cross section as an irregular oval, with the tumor located to the left and down a bit. Our problem is then to figure out what the correct dose will be if it is to be delivered along two or more of the arrows drawn on Figure 14.10. We also have the option of rotating the beam continuously around all or part of the patient. The solution is to place the tumor at the isocenter of the instrument and measure how deep the tumor is at several angles around the patient, determine TMR (or TAR) for each angle, and average the results. Note that with the tumor at the isocenter, SAD remains constant, while SSD varies. Moving field calculations are greatly simplified by using tissue maximum ratios rather than percent depth dose values.

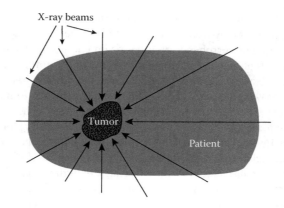

X-ray beams

Tumor

Patient

FIGURE 14.10 Isocentric therapy.

To minimize dose to healthy tissue, treatments are planned for a whole number of rotations for a specific therapy session. If this isn't possible, you can pick up where you left off at the next session.

14.6 PROTON AND ELECTRON BEAM DOSIMETRY

As mentioned in Section 8.2, electron beams are pretty easy to come by if you already have a linear accelerator X-ray therapy unit. Looking at Figures 13.13 and 13.14, you should realize that if the target and the flattener are removed, an electron beam hits the patient instead of X-rays. The electron beam produced in these linear accelerators is very narrow and needs to be broadened and flattened for therapy. This is done by placing two scattering foils in the beam—the first spreads out the beam, and the second flattens it. These foils are thin pieces of metal—remember that electrons have a higher probability of interacting with matter than X-ray photons. If the foils are made too thick, they'd be no different than the target, and they'd just make X-rays.

Because of the electrons' tendency to scatter, additional collimation (an **electron applicator** or **cone**) must be placed between the jaws (photon beam collimator) and the patient. This collimator is made from low Z material, which minimizes bremsstrahlung as the stray electrons slam into it. Because of its use, patients are typically treated with only one field (machine position relative to the patient) per day—isocentric treatment can still be done, albeit with separate fields on different days. To minimize air shielding, SSD values for electron therapy are kept low, typically less than 100 cm.

Blocking can further shape electron beams. Let's say the cone produces a 10×10 cm^2 square field at the surface of the patient, but you need a 5-cm diameter circular field. Take a 10×10 cm^2 sheet of lead, and cut out a 5-cm diameter circle. This can then be placed directly on the patient's skin. A material called cerrobend (an alloy of Bi, Pb, Sn, and Cd) is sometimes used instead of pure Pb. How thick does the lead need to be? The electrons hitting the metal will create X-rays, so it needs to be thick enough to absorb most of them. As it turns out, all you need to do is divide the electron beam energy by 2 MeV/mm—that gives the shielding thickness necessary in millimeters. For cerrobend, multiply beam energy by 0.6 mm/MeV. Notice that these two different mathematical operations are quite similar in their results.

The beam can also be shaped at the patient's skin with a **bolus**. A bolus is made from material that will interact with ionizing radiation similar to human tissue and is typically pliable. There are two reasons to use a bolus: (1) The treatment volume is a little shallow for the electron beam energy available—adding a bolus effectively makes it deeper. In other words, adding a bolus brings the isodose curves closer to the patient's skin, delivering more dose closer to the surface. (2) The surface of the patient is uneven. Because electrons have a relatively high probability of interacting with matter, surface irregularities will create inhomogeneities in the doses delivered at depth. The bolus can help straighten out the isodose curves by providing a flat entry surface. Note that a bolus can also be used with high-energy X-ray beam therapy for the same reasons listed here.

Electrons are charged particles and, unlike photons, will deliver dose in the patient's body up until a point, then not much after that (Bragg peak, see Section 6.2). In other words, they have a definable range in matter. Range (R) in humans can be estimated as the energy of the electron beam (E) divided by 2.0 MeV/cm (Equation 6.10).

Some dose is delivered after this point, but this is largely due to X-rays created by the electron beam hitting things such as the collimators and scattering foils as the beam emerges from the accelerator. This is often referred to as "X-ray contamination" and is inevitable when using electron beams. About 5% of the patient's total dose during electron beam therapy is due to X-rays.

What do isodose curves look like for electron beams? First, they are shallower and more compressed than photon beams of "equivalent" energy. This is because electrons have a much higher probability of interacting with matter and because electrons tend to deliver more of their dose at the Bragg peak. Electron isodose curves also "balloon" out at the sides because electrons tend to scatter through larger angles than high-energy X-rays. Percent depth dose (PDD) can be estimated for electron beams using simple (approximate!) formulas. Most electron beam therapy is performed at depths corresponding to PDD values of 90% and 80%. Ninety percent of the maximum dose will be delivered at a depth ($d_{90\%}$) equivalent to the energy of the electron beam (E) divided by 4.0 MeV/cm (Equation 14.7):

$$d_{90\%} = \frac{E}{4.0\,\text{MeV/cm}} \tag{14.7}$$

while 80% of the maximum dose is delivered at a depth corresponding to the beam energy divided by 3.0 MeV/cm (Equation 14.8):

$$d_{80\%} = \frac{E}{3.0\,\text{MeV/cm}} \tag{14.8}$$

Example 14.9

Electrons in a beam have an energy of 10 MeV. At what depth will the dose be 90% of the maximum?

$$d_{90\%} = \frac{10 \text{ MeV}}{4.0 \text{ MeV/cm}} = 2.5 \text{ cm}$$

What is the range, in tissue, of these electrons?

$$R_{e^-} = \frac{10 \text{ MeV}}{2.0 \text{ MeV/cm}} = 5.0 \text{ cm}$$

Like photons, increasing the energy of the electron beam will deliver more dose deeper, about 1 cm for every 2 MeV according to Equation 6.10. Even so, the depths where significant doses can be delivered remain fairly shallow for electron therapy, meaning it is only useful for relatively superficial treatments. Its most common applications today are for relatively shallow lesions with critical stuff behind it. An example would be treating the lymph nodes near the spinal column. The spinal column can't tolerate much of a radiation dose, making photon therapy less attractive.

Electron therapy is most commonly used to augment X-ray treatment of breast cancer. Photon therapy is generally performed tangential to the chest to minimize dose to the lungs. Because electron beams can't penetrate all the way to the lungs, they can be done *en face* (French for "straight out"). Since electron therapy is done as a complement to photon therapy when treating breast cancer, it is referred to as a "breast boost."

Electron therapy can also be used for skin, lip, head, and neck cancers. Total skin irradiation (TSI, six low-dose fields designed to irradiate all skin evenly) can be used in the treatment of some cases of skin cancer. During TSI, lead shielding is placed over the eyes and nail beds, which are more strongly affected by electron beam therapy. Also, a plastic sheet is placed in the beam to provide greater scattering.

Treatment times for electron beam therapy are calculated using a formula (Equation 14.9) similar to one used for X-rays (compare with Equation 14.4):

$$N_{mu} = \frac{\text{Tumor Dose}}{O_f \times \dfrac{D_n}{D_{max}} \times C_{fs}} \tag{14.9}$$

where O_f is the **output factor**, which is a calibrated dose rate for the machine. It is pretty much the same as C_{cal}, and usually has a value of 1.00 cGy/mu. The field size correction factor (C_{fs}) also includes any blocking.

Example 14.10

Using information from Example 14.9, how long will it take to deliver a 180 cGy dose at 2.5 cm if the output factor is 1.00 cGy/mu and the field size correction is 0.987?

The depth dose is 0.90 (from Example 14.9). Plug into Equation 14.9:

$$N_{mu} = \frac{180 \text{ cGy}}{1.00 \text{ cGy/mu} \times 0.90 \times 0.987} = 203 \text{ mu}$$

Like electrons, protons are charged particles, but they are much more massive. As a result, they have a much shorter range (Table 6.2) and don't scatter very much. Because they scatter less, they also have a much sharper Bragg peak than electrons, making them an excellent choice for radiation therapy, so long as the tumor is located at or very near the Bragg peak depth. That depth can be varied with the energy of the protons—something that can even be done during treatment! The peak also drops off very quickly shortly after delivering the bulk of the dose. No worries about an exit dose here—in fact, there's very little dose delivered beyond the peak (no pesky X-rays as with electrons). Because of the sharp Bragg peak, multifield treatments with proton beams can yield very tight dose distributions—much better than photons.

It might be a good idea to remember how protons interact with matter (Section 6.2). Primarily, they interact with electrons—knocking them loose from their orbitals. These secondary electrons have a very short range and do their damage locally. Protons can also interact with nuclei, but typically this doesn't happen until they slow down quite a bit (in or near the Bragg peak). When protons crash into a nucleus, a nuclear reaction takes place, and fragments are scattered—again, typically doing biological damage very close to where the nucleus was originally.

Protons can also be scattered by nuclei through Coulomb repulsions. This causes the beam to widen as it penetrates the matter—creating penumbra. The amount of scattering depends on the energy of the proton beam. Higher-energy protons are more likely to scatter and give larger penumbra, while lower-energy protons have smaller penumbra. We won't worry about why this is so.

Because protons have a much higher probability of interacting with matter than photons, tissue inhomogeneities (e.g., bones) are a bigger deal. It is important to have good anatomical data (computed tomography scans) before beginning treatment. With these data in hand, **compensators** (beam filters) are designed for each patient (Figure 14.11) to properly "sculpt" the beam.

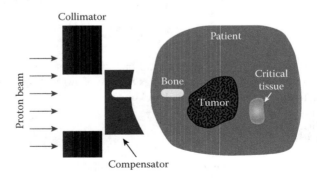

FIGURE 14.11   Sculpting a proton beam.

The proton beam is first shaped by the collimators, which set the size of the beam to the approximate size of the tumor. The compensator then adjusts the proton flux according to its thickness—the thicker parts attenuate the proton beam to a greater degree. Notice how the right surface of the compensator generally mirrors that of the patient's left side. This way almost all protons in the beam travel the same distance through compensator and tissue. This places the Bragg peak at the same depth at every point in the beam. In other words, this makes for a relatively flat isodose curve on the backside of the tumor, protecting the critical tissue beyond. As you might guess, compensators are made from low Z materials to mimic proton interactions in tissues. A thicker compensator at the beam edges also minimizes penumbra by minimizing proton flux at the beam edges. Scattering from protons initially traveling closer to the center of the beam tends to even the dose inside the patient.

By taking a chunk out of the compensator for the part of the beam that must pass through the bone, we can intensify the flux enough to even out the dose through that part of the tumor. Note that the removed chunk needs to be just the right size and perfectly aligned to properly dose the tumor and avoid excessive dose to healthy tissues beyond the tumor.

Like the photon and electron beams produced by linear accelerator beams, proton beams need to be broadened. Traditionally, this was done using scattering foils (just like broadening electron beams). These foils are thin pieces of high Z metals. This method is called **passive scattering**. Once broadened, a collimator trims the beam, and the compensator shapes its intensity.

Before scattering, the energy of the protons is varied using a **range shifter** (also known as range modulator). By varying the beam energy, the depth of the Bragg peak can be moved forward and backward (along the beam axis—left and right on Figure 14.11) through the tumor. A problem with this setup is that beam modulation is only one-dimensional. By shaping the compensator properly, we can get a very nice dose cutoff on the backside of the tumor, but using a range shifter means that the shape of the isodose curve on the nearside of the tumor will have the same shape as on the backside. If the tumor lacks this symmetry (very likely!), a significant dose will be delivered to healthy tissue on the nearside.

The solution to this problem is **active scattering**, as illustrated in Figure 14.12. Passing a beam of charged particles through a magnetic field will cause the beam to deflect. The degree of deflection depends on the size and charge of the particle and

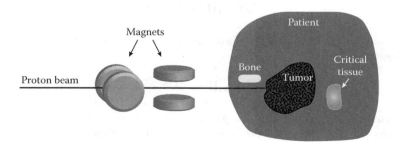

FIGURE 14.12   Active scattering of a proton beam.

the strength of the magnetic field. Two pairs of parallel magnets, offset 90° from each other, are used to move the Bragg peak in and out of the page (the magnets on the left) and up and down on the page (the magnets on the right). The strength of the magnetic fields is varied to allow continuous movement through the tumor in these two dimensions. Further upstream, the energy of the beam is dynamically varied, so the Bragg peak can be moved right and left inside of the tumor as pictured. This allows placement of the Bragg peak at any point, in three dimensions, within the tumor and minimizes dose outside. If both sets of magnets are switched off, the proton beam will travel straight through. Notice also that no compensator, and generally no collimator, is needed with active scattering, just a lot of skill in changing proton energies and magnetic strengths in ways that provide an even dose throughout the tumor while minimizing dose to healthy tissues.

Proton therapy has proven especially effective in treating cancer in the eye, skull base sarcomas, and prostate cancer. Like electron therapy, it can sometimes be used effectively in conjunction with photon therapy.

It sounds like proton therapy is pretty good stuff, yet there are only a couple dozen treatment centers worldwide. Why isn't it more commonly used? First, it can only treat stuff that's fairly close to the surface (protons like to interact with matter). Also, as mentioned in Section 8.2, the equipment needed takes up a lot of space and requires some serious geeks to operate and maintain. The high-energy protons are generated in a cyclotron, which requires a fair bit of shielding. These protons have a lot of velocity, and it is hard to get the beam to turn. Really big magnets are necessary (don't take your credit card!). Because of the size of the cumbersome magnets, rotation of the gantry is limited to 180° or less (take a look at Figure 8.4). Like linear accelerators, this gantry must have a great deal of precision in its movements. Typically beam variances at the isocenter must be less than 0.5 mm. Engineering this level of precision into an instrument this large does not come cheaply.

Because of the size requirements, a fair amount of space is required for a treatment facility. In addition to the cyclotron and its shielding, the beam lines and treatment rooms need to be heavily shielded. Constructing a proton therapy unit is roughly three times more expensive than a conventional photon therapy wing. Also, some radioactive contamination will develop wherever protons are used, from incidental nuclear reactions caused by the energetic protons. For example, a high-energy proton slamming into a ^{14}N atom (common in the air) creates ^{14}O and a neutron:

$$^{14}_{7}N + {}^{1}_{1}H \rightarrow {}^{14}_{8}O + {}^{1}n$$

Likewise, protons can react with ^{56}Fe (commonly used in construction) to create ^{56}Co and a neutron:

$$^{56}_{26}Fe + {}^{1}_{1}H \rightarrow {}^{56}_{27}Co + {}^{1}n$$

The neutrons created in these two p,n reactions can go on to cause other nuclear reactions. Note also that the products in these two reactions are unstable, decaying via positron emission and electron capture, which is expected, since both are proton-rich.

QUESTIONS

14.1 The PDD is 70.7% at 9.0 cm for an X-ray beam from a 6 MV linac. What dose is required at d_{max} (1.5 cm) to deliver 250 cGy to a tumor at 9 cm?

14.2 The d_{max} for a 10 MV linear accelerator is only 2.5 cm. How can we treat tumors at greater depth without damaging the healthy tissue surrounding it?

14.3 Determine the percent contributions of primary and secondary radiation at 19 cm depth for a 20 MV, 15×15 cm^2 field at 100 cm SSD. Percent depth doses are 53.5% for a "0" field and 55.0% for a 15×15 cm^2 field. Repeat for a 6 MV machine. PDD values are now 39.2% and 43.9%, respectively. Can you explain the difference between the two linear accelerators?

14.4 Calculate the percent depth dose at 7 cm depth for an X-ray beam from a 10 MV linear accelerator at 125 cm SSD. For a 10 MV beam $d_{max} = 2.5$ cm and the PDD at 100.0 cm SSD = 84.0%.

14.5 A patient, 23 cm thick at the area of treatment, needs a 215 cGy dose delivered to a tumor 5.5 cm below his skin. Using a 20 MV machine at 100 cm SSD and a 12×12 cm^2 field, determine the dose at d_{max} and the exit dose. How deep is d_{max} for this machine? The PDD values in Table 14.1 apply only to the machine as configured in this problem.

14.6 Using a PDD, determine the treatment time (in mu) to deliver 200 cGy to a tumor 8 cm below the surface of a patient exposed to a 7×7 cm^2 6 MV field at 150 cm SSD. This accelerator is calibrated to deliver 0.998 cGy/mu, has a field size correction factor of 0.978, an equipment attenuation factor of 0.95, and a percent depth dose of 74.6% at 8 cm depth and 100 cm SSD. The maximum dose is delivered at 1.5 cm.

14.7 What is the relationship between TAR and BSF? What is the relationship between TAR and TMR?

14.8 What happens to TAR and TMR values when SSD is decreased? Briefly explain.

14.9 Calculate the number of monitor units necessary to deliver 245 cGy to a tumor 6.5 cm below the patient's skin using a 15×15 cm^2 field on a 10 MV machine at 100 cm SAD. This accelerator is calibrated to deliver 1.02 cGy/mu at 100 cm SAD and has a TMR value of 0.932 under these conditions. The 15×15 cm^2 field size correction factor for this machine is 1.034.

14.10 Calculate the dose at d_{max} (2.5 cm) and the exit dose (at 15.0 cm, TMR = 0.741) for the previous problem.

14.11 Determine the treatment time (in mu) required to deliver a 225 cGy dose to a tumor 7.3 cm below the skin exposed to an 7×7 cm^2 6 MV field at 110 cm SSD. This accelerator is calibrated at SAD = 100 cm to deliver 1.001 cGy/mu under these conditions. The 7×7 cm^2 field size correction factor for

TABLE 14.1
Percent Depth Dose Values for Question 14.5

Depth (cm)	3.0	5.0	6.0	20.0	30.0
PDD	100.0	94.7	91.5	52.7	35.2

this machine is 0.978. The TMR values are 0.871 (at 7 cm depth) and 0.838 (at 8 cm). A 0.6 cm poly(methyl methacrylate) tray is also placed in the beam.

14.12 Define the following: secondary radiation, field symmetry, SAD, tissue–air ratio, and TMR.

14.13 Why are high Z metals used for blocking electron beams?

14.14 Why is X-ray production "inevitable" when using electron beams?

14.15 Give one reason why electron therapy would be preferred to photon therapy. Also give a reason why electron therapy might not be used.

14.16 Briefly explain the differences observed in the electron and photon isodose curves produced by the same MV machine. In terms of energy, what is the fundamental difference between these two beams?

14.17 What is the range, in tissue, of 12 MeV electrons? What treatment time (mu) will be required for a 170 cGy dose to be delivered 3 cm below the surface with a 12 MeV electron beam? The output factor is 1.00 cGy/mu, and the field size correction is 0.985 for this beam.

14.18 Why doesn't C_{attn} appear in Equation 14.9?

14.19 What would be the effect of taking too big of a chunk out of the compensator to adjust proton flux for a bone between the skin and the tumor along the proton beam path? What if the chunk were properly sized but not properly aligned?

14.20 Give one reason why proton therapy would be preferred to photon therapy. Also give a reason why proton therapy might not be used.

14.21 Why is it important to sculpt a proton beam for treatment?

Appendix A: Nuclide Data

Element	Z	N	Nuclide Symbol	Mass (u)[a]	Percent Abundance[b]	Half-Life	Decay Mode[c]	Particle Energy (MeV)[d]	Gamma Energy (keV)[e]	Total Energy of Decay (MeV)
Hydrogen	1	0	^{1}H	1.007825	99.9885	Stable				
		1	^{2}H	2.014102	0.0115	Stable				
		2	^{3}H	3.016049	None	12.32 a	β−	0.018591		0.018591
		3	^{4}H	4.0278	None	1.40E−22 s	n			2.91
Helium	2	1	^{3}He	3.016029	0.000134	Stable				
		2	^{4}He	4.002603	99.999866	Stable				
		3	^{5}He	5.0122	None	7.0E−22 s	n			
		4	^{6}He	6.018888	None	807 ms	β−	3.510		3.508
Lithium	3	3	^{6}Li	6.015123	7.59	Stable				
		4	^{7}Li	7.016005	92.41	Stable				
		5	^{8}Li	8.022487	None	0.840 s	β−	13		16.0052
Beryllium	4	3	^{7}Be	7.016929	None	53.3 d	EC		478	0.8619
		4	^{8}Be	8.005305	None	7.0E−17 s	α	0.0461		
		5	^{9}Be	9.012182	100	Stable				
		6	^{10}Be	10.013534	None	1.5E+06 a	β−	0.556		0.556
Boron	5	4	^{9}B	9.013329	None	8E−19 s	αα	0.0461		
		5	^{10}B	10.012937	19.9	Stable				
		6	^{11}B	11.009305	80.1	Stable				
		7	^{12}B	12.014352	None	20.20 ms	β−	13.37	4439	13.369
		8	^{13}B	13.017780	None	17.4 ms	β−	13.4	3684	13.437
		9	^{14}B	14.02540	None	13 ms	β−	14	6092	20.64
Carbon	6	4	^{10}C	10.016853	None	19.308 s	β+	1.87	718	3.6480
		5	^{11}C	11.011434	None	20.36 min	β+	0.960		1.982
							EC			

(Continued)

Element	Z	N	Nuclide Symbol	Mass (u)[a]	Percent Abundance[b]	Half-Life	Decay Mode[c]	Particle Energy (MeV)[d]	Gamma Energy (keV)[e]	Total Energy of Decay (MeV)
	6	6	^{12}C	12.000000	98.93	Stable				
		7	^{13}C	13.003355	1.07	Stable				
		8	^{14}C	14.003242	None	5715 a	β−	0.157		0.156476
Nitrogen	7	6	^{13}N	13.005739	None	9.97 min	β+	1.190		2.2205
		7	^{14}N	14.003074	99.636	Stable				
		8	^{15}N	15.000109	0.364	Stable				
		9	^{16}N	16.006101	None	7.13 s	β−	4.27	6129	10.421
								10.44	7115	
		12	^{19}N	19.01702	None	0.27 s	β− + n			12.5
							β−	4.06	96	
Oxygen	8	4	^{12}O	12.03440	None	1E−21 s	2p			1.77
		7	^{15}O	15.003065	None	2.037 min	β+	1.72		2.7542
							EC			
		8	^{16}O	15.994915	99.757	Stable				
		9	^{17}O	16.999132	0.038	Stable				
		10	^{18}O	17.999161	0.205	Stable				
		11	^{19}O	19.003579	None	26.9 s	β−	3.3	197	4.822
								4.60	1357	
		12	^{20}O	20.004076	None	13.5 s	β−	2.75	1057	3.815
Fluorine	9	8	^{17}F	17.002095	None	1.08 min	β+	1.74		2.7605
		9	^{18}F	18.000938	None	1.8293 h	β+	0.635		1.655
							EC			

(Continued)

Element	Z	N	Nuclide Symbol	Mass (u)[a]	Percent Abundance[b]	Half-Life	Decay Mode[c]	Particle Energy (MeV)[d]	Gamma Energy (keV)[e]	Total Energy of Decay (MeV)
		10	^{19}F	18.998403	100	Stable				
		11	^{20}F	19.999981	None	11.1 s	β−	5.39	1634	7.0245
		16	^{25}F	25.01210	None	0.07 s	β−		1703	13.4
Neon	10	8	^{18}Ne	18.005697	None	1.667 s	β+	3.42	1041	4.443
		9	^{19}Ne	19.001880	None	17.22 s	β+	2.24		3.2388
							EC		110	
									1357	
		10	^{20}Ne	19.992440	90.48	Stable				
		11	^{21}Ne	20.993847	0.27	Stable				
		12	^{22}Ne	21.991385	9.25	Stable				
		13	^{23}Ne	22.994467	None	37.1 s	β−	4.38	440	4.3758
								3.95		
		15	^{25}Ne	24.99779	None	0.61 s	β−	7.2	90	7.25
								6.3	980	
Sodium	11	11	^{22}Na	21.994437	None	2.604 a	β+	0.546	1275	2.8423
							EC			
		12	^{23}Na	22.989769	100	Stable				
		13	^{24}Na	23.990963	None	14.97 h	β−	1.391	1369	5.5155
									2754	
		14	^{25}Na	24.989954	None	59.3 s	β−	3.8	975	3.835
									585	
									390	
									1612	

(Continued)

Element	Z	N	Nuclide Symbol	Mass (u)[a]	Percent Abundance[b]	Half-Life	Decay Mode[c]	Particle Energy (MeV)[d]	Gamma Energy (keV)[e]	Total Energy of Decay (MeV)
Magnesium	12	11	^{23}Mg	22.994125	None	11.32 s	β+	3.09	440	4.056
		12	^{24}Mg	23.985042	78.99	Stable				
		13	^{25}Mg	24.985837	10.00	Stable				
		14	^{26}Mg	25.982593	11.01	Stable				
		15	^{27}Mg	26.984341	None	9.45 min	β−	1.75 1.59	844 1014	2.6100
		16	^{28}Mg	27.983877	None	21.0 h	β−	0.459	31 1342	1.832
		28	^{40}Mg	27.983877	None	20.91 h	β−	1.832		21
Aluminum	13	12	^{25}Al	24.990429	None	7.17 s	β+	3.26	1612	4.2767
		13	^{26}Al	25.986892	None	7.1E+05 a	β+ EC	1.17	1809	4.0043
		14	^{27}Al	26.981539	100	Stable				
		15	^{28}Al	27.981910	None	2.25 min	β−	2.86	1779	4.6424
Silicon	14	11	^{25}Si	25.00411	None	220 ms	β+		452 493 1612 945	12.74
		13	^{27}Si	26.986705	None	4.14 s	β+	3.85	2210	4.8124
		14	^{28}Si	27.976927	92.223	Stable				
		15	^{29}Si	28.976495	4.685	Stable				
		16	^{30}Si	29.973770	3.092	Stable				
		17	^{31}Si	30.97536	None	2.62 h	β−	1.48	1266	1.4919

(Continued)

Element	Z	N	Nuclide Symbol	Mass (u)[a]	Percent Abundance[b]	Half-Life	Decay Mode[c]	Particle Energy (MeV)[d]	Gamma Energy (keV)[e]	Total Energy of Decay (MeV)
		18	32Si	31.974148	None	1.6E+02 a	β−	0.221		0.2243
Phosphorus	15	15	30P	29.978314	None	2.50 min	β+	3.24		4.2324
							EC		2235	
		16	31P	30.973762	100	Stable				
		17	32P	31.973907	None	14.28 d	β−	1.709		1.7105
		18	33P	32.971725	None	25.3 d	β−	0.249		0.249
Sulfur	16	15	31S	30.979554	None	2.56 s	β+	4.39	1266	5.396
		16	32S	31.972071	94.99	Stable				
		17	33S	32.971459	0.75	Stable				
		18	34S	33.967867	4.25	Stable				
		19	35S	34.969032	None	87.2 d	β−	0.1674		0.1672
		20	36S	35.967081	0.01	Stable				
		21	37S	36.971126	None	5.05 min	β−	1.76	3104	4.8652
Chlorine	17	17	34mCl	33.973919	None	32.2 min	β+	2.5	2128	5.638
								1.3	1176	
									3304	
							IT		146	0.146
		17	34Cl	33.973762	None	1.527 s	β+	4.47		5.4920
		18	35Cl	34.968853	75.76	Stable				
		19	36Cl	35.968307	None	3.01E+05 a	β−	0.709		0.7097
							EC			
							β+	0.12		1.1422
		20	37Cl	36.965903	24.24	Stable				

(Continued)

Element	Z	N	Nuclide Symbol	Mass (u)[a]	Percent Abundance[b]	Half–Life	Decay Mode[c]	Particle Energy (MeV)[d]	Gamma Energy (keV)[e]	Total Energy of Decay (MeV)
		21	38Cl	37.968011	None	37.2 min	β−	4.91	2168	4.9165
								1.11	1642	
Argon	18	17	35Ar	34.975257	None	1.775 s	β+	4.943	1219	5.966
									1763	
									2694	
		18	36Ar	35.967545	0.3365	Stable				
		19	37Ar	36.966776	None	35.0 d	EC			0.8139
		20	38Ar	37.962732	0.0632	Stable				
		21	39Ar	38.964313	None	269 a	β−	0.565		0.565
		22	40Ar	39.962383	99.6003	Stable				
		23	41Ar	40.964501	None	1.83 h	β−	1.198	1294	2.4916
Potassium	19	18	37K	36.973377	None	1.23 s	β+	5.13	2796	6.1475
		19	38K	37.969080	None	7.63 min	β+	2.68	2168	5.9139
		20	39K	38.963707	93.2581	Stable				
		21	40K	39.963999	0.0117	1.25E+09 a	β−	1.33	1461	1.3111
							EC			1.5047
							β+			
		22	41K	40.961826	6.7302	Stable				
		23	42K	41.962403	None	12.36 h	β−	3.52	1525	3.5255
		24	43K	42.960716	None	22.3 h	β−	0.83	373	1.82
									618	
Calcium	20	20	40Ca	39.962591	96.941	Stable				
		21	41Ca	40.962278	None	1.03E+05 a	EC			0.4213

(Continued)

Element	Z	N	Nuclide Symbol	Mass (u)a	Percent Abundanceb	Half-Life	Decay Modec	Particle Energy (MeV)d	Gamma Energy (keV)e	Total Energy of Decay (MeV)
		22	^{42}Ca	41.958618	0.647	Stable				
		23	^{43}Ca	42.958767	0.135	Stable				
		24	^{44}Ca	43.955482	2.086	Stable				
		25	^{45}Ca	44.956186	None	162.7 d	β^-	0.258	12	0.256
		26	^{46}Ca	45.953693	0.004	Stable				
		27	^{47}Ca	46.954547	None	4.536 d	β^-	0.694 1.990	1297	1.992
		28	^{48}Ca	47.952534	0.187	4E+19 a	$\beta^-\beta^-$			4.272
		29	^{49}Ca	48.955673	None	8.72 min	β^-	2.18	3084	5.263
Scandium	21	23	^{44}Sc	43.959403	None	3.93 h	β^+ EC	1.47	1157	3.652
		24	^{45}Sc	44.955912	100	Stable				
		25	^{46}Sc	45.955170	None	83.81 d	β^-	0.357	1121 889	2.366
		26	^{47}Sc	46.952408	None	3.349 d	β^-	0.439 0.600	159	0.600
Titanium	22	22	^{44}Ti	43.9596902	None	59.2 a	EC		78 68	0.268
		23	^{45}Ti	44.958124	None	3.078 h	β^+ EC	1.04	720 1408	2.062
		24	^{46}Ti	45.952632	8.25	Stable				
		25	^{47}Ti	46.951763	7.44	Stable				

(Continued)

Element	Z	N	Nuclide Symbol	Mass (u)[a]	Percent Abundance[b]	Half-Life	Decay Mode[c]	Particle Energy (MeV)[d]	Gamma Energy (keV)[e]	Total Energy of Decay (MeV)
		26	^{48}Ti	47.947946	73.72	Stable				
		27	^{49}Ti	48.947870	5.41	Stable				
		28	^{50}Ti	49.944791	5.81	Stable				
		29	^{51}Ti	50.946616	None	5.76 min	β−	2.14	320	2.474
Vanadium	23	25	^{48}V	47.952255	None	15.98 d	β+	0.694	984	4.012
							EC		1312	
		26	^{49}V	48.948517	None	331 d	EC			0.602
		27	^{50}V	49.947158	0.250	1.4E+17 a	EC	1553.8		2.205
							β−		783	1.0379
		28	^{51}V	50.943960	99.750	Stable				
		29	^{52}V	51.944780	None	3.76 min	β−	2.47	1434	3.976
Chromium	24	24	^{48}Cr	47.954036	None	21.6 h	EC		308	1.66
							β+		112	
		25	^{49}Cr	48.951341	None	42.3 min	β+	1.39		2.626
							EC	1.45	91	
									153	
									62	
		26	^{50}Cr	49.946044	4.345	Stable				
		27	^{51}Cr	50.944772	None	27.702 d	EC		320	0.7526
		28	^{52}Cr	51.940507	83.789	Stable				

(Continued)

Element	Z	N	Nuclide Symbol	Mass (u)[a]	Percent Abundance[b]	Half-Life	Decay Mode[c]	Particle Energy (MeV)[d]	Gamma Energy (keV)[e]	Total Energy of Decay (MeV)
		29	^{53}Cr	52.940649	9.501	Stable				
		30	^{54}Cr	53.938880	2.365	Stable				
		31	^{55}Cr	54.940844	None	3.497 min	β−	2.49	1528	2.6031
Manganese	25	28	^{53}Mn	52.941295	None	3.7E+06 a	EC			0.5968
		29	^{54}Mn	53.940363	None	312.1 d	EC		835	1.377
							β−			0.697
							β+			1.377
		30	^{55}Mn	54.938045	100	Stable				
		31	^{56}Mn	55.938909	None	2.578 h	β−	2.84	847	3.6956
								1.04	1811	
									2113	
Iron	26	26	^{52}Fe	51.94812	None	8.28 h	β+	0.80	169	2.37
							EC		169	
		27	^{53}Fe	52.945312	None	8.51 min	β+	2.8		3.743
								2.4		
							EC		378	
		28	^{54}Fe	53.939611	5.845	Stable				
		29	^{55}Fe	54.938298	None	2.75 a	EC		126	0.2312
		30	^{56}Fe	55.934937	91.754	Stable				
		31	^{57}Fe	56.935394	2.119	Stable				
		32	^{58}Fe	57.933276	0.282	Stable				
		33	^{59}Fe	58.934880	None	44.50 d	β−	0.466	1099	1.565
								0.274	1292	

(Continued)

Element	Z	N	Nuclide Symbol	Mass (u)[a]	Percent Abundance[b]	Half-Life	Decay Mode[c]	Particle Energy (MeV)[d]	Gamma Energy (keV)[e]	Total Energy of Decay (MeV)
Cobalt	27	40	60Fe	59.934077	None	2.62E+06 a	β-	0.14	59	0.237
		30	57Co	56.936296	None	271.8 d	EC		122	0.836
									137	
									14	
		31	58Co	57.935758	None	70.88 d	EC			2.308
							β+	0.474	811	
		32	59Co	58.933195	100	Stable				
		33	60mCo	59.933885	None	10.47 min	IT		59	0.059
							β-	1.6	1333	2.883
		33	60Co	59.933822	None	5.271 a	β-	0.318	1333	2.8231
									1173	
Nickel	28	27	55Ni	54.95134	None	202 ms	β+	7.66		8.69
		28	56Ni	55.94214	None	5.9 d	EC		158	2.14
									812	
		29	57Ni	56.939800	None	35.6 h	EC		1378	3.262
							β+	0.85		
		30	58Ni	57.935343	68.0769	Stable				
		31	59Ni	58.934352	None	7.6E+04 a	EC			1.0728
							β+			
		32	60Ni	59.930786	26.2231	Stable				
		33	61Ni	60.931056	1.1399	Stable				
		34	62Ni	61.928345	3.6345	Stable				
		35	63Ni	62.929673	None	101 a	β-	0.0669		0.06698

(Continued)

Element	Z	N	Nuclide Symbol	Mass (u)[a]	Percent Abundance[b]	Half–Life	Decay Mode[c]	Particle Energy (MeV)[d]	Gamma Energy (keV)[e]	Total Energy of Decay (MeV)
		36	^{64}Ni	63.927966	0.9256	Stable				
		37	^{65}Ni	64.930088	None	2.517 h	β⁻	2.14	1482	2.138
								0.65	1116	
Copper	29	32	^{61}Cu	60.933462	None	3.35 h	β⁺	1.21		2.237
							EC		283	
									656	
		33	^{62}Cu	61.932587	None	9.74 min	β⁺	2.93		3.948
							EC		1173	
									876	
		34	^{63}Cu	62.929597	69.15	Stable				
		35	^{64}Cu	63.929768	None	12.701 h	EC		1346	1.675
							β⁻	0.578		0.579
							β⁺	0.651		1.6750
		36	^{65}Cu	64.927789	30.85	Stable				
		37	^{66}Cu	65.928873	None	5.10 min	β⁻	2.63	1039	2.641
Zinc	30	32	^{62}Zn	61.93433	None	9.22 h	EC			1.63
							β⁺	0.66	597	
									41	
									548	
									508	
		33	^{63}Zn	62.933216	None	38.5 min	β⁺	2.32		3.367
							EC		670	
									962	

(Continued)

Element	Z	N	Nuclide Symbol	Mass (u)[a]	Percent Abundance[b]	Half-Life	Decay Mode[c]	Particle Energy (MeV)[d]	Gamma Energy (keV)[e]	Total Energy of Decay (MeV)
		34	^{64}Zn	63.929142	48.268	Stable				
		35	^{65}Zn	64.929245	None	244.0 d	EC		1116	1.3521
							β+	0.325		
		36	^{66}Zn	65.926033	27.975	Stable				
		37	^{67}Zn	66.927127	4.102	Stable				
		38	^{68}Zn	67.924844	19.024	Stable				
		38	^{69m}Zn	68.927025	None	13.76 h	IT		439	0.439
							β−		574	1.345
		39	^{69}Zn	68.926554	None	56 min	β−	0.90	319	0.910
		40	^{70}Zn	69.925319	0.631	Stable				
		42	^{72}Zn	71.926861	None	46.5 h	β−	0.30	145	0.46
Gallium	31	36	^{67}Ga	66.928205	None	3.2611 d	EC		93	1.001
									185	
									300	
		37	^{68}Ga	67.927984	None	1.130 h	β+	1.899		2.921
							EC		1077	
		38	^{69}Ga	68.925574	60.108	Stable				
		39	^{70}Ga	69.926028	None	21.14 min	β−	1.65	1039	1.653
									176	
		40	^{71}Ga	70.924701	39.892	Stable	EC			0.655
		41	^{72}Ga	71.926370	None	14.10 h	β−	0.97	834	3.997
								0.74	2202	
									630	

(Continued)

Element	Z	N	Nuclide Symbol	Mass (u)[a]	Percent Abundance[b]	Half-Life	Decay Mode[c]	Particle Energy (MeV)[d]	Gamma Energy (keV)[e]	Total Energy of Decay (MeV)
		42	^{73}Ga	72.925170	None	4.87 h	β−	1.2	297	1.598
									326	
									53	
									13	
Germanium	32	32	^{64}Ge	63.9416	None	1.06 min	β+	3.0		4.48
							EC		427	
									667	
									128	
		36	^{68}Ge	67.928097	None	270.8 d	EC			0.11
		37	^{69}Ge	68.927972	None	1.63 d	EC			2.227
							β+	1.21	1107	
									574	
									872	
		38	^{70}Ge	69.924247	20.38	Stable				
		39	^{71}Ge	70.924954	None	11.4 d	EC			0.2325
		40	^{72}Ge	71.922076	27.31	Stable				
		41	^{73}Ge	72.923459	7.76	Stable				
		42	^{74}Ge	73.921178	36.72	Stable				
		43	^{75}Ge	74.922860	None	1.380 h	β−	1.19	265	1.176
		44	^{76}Ge	75.921403	7.83	1.3E+21 a	β−β−			2.04
		45	^{77}Ge	76.923549	None	11.22 h	β−	2.20	264	2.703
								1.38	211	
									216	

(Continued)

Element	Z	N	Nuclide Symbol	Mass (u)[a]	Percent Abundance[b]	Half-Life	Decay Mode[c]	Particle Energy (MeV)[d]	Gamma Energy (keV)[e]	Total Energy of Decay (MeV)
Arsenic	33	40	73As	72.923825	None	80.3 d	EC		53 / 13	0.341
		41	74As	73.923929	None	17.78 d	EC / β+ / β−	0.941 / 1.350 / 0.717	596 / 635	2.562 / 1.353
		42	75As	74.921596	100	Stable				
		43	76As	75.922394	None	26.3 h	β−	2.97 / 2.41	559	2.963
Selenium	34	38	72Se	71.92711	None	8.5 d	EC		46	0.34
		39	73Se	72.92677	None	7.1 h	β+ / EC	1.32	361 / 67	2.74
		40	74Se	73.922476	0.89	Stable				
		41	75Se	74.922524	None	119.78 d	EC		265 / 136 / 280	0.863
		42	76Se	75.919214	9.37	Stable				
		43	77Se	76.919914	7.63	Stable				
		44	78Se	77.917309	23.77	Stable				
		45	79Se	78.918500	None	3.5E+05 a	β−	0.16		0.151
		46	80Se	79.916521	49.61	Stable				

(Continued)

Element	Z	N	Nuclide Symbol	Mass (u)[a]	Percent Abundance[b]	Half-Life	Decay Mode[c]	Particle Energy (MeV)[d]	Gamma Energy (keV)[e]	Total Energy of Decay (MeV)
		47	^{81}Se	80.917993	None	18.5 min	β−	1.58	276 290	1.585
		48	^{82}Se	81.916699	8.73	9E+19 a	β−β−			2.995
		49	^{83}Se	82.919119	None	22.3 min	β−	1.0 1.8	357 510 225	3.67
Bromine	35	42	^{77m}Br	76.921494	None	4.3 min	IT		106	0.106
		42	^{77}Br	76.921380	None	2.376 d	EC		239 521	1.365
		43	^{78}Br	77.921146	None	6.45 min	β+ β+ EC β−	0.34 2.5	614	3.574 0.727
		44	^{79}Br	78.918337	50.69	Stable				
		45	^{80}Br	79.918530	None	17.66 min	β− EC	2.00 0.85	617 666	2.003 1.8705
		46	^{81}Br	80.916291	49.31	Stable				
		47	^{82}Br	81.916805	None	1.471 d	β−	0.444	777 554 619	3.093
Krypton	36	41	^{77}Kr	76.924668	None	1.24 h	β+	1.88 1.70		3.065

(Continued)

Element	Z	N	Nuclide Symbol	Mass (u)[a]	Percent Abundance[b]	Half-Life	Decay Mode[c]	Particle Energy (MeV)[d]	Gamma Energy (keV)[e]	Total Energy of Decay (MeV)
		42	78Kr	77.920365	0.355	Stable	EC		130 147	
		43	79Kr	78.920083	None	1.455 d	EC β+	0.60	2 261 398 606	1.626
		44	80Kr	79.916379	2.286	Stable				
		45	81Kr	80.916592	None	2.3E+05 a	EC		276	0.2808
		46	82Kr	81.913484	11.593	Stable				
		47	83Kr	82.914136	11.500	Stable				
		48	84Kr	83.911507	56.987	Stable				
		49	85Kr	84.912527	None	10.76 a	β-	0.687	514	0.687
		50	86Kr	85.910611	17.279	Stable				
		51	87Kr	86.913354	None	1.27 h	β-	3.5 3.9	403 2555	3.8884
		56	92Kr	91.92615	None	1.84 s	β-	6.0	142 1219	5.99
Rubidium	37	44	81Rb	80.918994	None	4.57 h	EC β+	1.05	190 446	2.24
		45	82Rb	81.918208	None	1.260 min	β+ EC	3.4	777	4.401

(Continued)

Element	Z	N	Nuclide Symbol	Mass (u)[a]	Percent Abundance[b]	Half-Life	Decay Mode[c]	Particle Energy (MeV)[d]	Gamma Energy (keV)[e]	Total Energy of Decay (MeV)
		46	^{83}Rb	82.915112	None	86.2 d	EC		520	0.91
									530	
									553	
		47	^{84}Rb	83.914385	None	33.2 d	EC			2.681
							β+	1.66	882	
								0.78		
							β-	0.894		0.894
		48	^{85}Rb	84.911790	72.17	Stable				
		49	^{86}Rb	85.911167	None	18.65 d	β-	1.775	1077	1.777
							EC			0.5186
		50	^{87}Rb	86.909181	27.83	4.9E+10 a	β-	5.31	1836	5.313
		51	^{88}Rb	87.911319	None	17.7 min			898	
		55	^{92}Rb	91.919725	None	4.48 s	β-	8.10	815	8.10
									2821	
									570	
		56	^{93}Rb	92.922033	None	5.85 s	β-	7.46	433	7.47
									986	
									213	
		57	^{94}Rb	93.926407	None	2.71 s	β-	9.5	837	10.29
									1578	
Strontium	38	44	^{82}Sr	81.918401	None	25.36 d	EC			0.18
		45	^{83}Sr	82.917555	None	1.350 d	EC			2.28

(Continued)

Element	Z	N	Nuclide Symbol	Mass (u)[a]	Percent Abundance[b]	Half-Life	Decay Mode[c]	Particle Energy (MeV)[d]	Gamma Energy (keV)[e]	Total Energy of Decay (MeV)
							β+	1.23	763	
									382	
		46	84Sr	83.913425	0.56	Stable				
		47	85Sr	84.912933	None	64.84 d	EC		514	1.065
		48	86Sr	85.909260	9.86	Stable				
		49	87Sr	86.908877	7.00	Stable				
		50	88Sr	87.905612	82.58	Stable				
		51	89Sr	88.907453	None	50.61 d	β−	1.488	909	1.493
		52	90Sr	89.907738	None	28.8 a	β−	0.546		0.546
		54	92Sr	91.911030	None	2.61 h	β−	0.54	1384	1.95
		56	94Sr	93.915360	None	1.25 min	β−	2.08	1428	3.51
		62	100Sr	99.9354	None	201 ms	β−		964	7.1
									899	
									66	
Yttrium	39	49	88Y	87.909503	None	106.63 d	EC		1836	3.623
							β+	0.76	898	
		50	89mY	88.905848	None	15.7 s	IT		909	0.9091
		50	89Y		100	Stable				
		51	90Y	89.907151	None	2.669 d	β−	2.281	2186	2.280
		53	92Y	91.90895	None	3.54 h	β−	3.64	935	3.64
									1405	
		56	94Y	93.911594	None	18.7 min	β−	4.92	919	4.92
									1139	
									551	

(Continued)

Element	Z	N	Nuclide Symbol	Mass (u)[a]	Percent Abundance[b]	Half-Life	Decay Mode[c]	Particle Energy (MeV)[d]	Gamma Energy (keV)[e]	Total Energy of Decay (MeV)
		63	102Y	101.93356	None	0.30 s	β−		152 1211	9.9
Zirconium	40	49	89Zr	88.908889	None	3.27 d	EC β+	0.90	909	2.833
		50	90Zr	89.904704	51.45	Stable				
		51	91Zr	90.905646	11.22	Stable				
		52	92Zr	91.905041	17.15	Stable				
		53	93Zr	92.906476	None	1.5E+06 a	β−	0.060	31	0.091
		54	94Zr	93.906315	17.38	Stable				
		55	95Zr	94.908043	None	64.02 d	β−	0.368 0.400	757 724	1.124
		56	96Zr	95.908273	2.80	2.0E+19 a	β−β−			3.35
		57	97Zr	96.910951	None	16.75 h	β−	1.92	743	2.659
		59	99Zr	98.91651	None	2.2 s	β−	3.54 3.50	469 546 594	4.56
		62	102Zr	101.92298	None	2.9 s	β−		600 535 65	4.61
Niobium	41	51	92Nb	91.907193	None	3.5E+07 a	EC		561	2.006

(Continued)

Element	Z	N	Nuclide Symbol	Mass (u)[a]	Percent Abundance[b]	Half-Life	Decay Mode[c]	Particle Energy (MeV)[d]	Gamma Energy (keV)[e]	Total Energy of Decay (MeV)
		52	^{93}Nb	92.906378	100	Stable			935	
		53	^{94}Nb	93.907284	None	2.0E+04 a	β−	0.473	871	2.045
									703	
		58	^{99}Nb	98.91162	None	15.0 s	β−	3.5	138	3.64
									98	
		63	^{104}Nb	103.9224	None	4.9 s	β−		192	8.1
Molybdenum	42	49	^{91}Mo	90.91175	None	15.5 min	β+ EC	3.44		4.43
		50	^{92}Mo	91.906811	14.77	Stable			1637	
		51	^{93}Mo	92.906812	None	3.5E+03 a	EC		1581	0.405
		52	^{94}Mo	93.905088	9.23	Stable			31	
		53	^{95}Mo	94.905842	15.90	Stable				
		54	^{96}Mo	95.904679	16.68	Stable				
		55	^{97}Mo	96.906021	9.56	Stable				
		56	^{98}Mo	97.905408	24.19	Stable				
		57	^{99}Mo	98.907712	None	2.7476 d	β−	1.214	740	1.357
									181	
									141	
		58	^{100}Mo	99.90748	9.67	7.0E+18 a	β−β−			3.034
		59	^{101}Mo	100.910347	None	14.61 min	β−	0.7	591	2.82
								2.23	192	
									1013	
									506	

(Continued)

Element	Z	N	Nuclide Symbol	Mass (u)[a]	Percent Abundance[b]	Half–Life	Decay Mode[c]	Particle Energy (MeV)[d]	Gamma Energy (keV)[e]	Total Energy of Decay (MeV)
		60	^{102}Mo	101.91030	None	11.3 min	β⁻	1.2	212, 148, 224	1.01
		61	^{103}Mo	102.91320	None	1.13 min	β⁻	3.7	83, 424, 46	3.8
		62	^{104}Mo	103.91376	None	1.00 min	β⁻	2.02	69, 70	2.16
Technetium	43	53	^{96}Tc	95.907871	None	4.3 d	EC		778, 850, 813	2.97
		54	^{97m}Tc	96.90647	None	91 d	IT, EC		97	0.097, 0.417
		54	^{97}Tc	96.906365	None	2.6E+06 a	EC			0.320
		55	^{98}Tc	97.907216	None	4.2E+06 a	β⁻	0.40	745, 652	1.80
		56	^{99m}Tc	98.906409	None	6.008 h	IT, β⁻	0.435	140.5, 322	0.143, 0.437
		56	^{99}Tc	98.906255	None	2.13E+05 a	β⁻	0.294	90	0.294
		61	^{104}Tc	103.91144	None	18.2 min	β⁻	5.3	358	5.60
Ruthenium	44	51	^{95}Ru	94.91041	None	1.64 h	EC, β⁺	1.20, 0.91	336, 1097, 627	2.57

(Continued)

Element	Z	N	Nuclide Symbol	Mass (u)[a]	Percent Abundance[b]	Half-Life	Decay Mode[c]	Particle Energy (MeV)[d]	Gamma Energy (keV)[e]	Total Energy of Decay (MeV)
		52	96Ru	95.90760	5.54	Stable				
		53	97Ru	96.90756	None	2.89 d	EC		216	1.11
									325	
		54	98Ru	97.90529	1.87	Stable				
		55	99Ru	98.905939	12.76	Stable				
		56	100Ru	99.904219	12.60	Stable				
		57	101Ru	100.905582	17.06	Stable				
		58	102Ru	101.904349	31.55	Stable				
		59	103Ru	102.906324	None	39.27 d	β−	0.223	497	0.763
		60	104Ru	103.905433	18.62	Stable				
		61	105Ru	104.907750	None	4.44 h	β−	1.187	724	1.918
								1.11	469	
								1.134	676	
Rhodium	45	54	99Rh	98.908132	None	16 d	EC		528	2.04
							β+	0.68	353	
								0.54	90	
								1.10		
		57	102Rh	101.906843	None	207 d	EC		475	2.323
							β+	1.30		
								0.82		
							β−	1.15	557	1.150

(Continued)

Element	Z	N	Nuclide Symbol	Mass (u)[a]	Percent Abundance[b]	Half-Life	Decay Mode[c]	Particle Energy (MeV)[d]	Gamma Energy (keV)[e]	Total Energy of Decay (MeV)
		58	^{103}Rh	102.905504	100	Stable				
		59	^{104}Rh	103.906655	None	42.3 s	β^-	2.44	556	2.440
							EC			1.139
		60	^{105}Rh	104.905692	None	35.4 h	β^-	0.566	319	0.567
								0.248		
Palladium	46	53	^{99}Pd	98.91177	None	21.4 min	β^+	2.18	136	3.39
							EC			
			^{101}Pd	100.90829	None	8.4 h	EC		296	1.980
							β^+	0.776	591	
		56	^{102}Pd	101.905609	1.02	Stable				
		57	^{103}Pd	102.906087	None	16.99 d	EC		40	0.543
									358	
		58	^{104}Pd	103.904036	11.14	Stable				
		59	^{105}Pd	104.905085	22.33	Stable				
		60	^{106}Pd	105.903486	27.33	Stable				
		61	^{107}Pd	106.905129	None	6.5E+06 a	β^-	0.04		0.034
		62	^{108}Pd	107.903892	26.46	Stable				
		63	^{109}Pd	108.905954	None	13.46 h	β^-	1.028	88	1.116
		64	^{110}Pd	109.90515	11.72	Stable				
		65	^{111}Pd	110.90764	None	23.4 min	β^-	2.1	580	2.22
									70	
		72	^{118}Pd	117.9190	None	2.1 s	β^-		125	4.1
									224	

(Continued)

Element	Z	N	Nuclide Symbol	Mass (u)[a]	Percent Abundance[b]	Half-Life	Decay Mode[c]	Particle Energy (MeV)[d]	Gamma Energy (keV)[e]	Total Energy of Decay (MeV)
Silver	47	52	^{99}Ag	98.9176	None	2.1 min	β+		265	5.4
							EC			
		57	^{104}Ag	103.908628	None	1.2 h	EC		556	4.279
			^{106}Ag	105.906666	None	24.0 min	EC	0.99	768	2.965
							β+	1.96	512	
		60	^{107}Ag	106.905097	51.839	Stable				
		61	^{108}Ag	107.905954	None	2.39 min	β−			0.20
							β−	1.65	633	1.65
							EC			1.92
		62	^{109}Ag	108.904752	48.161	Stable				
		63	^{110}Ag	109.906111	None	24.6 s	β−	0.88	434	
							EC		619	
							β−	2.891	658	2.891
							EC			0.89
Cadmium	48	56	^{104}Cd	103.90985	None	58 min	EC	0.29	84	1.14
							β+		709	
		57	^{105}Cd	104.90947	None	55.5 min	EC	1.69	962	2.738
							β+		347	
									1303	
		58	^{106}Cd	105.90646	1.25	Stable				

(Continued)

Element	Z	N	Nuclide Symbol	Mass (u)[a]	Percent Abundance[b]	Half-Life	Decay Mode[c]	Particle Energy (MeV)[d]	Gamma Energy (keV)[e]	Total Energy of Decay (MeV)
		59	107Cd	106.906614	None	6.52 h	EC		93	1.417
							β+	0.302	829	
		60	108Cd	107.90418	0.89	Stable				
		61	109Cd	108.904986	None	462 d	EC		88	0.214
		62	110Cd	109.903002	12.49	Stable				
		63	111Cd	110.904178	12.80	Stable				
		64	112Cd	111.902758	24.13	Stable				
		65	113Cd	112.904402	12.22	8.0E+15 a	β−			0.32
		66	114Cd	113.903359	28.73	Stable				
		67	115Cd	114.905431	None	2.228 d	β−	1.11	336	1.446
								0.593	528	
		68	116Cd	115.904756	7.49	3.0E+19 a	β−β−			2.81
		69	117Cd	116.907218	None	2.49 h	β−	0.67	315	2.52
								2.2	273	
									1303	
Indium	49	55	104In	103.9183	None	1.81 min	β+	4.30	658	7.9
							EC		834	
		62	111In	110.905111	None	2.8049 d	EC		245	0.862
									171	
		63	112In	111.905533	None	14.4 min	EC		617	2.584
							β+	1.56	607	

(Continued)

Element	Z	N	Nuclide Symbol	Mass (u)[a]	Percent Abundance[b]	Half-Life	Decay Mode[c]	Particle Energy (MeV)[d]	Gamma Energy (keV)[e]	Total Energy of Decay (MeV)
		64	113mIn	112.904479	None	1.658 h	β⁻	0.66		0.665
							IT		392	0.392
		64	113In	112.904058	4.29	Stable				
		65	114In	113.904917	None	1.198 min	β⁻	1.984	1300	1.989
							EC			1.449
							β⁺	0.40	558	
									576	
		66	115In	114.903878	95.71	4.4E+14 a	β⁻	0.49	498	0.499
		67	116In	115.905260	None	14.1 s	β⁻	3.3	1294	3.278
									463	
							EC			0.469
Tin	50	61	111Sn	110.907735	None	35 min	EC			2.45
							EC			
							β⁺	1.5	1153	
									1915	
									762	
									1611	
		62	112Sn	111.904818	0.97	Stable				
		63	113Sn	112.905173	None	115.1 d	EC		392	1.037
		64	114Sn	113.902779	0.66	Stable				
		65	115Sn	114.903342	0.34	Stable				
		66	116Sn	115.901741	14.54	Stable				
		67	117Sn	116.902952	7.68	Stable				
		68	118Sn	117.901603	24.22	Stable				

(Continued)

Element	Z	N	Nuclide Symbol	Mass (u)[a]	Percent Abundance[b]	Half-Life	Decay Mode[c]	Particle Energy (MeV)[d]	Gamma Energy (keV)[e]	Total Energy of Decay (MeV)
		69	^{119}Sn	118.903308	8.59	Stable				
		70	^{120}Sn	119.902195	32.58	Stable				
		71	^{121}Sn	120.904237	None	1.128 d	β−	0.383		0.391
		72	^{122}Sn	121.903439	4.63	Stable				
		73	^{123}Sn	122.905722	None	129.2 d	β−	1.42	1089	1.404
		74	^{124}Sn	123.905274	5.79	Stable				
		75	^{125}Sn	124.907785	None	9.64 d	β−	2.35	1067	2.357
									1089	
									822	
									916	
		78	^{128}Sn	127.91054	None	59.1 min	β−	0.63	482	1.27
									75	
		80	^{130}Sn	129.91385	None	3.72 min	β−	1.1	193	2.15
									780	
									70	
		82	^{132}Sn	131.91774	None	39.7 s	β−	1.76	86	3.12
									341	
									247	
									890	
									993	
		85	^{135}Sn	134.9347	None	0.52 s	β−		282	8.9
									732	
									923	

(Continued)

Element	Z	N	Nuclide Symbol	Mass (u)[a]	Percent Abundance[b]	Half-Life	Decay Mode[c]	Particle Energy (MeV)[d]	Gamma Energy (keV)[e]	Total Energy of Decay (MeV)
Antimony	51	69	^{120}Sb	119.905074	None	15.89 min	EC			2.68
							$\beta+$	1.72	1171	
		70	^{121}Sb	120.903816	57.21	Stable				
		71	^{122}Sb	121.905175	None	2.72 d	$\beta-$	1.414	564	1.984
								1.980		
							EC			1.616
							$\beta+$	0.570	1141	
		72	^{123}Sb	122.904214	42.79	Stable				
		73	^{124}Sb	123.905938	None	60.20 d	$\beta-$	0.61	603	2.904
								2.301	1691	
									723	
		75	^{126}Sb	125.90725	None	12.4 d	$\beta-$	1.9	666	3.67
									695	
									415	
		79	^{130}Sb	129.91155	None	39.5 min	$\beta-$	2.9	839	5.06
									794	
									331	
									182	
Tellurium	52	67	^{119}Te	118.906408	None	16.0 h	EC		644	2.293
							$\beta+$	0.627		
		68	^{120}Te	119.90402	0.09	Stable				
		69	^{121}Te	120.90493	None	19.1 d	EC		573	1.04
									508	

(Continued)

Element	Z	N	Nuclide Symbol	Mass (u)[a]	Percent Abundance[b]	Half–Life	Decay Mode[c]	Particle Energy (MeV)[d]	Gamma Energy (keV)[e]	Total Energy of Decay (MeV)
		70	^{122}Te	121.903044	2.55	Stable				
		71	^{123}Te	122.904270	0.89	9E+16 a	EC			0.052
		72	^{124}Te	123.902818	4.74	Stable				
		73	^{125m}Te	124.904587	None	58 d	IT		36 109	0.145
		73	^{125}Te	124.904431	7.07	Stable				
		74	^{126}Te	125.903312	18.84	Stable				
		75	^{127}Te	126.905217	None	9.4 h	β−	0.69	418 360	0.702
		76	^{128}Te	127.904463	31.74	4E+24 a	β−β−			0.87
		77	^{129}Te	128.906596	None	33.6 d	β−	1.45	460 28	1.500
		78	^{130}Te	129.906224	34.08	6E+20 a	β−β−			
		79	^{131}Te	130.908522	None	25.0 min	β−	2.1	150 452	2.235
		82	^{134}Te	133.91154	None	42.0 min	β−	0.6 0.7	767 211 278 80	1.51
		86	^{138}Te	137.9292	None	1.15 s	β−			6.4
Iodine	53	70	^{123}I	122.905598	None	13.221 h	EC		159	1.229
		72	^{125}I	124.904624	None	59.4 d	EC		35.5	0.1858
		73	^{126}I	125.905620	None	12.89 d	EC		666	2.154

(*Continued*)

Element	Z	N	Nuclide Symbol	Mass (u)[a]	Percent Abundance[b]	Half-Life	Decay Mode[c]	Particle Energy (MeV)[d]	Gamma Energy (keV)[e]	Total Energy of Decay (MeV)
							β^-	0.87	389	1.258
							β^+	1.13		2.154
		74	$^{127}\mathrm{I}$	126.904473	100	Stable				
		75	$^{128}\mathrm{I}$	127.905805	None	25.00 min	β^-	2.13	443	2.122
							EC		743	1.254
							β^+		743	
		76	$^{129}\mathrm{I}$	128.904988	None	1.57E+07 a	β^-	0.15	40	0.194
		77	$^{130}\mathrm{I}$	129.906674	None	12.36 h	β^-	1.04	536	2.949
							β^-	0.62	669	
									740	
		78	$^{131}\mathrm{I}$	130.906124	None	8.023 d	β^-	0.606	365	0.971
		84	$^{137}\mathrm{I}$	136.91787	None	24.5 s	β^-		1218	5.88
									601	
									1303	
									1220	
Xenon	54	66	$^{120}\mathrm{Xe}$	119.91215	None	46 min	EC			1.62
							β^+		25	
									73	
									178	
		69	$^{123}\mathrm{Xe}$	122.90847	None	2.00 h	EC		149	2.69
							β^+	1.51	178	
									330	

(Continued)

Element	Z	N	Nuclide Symbol	Mass (u)[a]	Percent Abundance[b]	Half-Life	Decay Mode[c]	Particle Energy (MeV)[d]	Gamma Energy (keV)[e]	Total Energy of Decay (MeV)
		70	124Xe	123.905893	0.0952	Stable				
		71	125mXe	124.906670	None	57 s	IT		140 / 112	0.253
		71	125Xe	124.906398	None	17.1 h	EC / β+	0.47	188	1.644
		72	126Xe	125.90427	0.0890	Stable				
		73	127Xe	126.905180	None	36.4 d	EC		203 / 172	0.662
		74	128Xe	127.903531	1.9102	Stable				
		75	129Xe	128.904779	26.4006	Stable				
		76	130Xe	129.903508	4.0710	Stable				
		77	131Xe	130.905082	21.2324	Stable				
		78	132Xe	131.904153	26.9086	Stable				
		79	133Xe	132.905906	None	5.243 d	β−	0.346	81	0.427
		80	134Xe	133.905394	10.4357	Stable				
		81	135Xe	134.90721	None	9.10 h	β−	0.91	250	1.165
		82	136Xe	135.90722	8.8573	Stable				
		83	137Xe	136.911563	None	3.82 min	β−	4.1 / 3.6	456	4.17
Cesium	55	75	130Cs	129.906706	None	29.21 min	EC / β+ / β−	1.98 / 0.44	536	2.98 / 0.36
		77	132Cs	131.906430	None	6.48 d	EC			2.125

(Continued)

Element	Z	N	Nuclide Symbol	Mass (u)[a]	Percent Abundance[b]	Half-Life	Decay Mode[c]	Particle Energy (MeV)[d]	Gamma Energy (keV)[e]	Total Energy of Decay (MeV)
		78	^{133}Cs	132.905452	100	Stable	β+ / β−	0.40	668, 465	1.279
		79	^{134}Cs	133.906713	None	2.065 a	β−	0.658	605, 796	2.0587
		80	^{135}Cs	134.905972	None	2.3E+06 a	EC	0.089		1.233
		81	^{136}Cs	135.907306	None	13.10 d	β−	0.21	819, 1048, 341	0.269
		82	^{137}Cs	136.907084	None	30.07 a	β−	0.341	662	2.548
		83	^{138}Cs	137.91101	None	32.2 min	β−	0.514	1436, 463, 1010	1.1756
							β−	2.9		5.37
		85	^{140}Cs	139.917277	None	1.06 min	β−	5.7, 6.21	602	6.22
Barium	56	73	^{129}Ba	128.90867	None	2.2 h	β+	1.42	214, 221, 129	2.44
		74	^{130}Ba	129.906321	0.106	2E+21 a	ECEC			2.62
		75	^{131}Ba	130.906931	None	11.53 d	EC / β+		496, 124, 216	1.38

(Continued)

Element	Z	N	Nuclide Symbol	Mass (u)[a]	Percent Abundance[b]	Half–Life	Decay Mode[c]	Particle Energy (MeV)[d]	Gamma Energy (keV)[e]	Total Energy of Decay (MeV)
		76	^{132}Ba	131.905061	0.101	Stable				
		77	^{133}Ba	132.906002	None	10.54 a	EC		356 81 303	0.517
		78	^{134}Ba	133.904508	2.417	Stable				
		79	^{135}Ba	134.905689	6.592	Stable				
		80	^{136}Ba	135.904576	7.854	Stable				
		81	^{137m}Ba	136.906638	None	2.552 min	IT		662	0.662
		81	^{137}Ba	136.905827	11.232	Stable				
		82	^{138}Ba	137.905247	71.698	Stable				
		83	^{139}Ba	138.908835	None	1.396 h	β–	2.27 2.14	166	2.318
		84	^{140}Ba	139.910600	None	12.75 d	β–	1.0 0.48 1.02	537 30	1.05
		85	^{141}Ba	140.914406	None	18.3 min	β–	2.59 2.73	190 304 277 344	3.21
Lanthanum	57	73	^{130}La	129.9123	None	8.7 min	β+ EC	3.7 4.3	357 551 908 545	5.63

(Continued)

Element	Z	N	Nuclide Symbol	Mass (u)[a]	Percent Abundance[b]	Half-Life	Decay Mode[c]	Particle Energy (MeV)[d]	Gamma Energy (keV)[e]	Total Energy of Decay (MeV)
		80	137La	136.90647	None	6E+04 a	EC			0.62
		81	138La	137.907112	0.090	1.05E+11 a	EC		1436	1.737
							β−	0.25	789	1.04
		82	139La	138.906353	99.910	Stable				
		83	140La	139.909473	None	1.678 d	β−	1.35	1596	3.762
								1.24	487	
								1.67	816	
									329	
		84	141La	140.910957	None	3.90 h	β−	2.43	1355	2.502
Cerium	58	72	130Ce	129.9147	None	23 min	EC		131	2.21
							β+			
		75	133mCe	132.9116	None	5.4 h	EC		477	2.937
							β+	1.3	510	
									58	
		77	135Ce	134.90915	None	17.7 h	EC		266	2.026
							β+	0.8	300	
									607	
		78	136Ce	135.90717	0.185	Stable				
		79	137Ce	136.90778	None	9.0 h	EC		447	1.222
							β+			
		80	138Ce	137.90599	0.251	Stable				

(Continued)

Element	Z	N	Nuclide Symbol	Mass (u)[a]	Percent Abundance[b]	Half-Life	Decay Mode[c]	Particle Energy (MeV)[d]	Gamma Energy (keV)[e]	Total Energy of Decay (MeV)
		81	^{139}Ce	138.906647	None	137.6 d	EC		166	0.28
		82	^{140}Ce	139.905439	88.450	Stable				
		83	^{141}Ce	140.908271	None	32.50 d	β−	0.436 0.581	145	0.581
		84	^{142}Ce	141.909244	11.114	Stable				
		85	^{143}Ce	142.912381	None	1.377 d	β−	1.110 1.404	293 57	1.461
Praseodymium	59	81	^{140}Pr	139.909071	None	3.39 min	EC β+	2.37	1596 307	3.39
		82	^{141}Pr	140.907653	100	Stable				
		83	^{142}Pr	141.910040	None	19.12 h	β− EC	2.162	1576 641	2.162 0.746
		84	^{143}Pr	142.910812	None	13.57 d	β−	0.933	742	0.934
Neodymium	60	81	^{141}Nd	140.909605	None	2.49 h	EC β+	0.802	1127 1293 1147	1.823
		82	^{142}Nd	141.907723	27.2	Stable				
		83	^{143}Nd	142.909814	12.2	Stable				
		84	^{144}Nd	143.910087	23.8	2.38E+15 a	α	1.83		1.905
		85	^{145}Nd	144.912574	8.3	Stable				

(Continued)

Element	Z	N	Nuclide Symbol	Mass (u)[a]	Percent Abundance[b]	Half-Life	Decay Mode[c]	Particle Energy (MeV)[d]	Gamma Energy (keV)[e]	Total Energy of Decay (MeV)
		86	^{146}Nd	145.913117	17.2	Stable				
		87	^{147}Nd	146.916096	None	10.98 d	β⁻	0.805	91	0.896
									531	
		88	^{148}Nd	147.916893	5.7	Stable				
		89	^{149}Nd	148.920144	None	1.73 h	β⁻	1.42	211	1.690
								1.13	114	
								1.03	270	
		90	^{150}Nd	149.920891	5.6	8E+18 a	β⁻β⁻			3.37
		91	^{151}Nd	150.923825	None	12.4 min	β⁻	1.2	117	2.442
									256	
									1181	
Promethium	61	84	^{145}Pm	144.912744	None	17.7 a	EC		73	0.163
									67	
		85	^{146}Pm	145.914692	None	5.53 a	EC		454	1.471
									736	
							β⁻	0.795	747	1.542
									633	
		86	^{147}Pm	146.915134	None	2.6234 a	β⁻	0.224	121	0.2241
		87	^{148}Pm	147.917468	None	5.37 d	β⁻	2.47	1465	2.47
								1.02	550	
		88	^{149}Pm	148.918330	None	2.212 d	β⁻	1.072	286	1.071
		89	^{150}Pm	149.9098	None	2.68 h	β⁻	2.3	334	3.45
								1.8	1325	

(Continued)

Element	Z	N	Nuclide Symbol	Mass (u)[a]	Percent Abundance[b]	Half–Life	Decay Mode[c]	Particle Energy (MeV)[d]	Gamma Energy (keV)[e]	Total Energy of Decay (MeV)
								1.6	1166 / 832	
		90	151Pm	150.921203	None	1.183 d	β–	0.84	340 / 168 / 275	1.187
Samarium	62	81	143Sm	142.914624	None	8.83 min	EC / β+	2.42	1057 / 1515	3.443
		82	144Sm	143.911999	3.07	Stable				
		83	145Sm	144.913406	None	340 d	EC		61	0.616
		84	146Sm	145.913041	None	1.03E+08 a	α	2.455		
		85	147Sm	146.914898	14.99	1.17E+11 a	α	2.23		2.31
		86	148Sm	147.914823	11.24	7E+15 a	α	1.96		1.99
		87	149Sm	148.917185	13.82	Stable				
		88	150Sm	149.917276	7.38	Stable				
		89	151Sm	150.919928	None	90 a	β–	0.076	22	0.077
		90	152Sm	151.919732	26.75	Stable				
		91	153Sm	152.922094	None	1.928 d	β–	0.69 / 0.64	103 / 70	0.808
		92	154Sm	153.922209	22.75	Stable				
		93	155Sm	154.924636	None	22.2 min	β–	1.52	104	1.627
Europium	63	87	150Eu	149.919698	None	36 a	EC		334 / 439 / 584	2.26
		88	151Eu	150.919850	47.81	Stable				

(Continued)

Element	Z	N	Nuclide Symbol	Mass (u)[a]	Percent Abundance[b]	Half-Life	Decay Mode[c]	Particle Energy (MeV)[d]	Gamma Energy (keV)[e]	Total Energy of Decay (MeV)
		89	152mEu	151.921898	None	1.60 h	IT		90	0.148
		89	152mEu	151.921788	None	9.29 h	β−	1.86		1.865
							EC		842	1.920
									963	
		89	152Eu	151.921740	None	13.54 a	EC		122	1.874
							β+	0.727	1408	
		90	153Eu	152.921230	52.19	Stable	β−	0.699	344	1.820
		91	154Eu	153.922975	None	8.60 a	β−	0.58	123	1.969
								0.27	1275	
							EC		82	0.717
									185	
Gadolinium	64	87	151Gd	150.920344	None	124 d	EC		154	0.464
							α	2.60	243	
		88	152Gd	151.919791	0.20	1.1E+14 a	α	2.14		2.205
		89	153Gd	152.921746	None	241.6 d	EC		97	0.484
									103	
		90	154Gd	153.920866	2.18	Stable				
		91	155Gd	154.922622	14.80	Stable				

(Continued)

Element	Z	N	Nuclide Symbol	Mass (u)[a]	Percent Abundance[b]	Half-Life	Decay Mode[c]	Particle Energy (MeV)[d]	Gamma Energy (keV)[e]	Total Energy of Decay (MeV)
		92	^{156}Gd	155.922123	20.47	Stable				
		93	^{157}Gd	156.923960	15.65	Stable				
		94	^{158}Gd	157.924104	24.84	Stable				
		95	^{159}Gd	158.926385	None	18.5 h	β−	0.96	364 58	0.971
		96	^{160}Gd	159.927054	21.86	Stable				
		97	^{161}Gd	160.929666	None	3.66 min	β−	1.56	361 315 102	1.955
Terbium	65	93	^{158}Tb	157.925410	None	1.8E+02 a	EC		944 962	1.220
		94	^{159}Tb	158.925347	100	Stable	β−	0.83	99	0.935
		95	^{160}Tb	159.927164	None	72.3 d	β−	0.57 0.86	879 299 966	1.835
		103	^{168}Tb	167.9434	None	8 s	β−			6
Dysprosium	66	88	^{154}Dy	153.92442	None	3E+06 a	α	2.87		
		89	^{155}Dy	154.92575	None	9.9 h	EC β+	0.845	227	2.095
		90	^{156}Dy	155.92428	0.056	Stable				
		91	^{157}Dy	156.925461	None	8.1 h	EC		326	1.34
		92	^{158}Dy	157.924409	0.095	Stable				

(Continued)

Element	Z	N	Nuclide Symbol	Mass (u)[a]	Percent Abundance[b]	Half-Life	Decay Mode[c]	Particle Energy (MeV)[d]	Gamma Energy (keV)[e]	Total Energy of Decay (MeV)
		93	159Dy	158.925736	None	144.4 d	EC		58	0.366
		94	160Dy	159.925198	2.329	Stable				
		95	161Dy	160.926933	18.889	Stable				
		96	162Dy	161.926798	25.475	Stable				
		97	163Dy	162.928731	24.896	Stable				
		98	164Dy	163.929175	28.260	Stable				
		99	165Dy	164.931700	None	2.33 h	β-	1.29	95	1.287
									362	
Holmium	67	102	168Dy	167.9371	None	8.7 min	β-	0.870	192	1.5
		96	163Ho	162.928730	None	4.57E+03 a	EC			0.00256
		97	164Ho	163.930231	None	29 min	EC	0.96	73	0.986
							β-	0.88	91	0.962
		98	165Ho	164.930322	100	Stable				
		99	166Ho	165.932281	None	1.118 d	β-	1.855	81	1.855
								1.773	1379	
Erbium	68	101	168Ho	167.9355	None	2.99 min	β-	1.90	741	2.93
		78	146Er	145.9521	None	2 s	EC			6.9
		93	161Er	160.93000	None	3.21 h	β+	0.82	827	1.99
							EC		211	
		94	162Er	161.928778	0.139	Stable				

(Continued)

Element	Z	N	Nuclide Symbol	Mass (u)[a]	Percent Abundance[b]	Half-Life	Decay Mode[c]	Particle Energy (MeV)[d]	Gamma Energy (keV)[e]	Total Energy of Decay (MeV)
		95	^{163}Er	162.930029	None	1.25 h	EC β^+	 0.19	1114 436 440	1.210
		96	^{164}Er	163.929200	1.601	Stable				
		97	^{165}Er	164.930723	None	10.36 h	EC			0.376
		98	^{166}Er	165.930293	33.503	Stable				
		99	^{167}Er	166.932048	22.869	Stable				
		100	^{168}Er	167.932370	26.978	Stable				
		101	^{169}Er	168.934588	None	9.39 d	β^-	0.34	8	0.351
		102	^{170}Er	169.935464	14.910	Stable				
		103	^{171}Er	170.938026	None	7.516 h	β^-	1.065	308 296 112	1.491
		104	^{172}Er	171.939352	None	2.05 d	β^-	0.36 0.28	610 407	0.891
Thulium	69	78	^{147}Tm	146.9610	None	0.58 s	EC p	1.051	81	10.7 1.061
		99	^{168}Tm	167.934170	None	93.1 d	EC β^+		198 816 448 184	1.679
							β^-			0.257
		100	^{169}Tm	168.934213	100	Stable				

(*Continued*)

Element	Z	N	Nuclide Symbol	Mass (u)[a]	Percent Abundance[b]	Half-Life	Decay Mode[c]	Particle Energy (MeV)[d]	Gamma Energy (keV)[e]	Total Energy of Decay (MeV)
		101	^{170}Tm	169.935798	None	128.6 d	β⁻	0.968 0.883	84	0.968
Ytterbium	70	97	^{167}Yb	166.934947	None	17.5 min	EC EC β⁺	0.64	79 113 106 176	0.314 1.954
		98	^{168}Yb	167.933897	0.13	Stable				
		99	^{169}Yb	168.935187	None	32.03 d	EC		63 198	0.910
		100	^{170}Yb	169.934762	3.04	Stable				
		101	^{171}Yb	170.936326	14.28	Stable				
		102	^{172}Yb	171.936381	21.83	Stable				
		103	^{173}Yb	172.938211	16.13	Stable				
		104	^{174}Yb	173.938862	31.83	Stable				
		105	^{175}Yb	174.941273	None	4.185 d	β⁻	0.466	396 283 114	0.470
		106	^{176}Yb	175.942572	12.76	Stable				
			^{177}Yb	176.945257	None	1.911 h	β⁻	1.40	150 1080	1.400
Lutetium	71	97	^{168}Lu	167.9387	None	5.5 min	EC		1484	4.51

(Continued)

Element	Z	N	Nuclide Symbol	Mass (u)[a]	Percent Abundance[b]	Half-Life	Decay Mode[c]	Particle Energy (MeV)[d]	Gamma Energy (keV)[e]	Total Energy of Decay (MeV)
		101	^{172m}Lu	171.939127	None	3.7 min	IT		42	0.0419
		103	^{174}Lu	173.940334	None	3.3 a	EC		1242	1.374
									77	
							β^+	0.38		
		104	^{175}Lu	174.940772	97.41	Stable				
		105	^{176}Lu	175.942686	2.59	3.76E+10 a	β^-	0.57	307	1.190
									202	
		106	^{177}Lu	176.943755	None	6.65 d	β^-	0.497	208	0.501
									113	
Hafnium	72	96	^{168}Hf	167.9406	None	26.0 min	EC		57	1.7
							β^+	0.236		
		100	^{172}Hf	171.93946	None	1.87 a	EC		24	0.34
									126	
									67	
									82	
		101	^{173}Hf	172.9407	None	23.6 h	EC		124	1.47
									297	
		102	^{174}Hf	173.940046	0.16	2.0E+15 a	α	2.50		
		103	^{175}Hf	174.941503	None	70 d	EC		343	0.687
		104	^{176}Hf	175.941409	5.26	Stable				
		105	^{177}Hf	176.943221	18.60	Stable				
		106	^{178}Hf	177.943699	27.28	Stable				

(Continued)

Element	Z	N	Nuclide Symbol	Mass (u)[a]	Percent Abundance[b]	Half-Life	Decay Mode[c]	Particle Energy (MeV)[d]	Gamma Energy (keV)[e]	Total Energy of Decay (MeV)
		107	^{179}Hf	178.945816	13.62	Stable				
		108	^{180}Hf	179.946550	35.08	Stable				
		109	^{181}Hf	180.949099	None	42.4 d	β−	0.405	482, 133, 346	1.030
		110	^{182}Hf	181.950553	None	8.9E+06 a	β−		270	0.37
Tantalum	73	95	^{168}Ta	167.9480	None	2.0 min	β+, EC	5.82	124	6.97
		106	^{179}Ta	178.945934	None	1.82 a	EC			0.1056
		107	^{180m}Ta	179.947546	0.012	7.1E+15 a				
		107	^{180}Ta	179.947465	None	8.15 h	β−, EC	0.71, 0.61	103, 93	0.708, 0.852
		108	^{181}Ta	180.947996	99.988	Stable				
		109	^{182}Ta	181.950152	None	114.43 d	β−	0.522, 0.25	68, 1121	1.814
		111	^{184}Ta	183.95401	None	8.7 h	β−	1.19	414, 253, 921	2.87
Tungsten	74	105	^{179}W	178.94707	None	37.8 min	EC		31	1.06
		106	^{180}W	179.946704	0.12	Stable				
		107	^{181}W	180.948198	None	121.2 d	EC		6	0.188
		108	^{182}W	181.948204	26.50	Stable				

(Continued)

Element	Z	N	Nuclide Symbol	Mass (u)[a]	Percent Abundance[b]	Half–Life	Decay Mode[c]	Particle Energy (MeV)[d]	Gamma Energy (keV)[e]	Total Energy of Decay (MeV)
		109	183W	182.950223	14.31	Stable				
		110	184W	183.950931	30.64	Stable				
		111	185W	184.953421	None	74.8 d	β-	0.433	125	0.432
		112	186W	185.954364	28.43	Stable				
		113	187W	186.957158	None	23.9 h	β-	0.622 1.312	686 480	1.311
		114	188W	187.958487	None	69.78 d	β-	0.349	291 227 64	0.349
Rhenium	75	109	184Re	183.952524	None	35 d	EC		903 792	1.481
		110	185Re	184.952955	37.40	Stable				
		111	186Re	185.954987	None	3.718 d	β-	1.071 0.933	137	1.069
							EC		123	0.579
		112	187Re	186.955753	62.60	4.12E+10 a	β-	0.002470		0.002469
		113	188Re	187.958112	None	17.004 h	β-	2.118 1.962	155	2.1203
Osmium	76	107	183Os	182.9531	None	13 h	EC β+		382 114 168	2.1
		108	184Os	183.952489	0.02	Stable				

(*Continued*)

Element	Z	N	Nuclide Symbol	Mass (u)[a]	Percent Abundance[b]	Half-Life	Decay Mode[c]	Particle Energy (MeV)[d]	Gamma Energy (keV)[e]	Total Energy of Decay (MeV)
		109	^{185}Os	184.954043	None	93.6 d	EC		646	1.0128
									875	
									880	
									717	
		110	^{186}Os	185.953838	1.59	2E+15 a	α	2.758		2.82
		111	^{187}Os	186.955750	1.96	Stable				
		112	^{188}Os	187.955838	13.24	Stable				
		113	^{189}Os	188.958147	16.15	Stable				
		114	^{190}Os	189.958447	26.26	Stable				
		115	^{191}Os	190.960928	None	15.4 d	β$^-$	0.143	129	0.313
		116	^{192}Os	191.961481	40.78	Stable				
		117	^{193}Os	192.964148	None	30.11 h	β$^-$	1.13	139	1.141
									461	
									73	
Iridium	77	113	^{190}Ir	189.9606	None	11.8 d	EC		187	1.955
									605	
									519	
		114	^{191}Ir	190.960594	37.3	Stable				
		115	^{192}Ir	191.962602	None	73.83 d	β$^-$	0.672	317	1.460
								0.535	468	
		116	^{193}Ir	192.962926	62.7	Stable	EC		485	1.047
		117	^{194}Ir	193.965076	None	19.3 h	β$^-$	2.251	329	2.234

(Continued)

Element	Z	N	Nuclide Symbol	Mass (u)[a]	Percent Abundance[b]	Half-Life	Decay Mode[c]	Particle Energy (MeV)[d]	Gamma Energy (keV)[e]	Total Energy of Decay (MeV)
Platinum	78	111	189Pt	188.96083	None	10.9 h	EC			1.97
							β+	0.89	721	
									608	
									94	
									569	
									244	
		112	190Pt	189.95993	0.014	6.5E+11 a	α	3.18		3.249
		113	191Pt	190.961685	None	2.88 d	EC		539	1.008
									410	
									360	
		114	192Pt	191.961038	0.782	Stable				
		115	193Pt	192.962985	None	50 a	EC			0.0568
		116	194Pt	193.962680	32.967	Stable				
		117	195Pt	194.964791	33.832	Stable				
		118	196Pt	195.964952	25.242	Stable				
		119	197Pt	196.967323	None	19.96 h	β−	0.642	77	0.719
								0.719	191	
		120	198Pt	197.967893	7.163	Stable				
		121	199Pt	198.970576	None	30.8 min	β−	1.69	543	1.703
									494	
									317	
Gold	79	116	195Au	194.965018	None	186.12 d	EC		99	0.227
		117	196Au	195.966551	None	6.167 d	EC		356	1.507

(Continued)

Element	Z	N	Nuclide Symbol	Mass (u)[a]	Percent Abundance[b]	Half-Life	Decay Mode[c]	Particle Energy (MeV)[d]	Gamma Energy (keV)[e]	Total Energy of Decay (MeV)
		118	197Au	196.966569	100	Stable	β-	0.259	333	0.687
							β+		426	1.507
		119	198Au	197.968225	None	2.6949 d	β-	0.961	412	1.372
Mercury	80	115	195Hg	194.96664	None	10.53 h	EC		780	1.57
									61	
		116	196Hg	195.965833	0.15	Stable				
		117	197Hg	196.967195	None	2.672 d	EC		77	0.600
		118	198Hg	197.966769	9.97	Stable				
		119	199Hg	198.968280	16.87	Stable				
		120	200Hg	199.968326	23.10	Stable				
		121	201Hg	200.970302	13.18	Stable				
		122	202Hg	201.970643	29.86	Stable				
		123	203Hg	202.972857	None	46.61 d	β-	0.213	279	0.492
		124	204Hg	203.973494	6.87	Stable				
		125	205Hg	204.976056	None	5.2 min	β-	1.6	204	1.533
		126	206Hg	205.97750	None	8.2 min	β-	0.94	305	1.31
Thallium	81	120	201Tl	200.97080	None	3.043 d	EC		167	0.48
									135	
		121	202Tl	201.97209	None	12.23 d	EC		440	1.36
							β+			
		122	203Tl	202.972344	29.52	Stable				

(*Continued*)

Element	Z	N	Nuclide Symbol	Mass (u)[a]	Percent Abundance[b]	Half-Life	Decay Mode[c]	Particle Energy (MeV)[d]	Gamma Energy (keV)[e]	Total Energy of Decay (MeV)
		123	204Tl	203.973849	None	3.78 a	β−	0.7634		0.7638
							EC			0.344
		124	205Tl	204.974428	70.48	Stable				
		125	206Tl	205.976095	None	4.20 min	β−	1.528	803	1.532
Lead	82	121	203Pb	202.973376	None	2.16 d	EC		279	0.97
		122	204Pb	203.973044	1.4	Stable				
		123	205Pb	204.974467	None	1.5E+07 a	EC			0.0505
		124	206Pb	205.974465	24.1	Stable				
		125	207Pb	206.975897	22.1	Stable				
		126	208Pb	207.976652	52.4	Stable				
		127	209Pb	208.981075	None	3.25 h	β−	0.645		0.644
		128	210Pb	209.984173	None	22.3 a	β−	0.017	47	0.0635
							α	0.061		
								3.72		
		129	211Pb	210.988732	None	36.1 min	β−	1.38	405	1.37
									832	
									427	
		130	212Pb	211.991888	None	10.64 h	β−	0.335	239	0.570
								0.569	300	
		131	213Pb	212.996500	None	10.2 min	β−	0.67	352	2.05
		132	214Pb	213.999798	None	27 min	β−	0.73	295	1.02
									242	
Bismuth	83	125	208Bi	207.979726	None	3.68E+05 a	EC		2614	2.878

(Continued)

Element	Z	N	Nuclide Symbol	Mass (u)[a]	Percent Abundance[b]	Half-Life	Decay Mode[c]	Particle Energy (MeV)[d]	Gamma Energy (keV)[e]	Total Energy of Decay (MeV)
		126	^{209}Bi	208.980399	100	2.0E+19 a	α	3.08		3.14
		127	^{210}Bi	209.984105	None	5.01 d	β−	1.162		1.161
							α	4.648	305	
								4.687	266	
Polonium	84	124	^{208}Po	207.981246	None	2.898 a	α	5.115	292	
							EC		570	
									602	
		125	^{209}Po	208.982430	None	102 a	α	4.880	261	
							EC		263	
									896	
		126	^{210}Po	209.982874	None	138.38 d	α	5.3044	803	
		128	^{212}Po	211.988868	None	0.298 μs	α	8.7844		
		130	^{214}Po	213.995186	None	163.7 μs	α	7.6869	799	
		134	^{218}Po	218.008966	None	3.10 min	α	6.0024	510	
							β−			
Astatine	85	124	^{209}At	208.986159	None	5.4 h	EC		545	3.49
							α	5.647	782	
									790	
		125	^{210}At	209.987131	None	8.1 h	EC		1181	3.98
									245	
									1483	
							α	5.524	83	

(Continued)

Element	Z	N	Nuclide Symbol	Mass (u)[a]	Percent Abundance[b]	Half–Life	Decay Mode[c]	Particle Energy (MeV)[d]	Gamma Energy (keV)[e]	Total Energy of Decay (MeV)
		126	^{211}At	210.987496	None	7.21 h	EC	5.442	106	0.785
							α	5.361	687	
								5.868	670	
Radon	86	133	^{219}Rn	219.009480	None	3.96 s	α	6.8193	271	
								6.553	402	
								6.425		
		135	^{221}Rn	221.01554	None	25 min	β^-	0.83	186	1.19
							α	6.037	254	
									265	
		136	^{222}Rn	222.017578	None	3.8235 d	α	5.4895	510	5.590
		137	^{223}Rn	223.0218	None	23 min	β^-		592	1.9
									635	
									416	
									654	
		138	^{224}Rn	224.0241	None	1.8 h	β^-		261	0.8
									266	
Francium	87	135	^{222}Fr	222.01754	None	14.3 min	β^-	1.72	206	2.03
								1.92	111	
		136	^{223}Fr	223.019731	None	21.8 min	β^-	1.15	50	1.149
									80	
									235	
							α	5.291	151	

(Continued)

Element	Z	N	Nuclide Symbol	Mass (u)[a]	Percent Abundance[b]	Half-Life	Decay Mode[c]	Particle Energy (MeV)[d]	Gamma Energy (keV)[e]	Total Energy of Decay (MeV)
								5.314	59 145	
		137	^{224}Fr	224.02324	None	3.0 min	β⁻	2.6 1.8	216 132 837	2.83
Radium	88	135	^{223}Ra	223.018502	None	11.435 d	α	5.7164 5.607	269 154 324	
		136	^{224}Ra	224.020212	None	3.63 d	α	5.6855 5.449	241	
		137	^{225}Ra	225.023605	None	14.9 d	β⁻ α	0.32 5.006 4.976	40	0.356
		138	^{226}Ra	226.025410	None	1599 a	α	4.7844 4.602	186	
		139	^{227}Ra	227.029171	None	42 min	β⁻	1.31 1.03	27 300 303 284	1.328
		140	^{228}Ra	228.031064	None	5.76 a	β⁻	0.039 0.015 0.026	14	0.046
Actinium	89	136	^{225}Ac	225.02323	None	10.0 d	α	5.829	100	

(Continued)

Element	Z	N	Nuclide Symbol	Mass (u)[a]	Percent Abundance[b]	Half-Life	Decay Mode[c]	Particle Energy (MeV)[d]	Gamma Energy (keV)[e]	Total Energy of Decay (MeV)
								5.793		
								5.731		
		137	^{226}Ac	226.026090	None	1.224 d	β−	0.89	230	1.113
								1.11	158	
							EC		254	0.641
									186	
							α	5.4		
		138	^{227}Ac	227.027747	None	21.772 a	β−	0.045	15	0.045
							α	4.9534	100	
								4.941		
		139	^{228}Ac	228.031015	None	6.15 h	β−	1.158	911	2.124
								1.731	969	
									338	
Thorium	90	135	^{225}Th	225.02395	None	8.72 min	α	4.27		
							α	6.479	322	0.67
								6.441	246	
								6.501	359	
		139	^{229}Th	229.031762	None	7.4E+03 a	α	4.845	194	
								4.901	86	
								4.814	211	
									32	
		140	^{230}Th	230.033134	None	7.56E+04 a	α	4.688	68	
								4.621	110	

(Continued)

Element	Z	N	Nuclide Symbol	Mass (u)[a]	Percent Abundance[b]	Half-Life	Decay Mode[c]	Particle Energy (MeV)[d]	Gamma Energy (keV)[e]	Total Energy of Decay (MeV)
		141	^{231}Th	231.036297	None	1.063 d	β−	0.305	620 / 26 / 84	0.392
		142	^{232}Th	232.038055	100	1.40E+10 a	α / SF	4.012 / 3.974	64 / 141	
		143	^{233}Th	233.041577	None	21.83 min	β−	1.245	87 / 29 / 459	1.243
		144	^{234}Th	234.043596	None	24.10 d	β−	0.198	63 / 92 / 93	0.273
Protactinium	91	139	^{230}Pa	230.034533	None	17.4 d	EC / β− / α	0.51 / 4.766 / 5.345	952 / 315	1.311 / 0.560
		140	^{231}Pa	231.035884	None	3.28E+04 a	α	5.013 / 4.950 / 5.029	27 / 300	
		141	^{232}Pa	232.038582	None	1.32 d	β−	0.31 / 0.29	969 / 894	1.34
		142	^{233}Pa	233.040240	None	26.967 d	EC / β−	0.256	312	0.50 / 0.570

(Continued)

Element	Z	N	Nuclide Symbol	Mass (u)[a]	Percent Abundance[b]	Half-Life	Decay Mode[c]	Particle Energy (MeV)[d]	Gamma Energy (keV)[e]	Total Energy of Decay (MeV)
		143	^{234}Pa	234.043302	None	6.69 h	β^-	0.15		2.195
								0.48	131	
								0.65	34	
									1938	
Uranium	92	140	^{232}U	232.037156	None	69.8 a	α	5.3203	58	
								5.2635		
		141	^{233}U	233.039635	None	1.592E+05 a	α	4.824	43	
								4.783	97	
									55	
		142	^{234}U	234.040952	0.0054	2.46E+05 a	SF			
							α	4.776	53	
								4.725	121	
		143	^{235}U	235.043930	0.7204	7.04E+08 a	α	4.398	186	
								4.366	144	
		144	^{236}U	236.045568	None	2.342E+07 a	SF			
							α	4.494	49	
								4.445	113	
		145	^{237}U	237.048724	None	6.752 d	SF			0.519
							β^-	0.24	60	
								0.25	208	
		146	^{238}U	238.050788	99.2742	4.468E+09 a	α	4.197	50	
								4.147		

(Continued)

Element	Z	N	Nuclide Symbol	Mass (u)[a]	Percent Abundance[b]	Half-Life	Decay Mode[c]	Particle Energy (MeV)[d]	Gamma Energy (keV)[e]	Total Energy of Decay (MeV)
		147	^{239}U	239.054288	None	23.47 min	SF			
							β-	1.21	75	1.261
								1.28	44	
		148	^{240}U	240.056586	None	14.1 h	β-	0.36	44	0.40
Neptunium	93	144	^{237}Np	237.048173	None	2.14E+06 a	α	4.788	29	
								4.771	87	
		145	^{238}Np	238.050941	None	2.103 d	β-	0.263	985	1.2915
								1.248	1029	
		146	^{239}Np	239.502931	None	2.356 d	β-	0.438	106	0.723
								0.341	278	
									228	
Plutonium	94	144	^{238}Pu	238.049560	None	87.7 a	α	5.4992	44	
								5.4565	100	
							SF			
		145	^{239}Pu	239.052163	None	2.410E+04 a	α	5.156	52	
								5.144	30	
								5.105	1057	
		146	^{240}Pu	240.053814	None	6.56E+03 a	SF			
							α	5.1685	45	
								5.1241	104	
		147	^{241}Pu	241.056845	None	14.29 a	SF			
							β-	0.0208	149	0.0208
							α	4.897		

(Continued)

Element	Z	N	Nuclide Symbol	Mass (u)[a]	Percent Abundance[b]	Half-Life	Decay Mode[c]	Particle Energy (MeV)[d]	Gamma Energy (keV)[e]	Total Energy of Decay (MeV)
		148	^{242}Pu	242.058743	None	3.75E+05 a	α / SF	4.853 / 4.901 / 4.856	104 / 45	
		149	^{243}Pu	243.061997	None	4.956 h	β−	0.58 / 0.49	84	0.579
		150	^{244}Pu	244.06420	None	8.11E+07 a	α / SF	4.589 / 4.546		
Americium	95	145	^{240}Am	240.05529	None	2.12 d	EC / α	5.378	988 / 889	1.38
		146	^{241}Am	241.056829	None	432.7 a	α	5.4857 / 5.4430	60 / 26 / 955	
		147	^{242}Am	242.059543	None	16.02 h	SF / β−	0.63 / 0.67	42	0.6645
		148	^{243}Am	243.061381	None	7.37E+03 a	EC / α	5.276 / 5.234	45 / 75 / 31	0.751
		149	^{244m}Am	244.064373	None	26 min	SF / β−	1.50	662 / 43	1.516

(Continued)

Element	Z	N	Nuclide Symbol[a]	Mass (u)[a]	Percent Abundance[b]	Half-Life	Decay Mode[c]	Particle Energy (MeV)[d]	Gamma Energy (keV)[e]	Total Energy of Decay (MeV)
		149	244mAm	244.064279	None	10.1 h	β−	0.387	744 154 898	1.427
Curium	96	145	241Cm	241.057647	None	32.8 d	EC α	5.939 5.929 5.884	472 146	0.767
		146	242Cm	242.058836	None	162.8 d	α SF	6.1127 6.0694	44	
		147	243Cm	243.061389	None	29.1 a	α EC SF	5.785 5.742	278 228	
		148	244Cm	244.062753	None	18.1 a	α SF	5.8048 5.7627	43	
		149	245Cm	245.065491	None	8.5E+03 a	α SF	5.362	175 133	
		150	246Cm	246.067224	None	4.77E+03 a	α SF	5.386 5.342	45	

(Continued)

Element	Z	N	Nuclide Symbol	Mass (u)[a]	Percent Abundance[b]	Half-Life	Decay Mode[c]	Particle Energy (MeV)[d]	Gamma Energy (keV)[e]	Total Energy of Decay (MeV)
		151	^{247}Cm	247.070354	None	1.56E+07 a	SF			
							α	4.870	403	
								5.267		
Berkelium	97	149	^{246}Bk	246.06867	None	1.80 d	EC		799	1.4
		150	^{247}Bk	247.07031	None	1.4E+03 a	α	5.532	84	
								5.711	268	
		151	^{248}Bk	248.07308	None	23.7 h	β–	0.86	551	0.87
							EC			0.72
		152	^{249}Bk	249.074980	None	330 d	β–	0.124	328	0.124
							α	5.417	308	
Californium	98	153	^{251}Cf	251.079587	None	9.0E+02 a	SF		177	
							α	5.677	227	
								5.852		
								6.014		
		154	^{252}Cf	252.08163	None	2.646 a	α	6.1181	43	
								6.0756	100	
							SF			
		155	^{253}Cf	253.085127	None	17.8 d	β–	0.27		0.29
							α	5.98		
								5.92		
Einsteinium	99	150	^{249}Es	249.07641	None	1.70 h	EC		380	1.45

(*Continued*)

Element	Z	N	Nuclide Symbol	Mass (u)[a]	Percent Abundance[b]	Half-Life	Decay Mode[c]	Particle Energy (MeV)[d]	Gamma Energy (keV)[e]	Total Energy of Decay (MeV)
		151	^{250}Es	250.0787	None	8.6 h	α	6.77	813	
							EC		829	2.1
									303	
									350	
		152	^{251}Es	251.079984	None	1.38 d	EC		178	0.38
									153	
		153	^{252}Es	252.0830	None	1.29 a	α	6.492		
								6.462		
							α	6.632	52	1.26
								6.562	64	
									418	
							EC		785	
									139	
		154	^{253}Es	253.084825	None	20.47 d	α	6.6327	42	
								6.5916	389	
							SF			
		155	^{254}Es	254.088022	None	276 d	α	6.429	64	
		156	^{255}Es	255.09027	None	40 d	β−			0.29
							α	6.300	33	
								6.260		
							SF			
Fermium	100	151	^{251}Fm	251.081567	None	5.3 h	EC		881	1.47

(Continued)

Element	Z	N	Nuclide Symbol	Mass (u)[a]	Percent Abundance[b]	Half-Life	Decay Mode[c]	Particle Energy (MeV)[d]	Gamma Energy (keV)[e]	Total Energy of Decay (MeV)
							α	6.833	453, 451, 425, 480, 358	
		152	^{252}Fm	252.08247	None	1.058 d	α	7.039, 6.998	96, 42	
		153	^{253}Fm	253.085185	None	3.0 d	SF, EC	6.943, 6.676	272	0.336
		154	^{254}Fm	254.086854	None	3.240 h	α, SF	7.192, 7.150	99, 43	
		155	^{255}Fm	255.08996	None	20.1 h	α, SF	7.022, 6.963	81, 58	
		156	^{256}Fm	256.09177	None	2.63 h	SF, SF			
		157	^{257}Fm	257.09510	None	100.5 d	α, SF	6.92, 6.519	241, 179	
Mendelevium	101	154	^{255}Md	255.09108	None	27 min	SF, EC		170	1.04

(Continued)

Element	Z	N	Nuclide Symbol	Mass (u)[a]	Percent Abundance[b]	Half-Life	Decay Mode[c]	Particle Energy (MeV)[d]	Gamma Energy (keV)[e]	Total Energy of Decay (MeV)
		155	256Md	256.0941	None	1.29 h	α	7.327	231	
									453	
									406	
							EC		634	2.1
									644	
		156	257Md	257.095541	None	5.5 h	α	7.206	409	
								7.142	379	
								7.074		
							EC			0.41
		157	258Md	258.098431	None	51.5 d	α	6.718	371	
								6.763		
		158	259Md	259.1005	None	1.64 h	α		368	
							SF			
Nobelium	102	150	252No	252.08898	None	2.45 s	α	8.42		
								8.37		
							SF			
		151	253No	253.0907	None	1.6 min	α	8.01	222	
								8.04	280	
								8.06	152	
							EC			3.2
		152	254No	254.09096	None	50 s	α	8.09		
							EC			1.2
							SF			
		153	255No	255.09324	None	3.1 min	α	8.12	187	2.01

(*Continued*)

Element	Z	N	Nuclide Symbol	Mass (u)[a]	Percent Abundance[b]	Half-Life	Decay Mode[c]	Particle Energy (MeV)[d]	Gamma Energy (keV)[e]	Total Energy of Decay (MeV)
							EC	7.93		
								8.08		
		154	^{256}No	256.09428	None	2.9 s	α	8.45		
							SF			
		155	^{257}No	257.09685	None	25 s	α	8.22	77	
								8.27	124	
								8.32	101	
		156	^{258}No	258.0982	None	1.2 ms	SF			
		157	^{259}No	259.1010	None	58 min	α	7.520		0.5
								7.551		
								7.581		
Lawrencium	103	152	^{255}Lr	255.0967	None	21 s	EC	8.37		
							α	8.42		
		153	^{256}Lr	256.0986	None	27 s	EC	8.43		3.2
							α	8.39		
								8.52		
		154	^{257}Lr	257.0996	None	0.65 s	EC	8.86		4.0
							SF	8.80		
							α			
		155	^{258}Lr	258.1018	None	4.1 s	EC	8.60		2.5
							α			

(Continued)

Element	Z	N	Nuclide Symbol	Mass (u)[a]	Percent Abundance[b]	Half-Life	Decay Mode[c]	Particle Energy (MeV)[d]	Gamma Energy (keV)[e]	Total Energy of Decay (MeV)
		156	^{259}Lr	259.1029	None	6.2 s	EC, α	8.62, 8.56		
		157	^{260}Lr	260.1055	None	3 min	α, SF	8.44		
		158	^{261}Lr	261.1069	None	40 min	α, EC	8.03		2.7
		159	^{262}Lr	262.1097	None	3.6 h	SF, EC			2.2
Rutherfordium	104	155	^{259}Rf	259.1056	None	2.5 s	α, SF, EC	8.77, 8.86		
		156	^{260}Rf	260.1064	None	20 ms	SF			
		157	^{261}Rf	261.10877	None	4.2 s	SF			
		158	^{262}Rf	262.1099	None	2.3 s	SF			
		159	^{263}Rf	263.1125	None	17 min	SF			
Dubnium	105	150	^{255}Db	255.1074	None	1.6 s	α, SF			
		151	^{256}Db	256.1081	None	1.6 s	α, SF, EC	9.01		
		152	^{257}Db	257.1077	None	1.5 s	α	8.97		6.5

(Continued)

Element	Z	N	Nuclide Symbol	Mass (u)[a]	Percent Abundance[b]	Half-Life	Decay Mode[c]	Particle Energy (MeV)[d]	Gamma Energy (keV)[e]	Total Energy of Decay (MeV)
							SF	9.07		
							EC			5.3
		153	258Db	258.1092	None	4.2 s	α	9.17		
								9.08		
		154	259Db	259.1096	None	0.5 s	EC	9.47		
							α			
							SF			
		155	260Db	260.1113	None	1.5 s	α	9.05		
								9.08		
								9.13		
		156	261Db	261.1121	None	1.8 s	SF			
							α	8.93		
							SF			
		157	262Db	262.1141	None	34 s	α	8.45		
								8.53		
								8.67		
		158	263Db	263.1150	None	27 s	SF			
							SF	8.35		
							α			
							EC			
Seaborgium	106	152	258Sg	258.1132	None	2.9 ms	SF			

(Continued)

Element	Z	N	Nuclide Symbol	Mass (u)ᵃ	Percent Abundanceᵇ	Half-Life	Decay Modeᶜ	Particle Energy (MeV)ᵈ	Gamma Energy (keV)ᵉ	Total Energy of Decay (MeV)
		153	^{259}Sg	259.1145	None	0.5 s	α SF	9.62		
		154	^{260}Sg	260.11442	None	4 ms	α	9.77 9.72		
		155	^{261}Sg	261.1161	None	0.23 s	SF α	9.56		
		156	^{262}Sg	262.1164	None	12 ms	SF			
		157	^{263}Sg	263.1183	None	1.0 s	SF α	9.06		
		158	^{264}Sg	264.1189	None	0.05 s	SF α	9.25		
		159	^{265}Sg	265.1211	None	8 s	α	8.84 8.94		
		160	^{266}Sg	266.1221	None	21 s	SF α	8.77 8.52		
		165	^{271}Sg	271.133	None	3 min	α SF	8.5		
Bohrium	107	153	^{260}Bh	260.122	None	35 ms	α	10.2		
		157	^{264}Bh	264.1246	None	1.0 s	α	9.49 9.62		

(Continued)

Element	Z	N	Nuclide Symbol	Mass (u)ᵃ	Percent Abundanceᵇ	Half–Life	Decay Modeᶜ	Particle Energy (MeV)ᵈ	Gamma Energy (keV)ᵉ	Total Energy of Decay (MeV)
		159	^{266}Bh	266.1269	None	2 s	α	9.2		
		165	^{272}Bh	272.138	None	9.8 s	α	9.02		
Hassium	108	157	^{264}Hs	264.12839	None	0.08 ms	α / SF	10.43		
		162	^{270}Hs	270.1347	None	3.6 s	α	9.2		
		167	^{275}Hs	275.146	None	0.2 s	α	9.4		
Meitnerium	109	157	^{266}Mt	266.1373	None	2.6 ms	α	11.05		
		158	^{267}Mt	267.1375	None					
		159	^{268}Mt	268.1387	None	0.02 s	α	10.10 / 10.24		
		165	^{274}Mt	274.147	None	440 ms	α	10		
		167	^{276}Mt	276.151	None	0.72 s	α	9.7		
Darmstadtium	110	157	^{267}Ds	267.1443	None	3 μs	α	11.60		
		169	^{279}Ds	279.159	None	0.19 s	SF / α	9.70		
		171	^{281}Ds	281.162	None	9.6 s	SF			
Roentgenium	111	161	^{272}Rg	272.1536	None	2 ms	α	10.99		
		167	^{278}Rg	278.162	None	4.2 ms	α	10.7		
		169	^{280}Rg	280.164	None	3.6 s	α	9.7		
Copernicium	112	165	^{277}Cn	277.1639	None	0.69 ms	α	11.45 / 11.31		
		170	^{282}Cn	282.170	None	0.7 ms	SF			
		173	^{285}Cn	285.174	None	34 s	α	9.2		

(Continued)

Element	Z	N	Nuclide Symbol	Mass (u)[a]	Percent Abundance[b]	Half–Life	Decay Mode[c]	Particle Energy (MeV)[d]	Gamma Energy (keV)[e]	Total Energy of Decay (MeV)
Nihonium	113	170	283Nh	283.176	None	100 ms	α	10.1		
		171	284Nh	284.178	None	0.48 s	α	10.0		
Flerovium	114	174	288Fl	288.186	None	0.8 s	α	9.9		
		175	289Fl	289.187	None	2.7 s	α	9.8		
Moscovium	115	172	287Mc	287.191	None	32 ms	α	10.6		
		173	288Mc	288.192	None	87 ms	α	10.5		
Livermorium	116	174	290Lv	290.199	None	8 ms	α	10.9		
		175	291Lv	291.200	None	18 ms	α	10.7		
		176	292Lv	292.200	None	18 ms	α	10.7		
Tennessine	117	176	293Ts	293.208	None	18 ms	α	11.0		
		177	294Ts	294.210	None	290 ms	α	10.8		
Oganesson	118	176	294Og	294.214	None	1.4 ms	α	11.7		

a   The atomic mass of a nuclide includes the mass of Z electrons.

b   The percent abundance of a nuclide is its atomic percent present in nature.

c   Not all decay modes are listed for every nuclide; some relatively minor modes may be omitted.

d   Only the most commonly observed energies are listed. Values for β− and β+ are maximum energy values (E_{max}).

e   Only the most commonly observed energies, associated with each decay mode, are listed. Gamma energies listed after EC/positron could be associated with one or both.

Appendix B: Table of Chi-Squared

Degrees of Freedom	Probability of Observing an Outcome at Least as Large as the Chi-Squared Value						
	0.99	0.95	0.90	0.50	0.10	0.05	0.01
2	0.020	0.103	0.211	1.386	4.605	5.991	9.210
3	0.115	0.352	0.584	2.366	6.251	7.815	11.35
4	0.297	0.711	1.064	3.357	7.779	9.488	13.28
5	0.554	1.145	1.610	4.351	9.236	11.07	15.09
6	0.872	1.635	2.204	5.348	10.65	12.59	16.81
7	1.239	2.167	2.833	6.346	12.02	14.07	18.48
8	1.646	2.733	3.490	7.344	13.36	15.51	20.09
9	2.088	3.325	4.168	8.343	14.68	16.92	21.67
10	2.558	3.940	4.865	9.342	15.99	18.31	23.21
11	3.053	4.575	5.578	10.34	17.28	19.68	24.73
12	3.571	5.226	6.304	11.34	18.55	21.03	26.22
13	4.107	5.892	7.042	12.34	19.81	22.36	27.69
14	4.660	6.571	7.790	13.34	21.06	23.69	29.14
15	5.229	7.261	8.547	14.34	22.31	25.00	30.58
16	5.812	7.962	9.312	15.34	23.54	26.30	32.00
17	6.408	8.672	10.09	16.34	24.77	27.59	33.41
18	7.015	9.390	10.87	17.34	25.99	28.87	34.81
19	7.633	10.12	11.65	18.34	27.20	30.14	36.19
20	8.260	10.85	12.44	19.34	28.41	31.41	37.57
21	8.897	11.59	13.24	20.34	29.62	32.67	38.93
22	9.542	12.34	14.04	21.34	30.81	33.92	40.29
23	10.20	13.09	14.85	22.34	32.01	35.17	41.64
24	10.86	13.85	15.66	23.34	33.20	36.42	42.98
25	11.52	14.61	16.47	24.34	34.38	37.65	44.31
26	12.20	15.38	17.29	25.34	35.56	38.89	45.64
27	12.88	16.15	18.11	26.34	36.74	40.11	46.96
28	13.57	16.93	18.94	27.34	37.92	41.34	48.28
29	14.26	17.71	19.77	28.34	39.09	42.56	49.59
30	14.95	18.49	20.60	29.34	40.26	43.77	50.89
40	22.16	26.51	29.05	39.34	51.81	55.76	63.69
50	29.71	34.76	37.69	49.33	63.17	67.51	76.15
60	37.49	43.19	46.46	59.33	74.40	79.08	88.38
70	45.44	51.74	55.33	69.33	85.53	90.53	100.4
80	53.54	60.39	64.28	79.33	96.58	101.88	112.3
90	61.75	69.13	73.29	89.33	107.6	113.1	124.1
100	70.07	77.93	82.36	99.33	118.5	124.3	135.8

Appendix C: Useful Constants, Conversion Factors, SI Prefixes, and Formulas

C.1 CONSTANTS

Speed of light	$c = 2.998 \times 10^8$ m/s
Planck's constant	$h = 6.626 \times 10^{-34}$ J $\cdot$ s
Avogadro's number	6.022×10^{23} things/mole
W-quantity in air	33.85 eV/IP
Proton mass	1.673×10^{-27} kg $= 1.007276$ u
Neutron mass	1.675×10^{-27} kg $= 1.008665$ u
Electron mass	9.109×10^{-31} kg $= 5.486 \times 10^{-4}$ u
Boltzmann's constant	1.3806×10^{-23} J/K $= 8.63 \times 10^{-5}$ eV/K

C.2 CONVERSION FACTORS

$1\ u = 1.66054 \times 10^{-24}$ g	$1\ Ci = 3.7 \times 10^{10}$ dps	$1\ \mu Ci = 2.22 \times 10^6$ dpm
$1\ Bq = 1$ dps	5.259×10^5 min $= 1\ a = 365.24$ d	$1\ J = 6.242 \times 10^{18}$ eV
$1\ J = 1\dfrac{kg \cdot m^2}{s^2}$	$1R = \dfrac{2.58 \times 10^{-4}\ C}{kg}$	$1\dfrac{J}{kg} = 1\ Gy$
$1\ rad = \dfrac{100\ erg}{g} = \dfrac{10^{-2}\ J}{kg} = 1\ cGy$	$931.5\ MeV = 1\ u$	$1\ rem = 10^{-2}\ Sv = 1\ cSv = 10\ mSv$

C.3 SI PREFIXES

Prefix	Symbol	Factor
tera	T	10^{12}
giga	G	10^9
mega	M	10^6
kilo	k	10^3
centi	c	10^{-2}
milli	m	10^{-3}

(*Continued*)

Prefix	Symbol	Factor
micro	μ	10^{-6}
nano	n	10^{-9}
pico	p	10^{-12}
femto	f	10^{-15}

C.4 FORMULAS

$c = \lambda \nu$ $\qquad\qquad$ $E = h\nu$ $\qquad\qquad$ $A = Z + N$

$\ln \dfrac{A_1}{A_2} = \lambda t$ $\qquad\qquad$ $A_2 = A_1 e^{-\lambda t}$ $\qquad\qquad$ $\lambda = \dfrac{\ln 2}{t_{1/2}}$

$\ln \dfrac{N_1}{N_2} = \lambda t$ $\qquad\qquad$ $N_2 = N_1 e^{-\lambda t}$ $\qquad\qquad$ $A = \lambda N$

$\% \text{ efficiency} = \dfrac{\text{cpm}}{\text{dpm}} \times 100\%$ $\qquad\qquad$ $\text{specific activity} = \dfrac{\text{activity}}{\text{mass}}$

Secular equilibrium: $\qquad\qquad$ $N_B = \dfrac{\lambda_A}{\lambda_B} N_A (1 - e^{-\lambda_B t})$ $\qquad$ $A_B = A_A (1 - e^{-\lambda_B t})$

at equilibrium: $\qquad\qquad$ $A_B = A_A \dfrac{N_B}{N_A} = \dfrac{\lambda_A}{\lambda_B} = \dfrac{t_{1/2(B)}}{t_{1/2(A)}}$ $\quad$ or $\quad$ $N_A \lambda_A = N_B \lambda_B$

Transient equilibrium: $\quad$ $\dfrac{N_B}{N_A} = \dfrac{t_{1/2(B)}}{t_{1/2(A)} - t_{1/2(B)}}$ $\quad$ $\dfrac{A_A}{A_B} = 1 - \dfrac{t_{1/2(B)}}{t_{1/2(A)}}$ $\quad$ $t_{max} = \left[\dfrac{1.44 t_{1/2(A)} t_{1/2(B)}}{(t_{1/2(A)} - t_{1/2(B)})} \right] \times \ln \dfrac{t_{1/2(A)}}{t_{1/2(B)}}$

$\sigma = \sqrt{\bar{x}}$ $\qquad\qquad$ $\text{RSD} = \dfrac{\sigma}{\bar{x}}$ $\qquad\qquad$ $\sigma_{mean} = \dfrac{\sigma}{\sqrt{N}} = \dfrac{\sigma}{\sqrt{t}}$

$\chi^2 = \dfrac{1}{\bar{x}} \sum (\bar{x} - x)^2$ $\qquad\qquad$ $E = mc^2$

$K_\alpha = K_T \left(\dfrac{M}{M + m} \right)$ $\qquad\qquad$ $K_F = K_T \left(\dfrac{m}{M + m} \right)$ $\qquad\qquad$ $\alpha_T = \dfrac{I_e}{I_\gamma}$

$\text{KE} = \frac{1}{2} m v^2$ $\qquad\qquad$ $\text{KE} = \left(\dfrac{1}{\sqrt{1 - \dfrac{v^2}{c^2}}} - 1 \right) mc^2$ $\qquad\qquad$ $v = c \sqrt{1 - \left(\dfrac{mc^2}{\text{KE} + mc^2} \right)^2}$

$\text{LET} = \text{SI} \times W$ $\qquad\qquad$ $\text{Range} = R = \dfrac{E}{\text{LET}}$ $\qquad\qquad$ $R_{\alpha-\text{air}} \cong 0.31 E_\alpha^{3/2}$

$\text{RSP} = \dfrac{R_{\text{air}}}{R_{\text{abs}}}$ $\qquad\qquad$ $\text{SI} = \dfrac{4500 \dfrac{\text{IP}}{\text{m}}}{\dfrac{v^2}{c^2}}$ $\qquad\qquad$ $R_{e^-} = \dfrac{E}{2.0 \text{ MeV/cm}}$

(Continued)

$$E_{SC} = \frac{E_0}{1 + \left(\frac{E_0}{0.511}\right) \times (1 - \cos\theta)} \qquad I = I_0 e^{-\mu x} \qquad \ln\frac{I}{I_0} = -\mu x$$

$$HVL = \frac{\ln 2}{\mu} \qquad TVL = \frac{\ln 10}{\mu} \qquad \mu_m = \frac{\mu}{\rho}$$

$$\mu_a = \frac{\mu \times \text{atomic mass}}{\rho \times \text{Avogadro's \#}} \qquad \mu_{\text{eff}} = \frac{\ln\frac{I_0}{I}}{x} \qquad \mu_{\text{eff}} = \frac{\ln 2}{HVL}$$

$$E_{ED} = \frac{E_0^2}{E_0 + 0.2555} \qquad E_{BS} = \frac{E_0}{1 + (3.91 \times E_0)} \qquad E_0 = E_{ED} + E_{BS}$$

$$E_{tr} = -Q \times \left(\frac{A_A + A_X}{A_A}\right) \qquad E_{ecb} \approx 1.11 \times \left(\frac{A_A + A_X}{A_A}\right) \times \left(\frac{Z_A Z_X}{A_A^{1/3} + A_X^{1/3}}\right) \qquad E_{KC} \approx E_{KX} \times \left(\frac{A_X}{A_C}\right)$$

$$N_B = \sigma\Phi N_A t \qquad N_B = \frac{\sigma\Phi N_A}{\lambda}\left(1 - e^{-\lambda t}\right) \qquad A_0 = \sigma\Phi N_A \left(1 - e^{-\lambda t}\right)$$

$$A_{\text{sat}} = \sigma\Phi N_A \qquad A_0 = \sigma\Phi N_A \lambda t$$

$$K_T = 0.80 \times \frac{Z_1 Z_2}{\sqrt[3]{A_1} + \sqrt[3]{A_2}} \qquad K_1 = K_T\left(\frac{A_2}{A_1 + A_2}\right) \qquad K_2 = K_T\left(\frac{A_1}{A_1 + A_2}\right)$$

$$\frac{\text{rate}_A}{\text{rate}_B} = \sqrt{\frac{MW_B}{MW_A}} \qquad H = \text{dose} \times Q \qquad f = \frac{\text{dose in medium}}{\text{dose in air}}$$

$$X = \frac{\Gamma \cdot A \cdot t}{d^2} \qquad X = \frac{\Gamma \cdot A \cdot t}{d^2} B \qquad B = \frac{I}{I_0}$$

$$f = 3.5 \times 10^{-4} \, ZE \qquad \%D_n = \frac{D_n}{D_{\text{max}}} \times 100 \qquad \text{Dose Rate Change} = \frac{1}{\left(\frac{d_2}{d_1}\right)^2}$$

$$N_{\text{mu}} = \frac{\text{Tumor Dose}}{C_{\text{cal}} \times \dfrac{D_n}{D_{\text{max}}} \times C_{\text{fs}} \times C_{\text{attn}}} \qquad N_{\text{mu}} = \frac{\text{Tumor Dose}}{C_{\text{cal}} \times C_{\text{fs}} \times C_{\text{attn}} \times TMR} \times \left(\frac{SAD_{\text{treat}}}{SAD_{\text{cal}}}\right)^2$$

$$d_{90\%} = \frac{E}{4.0\,\text{MeV/cm}} \qquad d_{80\%} = \frac{E}{3.0\,\text{MeV/cm}} \qquad N_{\text{mu}} = \frac{\text{Tumor Dose}}{O_f \times \dfrac{D_n}{D_{\text{max}}} \times C_{\text{fs}}}$$

Appendix D: Periodic Table of the Elements

1A	2A	3B	4B	5B	6B	7B	8B	8B	8B	1B	2B	3A	4A	5A	6A	7A	8A
1 H 1.008																	2 He 4.003
3 Li 7.0	4 Be 9.012											5 B 10.8	6 C 12.01	7 N 14.01	8 O 16.00	9 F 19.00	10 Ne 20.18
11 Na 22.99	12 Mg 24.30											13 Al 26.98	14 Si 28.08	15 P 30.97	16 S 32.1	17 Cl 35.45	18 Ar 39.95
19 K 39.10	20 Ca 40.08	21 Sc 44.96	22 Ti 47.87	23 V 50.94	24 Cr 52.00	25 Mn 54.94	26 Fe 55.84	27 Co 58.93	28 Ni 58.69	29 Cu 63.55	30 Zn 65.38	31 Ga 69.72	32 Ge 72.63	33 As 74.92	34 Se 78.97	35 Br 79.90	36 Kr 83.80
37 Rb 85.47	38 Sr 87.62	39 Y 88.90	40 Zr 91.22	41 Nb 92.91	42 Mo 95.95	43 Tc (99)	44 Ru 101.1	45 Rh 102.9	46 Pd 106.4	47 Ag 107.9	48 Cd 112.4	49 In 114.8	50 Sn 118.7	51 Sb 121.8	52 Te 127.6	53 I 126.9	54 Xe 131.3
55 Cs 132.9	56 Ba 137.3	71 Lu 175.0	72 Hf 178.5	73 Ta 180.9	74 W 183.8	75 Re 186.2	76 Os 190.2	77 Ir 192.2	78 Pt 195.1	79 Au 197.0	80 Hg 200.6	81 Tl 204.4	82 Pb 207.2	83 Bi 209.0	84 Po (209)	85 At (210)	86 Rn (222)
87 Fr (223)	88 Ra (226)	103 Lr (261)	104 Rf (263)	105 Db (268)	106 Sg (269)	107 Bh (270)	108 Hs (269)	109 Mt (278)	110 Ds (280)	111 Rg (281)	112 Cn (283)	113 Nh (285)	114 Fl (288)	115 Mc (289)	116 Lv (291)	117 Ts (294)	118 Og (294)

57 La 138.9	58 Ce 140.1	59 Pr 140.9	60 Nd 144.2	61 Pm (145)	62 Sm 150.4	63 Eu 152.0	64 Gd 157.2	65 Tb 158.9	66 Dy 162.5	67 Ho 164.9	68 Er 167.3	69 Tm 168.9	70 Yb 173.0
89 Ac (227)	90 Th 232.0	91 Pa 231.0	92 U 238.0	93 Np (237)	94 Pu (244)	95 Am (243)	96 Cm (247)	97 Bk (247)	98 Cf (251)	99 Es (252)	100 Fm (257)	101 Md (258)	102 No (259)

Bibliography

Acton, J. M. and Hibbs, M., *Why Fukushima Was Preventable*. The Carnegie Papers. Washington DC: Carnegie Endowment, 2012.

Alley, W. M. and Alley, R., *Too Hot to Touch* (New York: Cambridge University Press), 2013.

Allison, W., *Radiation and Reason: The Impact of Science on a Culture of Fear* (York, UK: York), 2009.

American Nuclear Society., *Fukushima Daiichi: ANS Committee Report*, 2012. http://www. ans.org. Accessed on May 25, 2012.

Baum, E. M., Ernesti, M. C., Knox, H. D., Miller, T. R., and Watson, A. M., *Nuclides and Isotopes: Chart of the Nuclides*, 17th ed. (Schenectady, NY: Bechtel Marine Propulsion Corporation), 2010.

Bertulani, C. A. and Schechter, H., *Introduction to Nuclear Physics* (New York: Nova Science), 2002.

Chandra, R., *Nuclear Medicine Physics: The Basics*, 7th ed. (Philadelphia, PA: Lippincott, Williams & Wilkins), 2011.

Chase, G. D. and Rabinowitz, J. L., *Principles of Radioisotope Methodology*, 3rd ed. (Minneapolis, MN: Burgess), 1967.

Cherry, S. R., Sorenson, J. A., and Phelps, M. E., *Physics in Nuclear Medicine*, 4th ed. (Philadelphia, PA: Saunders), 2012.

Choppin, G., Liljenzin, J. O., Rydberg, J., and Ekberg, C., *Radiochemistry and Nuclear Chemistry*, 4th ed. (Oxford: Academic Press), 2013.

Clark, C., *Radium Girls: Women and Industrial Health Reform 1910–1935* (Chapel Hill, NC: University of North Carolina Press), 1997.

Cravins, G., *Power to Save the World: The Truth about Nuclear Energy* (New York: Knopf), 2007.

De Sanctis, E., Monti, S., and Ripani, M., *Energy from Nuclear Fission: An Introduction* (Switzerland: Springer), 2016.

Dunlap, R. A., *An Introduction to the Physics of Nuclei and Particles* (Belmont, CA: Brooks/Cole), 2004.

Ehmann, W. D. and Vance, D. E., *Radiochemistry and Nuclear Methods of Analysis* (New York: Wiley), 1991.

Ferguson, C. D., *Nuclear Energy: What Everyone Needs to Know* (New York: Oxford), 2011.

Friedlander, G., Kennedy, J. W., and Miller, J. M., *Nuclear and Radiochemistry*, 3rd ed. (New York: Wiley), 1981.

Gale, R. P. and Lax, E., *Radiation: What It Is, What You Need to Know* (New York: Random House), 2013.

Graham, D. T., Cloke, P., and Vesper, M., *Principles and Applications of Radiological Physics*, 6th ed. (Edinburgh: Churchill Livingstone), 2011.

Heath, R. L., *Scintillation Spectrometry: Gamma-Ray Spectrum Catalogue*. Vol. 1., 2nd ed. (Idaho National Laboratory), 1997. http://www.inl.gov/gammaray/. Accessed 11/12/2005.

Hendee, W. R., Ibbott, G. S., and Hendee, E. G., *Radiation Therapy Physics*, 3rd ed. (New York: Wiley), 2005.

Hendee, W. R. and Russell Ritenour, E., *Medical Imaging Physics*, 4th ed. (New York: Wiley), 2002.

Hoppe, M. E. and Schmidt, T. G. 2012. Estimation of organ and effective dose due to Compton backscatter security scans. *Med. Phys.* 39(6), 3396–3403.

Hála, J. and Navratil, J. D., *Radioactivity, Ionizing Radiation and Nuclear Energy*, 2nd ed. (Berkova, Czech Republic: Konvoj), 2003.

Khan, F. M., *The Physics of Radiation Therapy*, 5th ed. (Baltimore, MD: Lippincott, Williams & Wilkins), 2014.

Knoll, G. F., *Radiation Detection and Measurement*, 4th ed. (New York: Wiley), 2010.

Kondo, S., *Health Effects of Low-Level Radiation* (Osaka, Japan: Kinki University Press), 2010.

Kröner, M., Weber, C. H., Wirth, S., Pfeifer, K. J., Eriser, M. F., and Terrill, M. 2007. Advances in digital radiography: physical principles and system overview. *RadioGraphics* 27, 675–686.

Lawrence Berkeley Laboratory, *The ABC's of Nuclear Science* (Berkeley, CA: The Regents of the University of California), 2009. http://www.lbl.gov/abc/. Accessed 08/04/2017.

Lieser, K. H., *Nuclear and Radiochemistry: Fundamentals and Applications*, 2nd ed. (Berlin: Wiley-VCH), 2001.

Loveland, W., Morrissey, D. J., and Seaborg, G. T., *Modern Nuclear Chemistry*, 2nd ed. (New York: Wiley), 2017.

Malley, M. C., *Radioactivity: A History of a Mysterious Science* (New York: Oxford), 2011.

Mann, R., *An Introduction to Particle Physics and the Standard Model* (Boca Raton, FL: CRC Press), 2010.

Moody, K. J., Hutcheon, I. D., and Grant, P. M., *Nuclear Forensic Analysis*, 2nd ed. (Boca Raton, FL: Taylor & Francis), 2014.

Muller, R. A., *Physics for Future Presidents: The Science Behind the Headlines* (New York: W. W. Norton), 2008.

Nero, A. V., *A Guidebook to Nuclear Reactors* (Berkeley, CA: University of California Press), 1979.

Prekeges, J., *Nuclear Medicine Instrumentation*, 2nd ed. (Boston, MA: Jones and Bartlett), 2012.

Saha, G. B., *Fundamentals of Nuclear Pharmacy*, 6th ed. (New York: Springer), 2010.

Saha, G. B., *Physics and Radiobiology of Nuclear Medicine*, 4th ed. (New York: Springer), 2012.

Semat, H. and Albright, J. R., *Introduction to Atomic and Nuclear Physics*, 5th ed. (New York: Holt, Rinehart & Winston), 1972.

Settle, F. A., Alsos Digital Library for Nuclear Issues (Washington and Lee University), 2011. http://alsos.wlu.edu. Accessed 08/05/2017.

Stanton, R. and Stinson, D., *Applied Physics for Radiation Oncology* (Madison, WI: Medical Physics), 1996.

Tammemagi, H. and Jackson, D., *Half-Lives: A Guide to Nuclear Technology in Canada* (Don Mills, Ontario: Oxford), 2009.

Walker, J. S., *Three Mile Island: A Nuclear Crisis in Historical Perspective* (Berkeley, CA: University of California Press), 2004.

Weart, S. R., *The Rise of Nuclear Fear* (Cambridge, MA: Harvard University Press), 2012.

Webb, S., Ed., *The Physics of Medical Imaging* (Bristol, MA: Adam Hilger), 1988.

Wolfson, R., *Nuclear Choices: A Citizen's Guide to Nuclear Technology* (Cambridge, MA: MIT Press), 1991.

Index

A

Abundances, percent, 19, 23, 221–222, 240–241
Accelerators, 183–184, 189, 219–222, 258–259, 291
  cyclotrons, 185, 189–190, 220–222, 241, 330
  linear, 219–220
  linear, medical applications, 183–184, 296, 300–306, 310, 325
  synchrotrons, 221–222
Activities, 24–28, 137, 274
  in branched decay, 32–34
  equilibria, *see* Equilibria
  high, 171–173, 185
  low, detection of, 162
  production of radioactive nuclides, 216–218
  in radionuclide generators, 190–191
  specific, 29–31, 87, 92
  units, 24, 26
ALARA, xiv, 14–15, 180, 284
Alpha decay, *see* Decay
Alpha particles, 6, 97–98, 101, 131, 237
  beams, 186, 189, 203
  delayed, 126
  effective Coulomb barrier with, 209–210
  interactions with matter, 13, 131–132, 135–139, 273–275
  maximum energy, 99, 101
  medical applications, 183, 197–198
  in neutron production, 87–88
  in Rutherford's reaction, 205
  velocity of, 133–135
Amino acids, 196
Annihilation photons, 106–107, 146, 171, 173, 194, 200
Antineutrinos, 102–104, 106, 126
Antiparticles, 7, 102–103, 106, 127
Attenuation coefficients
  atomic, 150
  effective, 151
  linear, 148–150, 286
  mass, 149–150, 275–276
Atomic masses, 23, 28, 51–52, 59–61, 104, 107
  average, 23, 150
  unified, 4, 23, 51
Atomic numbers (Z), 5–7, 19, 57, 221, 293
  Coulomb barrier, 206, 209–210, 227–228, 243
  detector material, 160, 163–164
  shielding material, 142, 199, 286
  trends involving, 115–117, 141, 243, 293
    electron binding energy, 111
  fluorescent yield, 113, 121
  photon interactions, 146–147
Atomic structure, 3–7, 109–110
Auger effect, 112–113, 138, 145, 197–198
Auger electron, *see* Electron, Auger
Avalanches, *see* Cascades
Avogadro's number, 27, 37, 150, 405

B

Background corrections, 45, 159
Background radiation, 14–17, 277
  cosmogenic nuclides, 222–223
  variations in, 14, 280, 283
Backscatter factor (BSF), 316, 320
Backscattering, 142, 144, 168–169, 315
Becquerel, Henri, 26
Belt of stability, 9–12, 101, 114–119; *see also* Valley of beta stability
Bending magnets, 302–304
  achromatic, 304
  chromatic, 304
  proton therapy, 185, 329–330
Beta decay, *see* Decay
Beta energy, maximum (E_{max}), 103–104, 107–108, 139
Beta particles, 6–7, 102–104, 122, 125, 236–237
  detection of, 157–158, 162–163, 174
  energy of, 57, 60–61, 103–105
  interactions with matter, 13, 131, 138–142
Bhagavad Gita, 89
Binding energies
  electron, 111, 113, 131–133, 143, 145
    electron capture, 111–112
    internal conversion, 121–122
    W–quantity, 135
  nuclear, 51–52
    per nucleon, 53–54, 123, 237, 240
    thermal neutron, 233–234
Blocking, 325–327
Bohr model, 4, 8, 110
Bohr, Niels, 110
Boltzmann's constant, 239–240, 405
Bolus, 326
Bragg peaks, 137, 185–186, 298, 326, 328–330
Branched decay, *see* Decay, branched
Branch ratios, 32–34, 114, 124
  in gamma spectra, 169, 174
  in secular equilibria, 37
  spontaneous fission and cluster decay, 122–124, 227, 229

Breast boost, 327
Bremsstrahlung, 111, 141, 275, 286, 292–293;
          see also X–rays, continuous
     in gamma spectra, 173–174
     minimizing, 162, 286, 325
Brookhaven National Laboratory, 32, 214

C

Calibration factor (C_{cal}), 317, 327
Carbon dating, see Dating
Carrier, 191–192, 221
Cascades, 157–158
Cathode ray tube (CRT), 301–302
Centrifuges, gas, 72–73
Cesium-137, 28, 94, 187, 285–286
     gamma spectrum, 165–169, 175
     significant releases, 78, 262
Chart of the nuclides, 18–19, 104, 115–116, 124,
          210–211, 215
     online, 19, 105, 215
     printed, 19, 109, 119
Chelates, 195
Chemical vs. nuclear processes, 56, 67–68,
          187, 203
     equilibria, 40
     reactions, 6, 203, 213
Chemical vs. nuclear stability, 10
Cherenkov radiation, 142
Chernobyl, 75–79, 268
     cause, 76–78, 254, 264–266
     compared to Fukushima, 78–79, 266
     health effects, 76, 79, 266, 281–283
     radiation release, 16, 77, 262
China Syndrome, The, 77
Chi-squared, 45–46, 403
Cobalt-60, 187, 240, 250, 269
     gamma spectrum, 163–164, 171
     in radiation therapy, 183–185, 296–298, 321
     in weapons, 92
Coincidence, 158
Cold War, 74, 83, 87
Collimators, 173, 312–314, 316–318
     in CT scans, 181
     on gamma cameras, 199–200
     in radiation therapy, 298–299, 302–303, 305,
          311–312
          with electron beams, 325–326
          with proton beams, 328–330
     in X-ray production, 181, 295
Colloids, 194–195
Compensators, 328–330
Compound nuclei, 203, 207–209, 213–214
     energy, 207–209
     in fission, 230–232, 234–235, 243
Compton continua, 167–169, 174–175

Compton edge, 167–16
Compton electron, 143–145, 168, 298, 311
Compton scattering, 143–145, 298
     in gamma spectra, 167–168, 174–175
     in medical applications, 181, 200, 315–316
     relative probability, 146–147, 166–167,
          274, 298
Computed tomography (CT), 17, 181–182,
          196–197, 279, 328
Condoms, 187
Cone, see Electron, applicator
Conservation of angular momentum, 102, 104, 207
Conservation of energy, 51, 141, 142, 143
     in alpha decay, 98–99, 101
     in beta decay, 102–103
Conservation of momentum, 98, 205–206, 228
Conservation of particles, 102
Conversion, 255
Conversion coefficients (α_T), 122, 168–169, 193
Conversion electron, see Electron, conversion
Cosmic radiation, 15, 30, 222–224
     dose from, 17, 273, 280
Cosmogenic nuclides, 15, 222–224
Coulomb barriers (E_{cb}), 206, 227–228, 233, 237,
          239, 243, 269
Coulomb forces, 5–6, 53–54, 124, 328
     in accelerators, 219–221
     between nuclei, 205–207, 221, 227–228, 243
     in electron scattering, 140, 142
     in large nuclei, 10, 53–54, 242
Count rates, 24–25, 27, 32, 41–45, 156–159
Criticality, 249
Critical masses, 68, 248
Curie, Marie, 26
Cyclotrons, see Accelerators

D

Dating, 30–32, 223–224
Daughters, 34–35, 193
     in alpha decay, 56, 97–99, 101–102, 198
     in beta decay, 103
     in cluster decay, 123
     in decay diagrams, 57, 62
     in delayed particle emission decay,
          125–126, 237
     in double beta decay, 127
     in electron capture decay, 108–111, 113
     in equilibria, 34–41, 94, 190–191
     in isomeric transitions, 61–62
     in nuclear forensics, 94
     in positron decay, 107
Dead time, 45, 158–159
Decay, 6, 9, 24–28, 193, 203, 204
     alpha, 6, 97–102
     beta, 6–7, 58, 60–61, 102–106, 162, 230

branched, 16, 32–34, 37, 102, 113–115,
    123–124, 174, 227
cluster, 123–124
constant (k), 25, 27, 33, 35, 216
delayed particle emission, 124–126
diagrams, 57–62, 101, 104–105, 107–108,
    109, 116, 118
double beta, 126–127
electron capture, 7, 60, 108–113
gamma, 6, 8, 61–62
internal conversions, 121–122, 168
isomeric transitions, 119–122
isotropic nature, 24, 162
multiple modes, 32–34, 113–114
positron, 7, 59, 106–108, 115
proton/neutron emission, 124
rate, see Activities
series, 15–16, 31, 34, 36–37, 101–102, 183
spontaneous fission, 122, 227–229, 233
warming the Earth, 56–57
Dees, 219–221
Delayed particle emissions, 124–126, 237, 250
Delta rays, 132
Detectors, 24, 27, 46, 155–164
    charge-coupled device (CCD), 182
    gas-filled, 155–159
        Geiger-Müller tubes, 155, 157–159, 278
        ionization chambers, 155–156, 278,
            302–303
        proportional counters, 156–157
    scintillation, 159–163
        inorganic, 160–162, 165–168, 175, 182,
            199–201
        organic, 162–163
    semiconductor, 163–164, 175, 201
        CZT, 164, 201
    smoke, 136–137
    storage phosphor, 182
    well, 171–172
Diffusion, gaseous, 72–73, 94
Dirty bombs, 93–94
Doses, 1, 13–15, 273
    background, 15–17, 273, 280
    buildup region, 298
    depth, 309–310, 317, 318
    effective, see Effective dose equivalent
    effective limits, 277–278
    equivalent, 274–276
        rate, 288
    exit, 317–318, 328
    fractionation, 322
    nuclear medicine, 193
    nuclear plant accidents, 75–79, 265
    occupational, 74–75, 164–165, 179, 183,
        277–278, 284
    patient

    CT, 181–182
    radiation therapy, 296–297, 299–300,
        309, 325
    percentage depth (PDD), 296–297, 309–319, 326
    rate, 16, 284–285, 313–315, 323
        change, 313–315
    recommendations, 276–278
    risk, 14–15, 179–180, 279–284
    units, 274
Doomsday device, 92
Dosimeters
    pocket, 79, 156
    thermoluminescent (TLD), 164–165, 278

E

E. coli, 187
Effective dose equivalent, 276
Efficiencies, percent, 27, 33, 159
    organic scintillators, 162
    semiconductor detectors, 164
    well detectors, 171
    X-ray detectors, 182
Elastic collision (scattering), 132, 142
Electromagnetic
    fields, 221
    forces, see Coulomb forces
    radiation, 1–3, 132, 141, 157–158
    spectrum, 2–3, 131, 273
    waves, 220, 300–301
Electrons, 4, 8
    applicator, 306, 325
    Auger, 112–113, 145, 198
    beams
        dosimetry, 186, 325–328
        in food irradiation, 187
        generation, 185, 302, 306
        range in tissue, 139–140
        in X-ray production, 291–292, 294–295,
            302–304
    capture, see Decay
    conversion, 121–122, 168
    energy levels (shells), 4, 10
    equilibrium, 298, 311
    gun, 301–303
    incident, 139–141
    orbitals, 4, 7, 108–110, 293, 328
    scattering, 140–142, 186, 306, 325–326
        backscattering, 142
        elastic, 142
        inelastic, 140–141, 292
Electron volt, 51
Elements, 4, 10, 23, 122, 163
    artificial, synthesis of, 241–243
    symbols, 5
    transuranic, 241–243

Endoergic, 204–205, 207, 240
Eniwetok Island, 242
Equilibria, 34–41, 94–95
  no, 39–40
  secular, 35–37, 95
  transient, 37–39, 190–191
Equipment attenuation factor (C_{attn}), 316–317
Excitations, 1–2, 132–133
  detectors, in, 157, 159–160, 207
  energy, 133, 206, 208, 214, 229, 232,
    234–236, 243
Excited state, 8, 62, 203, 209
  in alpha decay, 97, 99–100
  electronic, 110, 132
  high-energy, 125
  in isomeric transitions, 119–122, 169–171
Exoergic, 57, 204, 240
Exothermic, *see* Exoergic
Exposure, 273–274
  calculating, 273
  from Chernobyl, 76–78, 254
  and dose, 17, 273, 275
  health effects, 14, 179–180, 273, 279
  managing, 17, 164–165, 277–278, 284–285
  rate constant, 285
  units (R), 273
  from uranium mining, 75

F

Fat Man, 89–90
Fertile materials, 74, 230, 247–248,
  255–256, 258
f factors, 275–276
Fields
  moving, 324–325
  size, 311, 312
  standard, 312
  symmetry, 312–313
Filtration, 151, 293
Fissile materials, 67–68, 87, 90–91, 94,
  247–248, 251
Fission, 53–55, 227–237
  asymmetric, 231–233
  barrier, 227–228, 233–235
  bombs, 87–90, 92
  chain reaction, 68–69, 230, 247–248
  cross section, 233–234, 247–249, 256
  energy, 55, 67–68, 227–229, 234–237
    delayed, 236–237
    prompt, 236
  neutron-induced, 67–69, 229–237, 247
  photo, 237
  products, 67, 77–80, 188, 227, 230–232, 236,
    256, 260, 262–263
  spontaneous, 90, 122–123, 227–229
  symmetric, 231–232

yield, 228, 230–231
  cumulative, 231
  independent, 230
Fissionable material, 230
Fizzle, 90
Flatteners, 303–305, 323, 325
Fluorescence yield (ω_r), 113, 193
Fluorodeoxyglugose–^{18}F (FDG), 196
Flux (φ), 89, 214–215, 240–242, 259–260,
  262, 263
Food irradiation, 187–188
Forensics, nuclear, 93–95
Fossil fuels, 30, 66, 69, 75–76, 85, 255
Frequency, 2–3
Fukushima
  ALARA, 15
  cause, 78, 266–268
  health effects, 76, 280, 283
  impact, 66, 79
  radiation released, 16, 78, 262
Fusion, 53–55, 237–240
  bombs, 90–92
  cold, 225, 240, 243
  D+T, 269–270
  heat of, 56
  ignition temperature, 239, 269
  reactors, 84, 240, 269–270
  stars, in, 237–240
  technologies in nuclear medicine, 197

G

Gamma decay, *see* Decay
Gamma rays, 2–3, 8, 111, 120, 131
  attenuation, 148–151, 286
  vs. conversion electrons, *see* Conversion
    coefficient
  detecting, 155, 160–161, 162, 165, 199
  interactions with matter, 13, 143–148,
    273–274
  prompt, 229, 236
  units, 62
  used in diagnostics, 193, 198
  used in food irradiation, 187
  used in therapeutics, 183, 197–198
Gamma spectroscopy, 155, 165–175
  annihilation peak, 171–172
  backscatter peak, 168
  Compton continuum, 167–169
  Compton edge, 167–169
  energy resolution, 161–164, 168, 175
  escape peaks, 173
  full width at half maximum (FWHM), 175
  iodine escape peak, 170
  lead X-ray peak, 173, 200
  peak intensity, 162–163
  peak-to-total ratio, 174–175

photopeak, 166
sum peak, 171
X-ray peak, 168–169
Gaussian distribution, 41–42, 45
Geometry
beam, 148, 150
sample, 159
Global climate change, 65–66, 76
Graham's Law, 72
Ground state, 8, 62, 104–105, 115, 207–209
in alpha decay, 99–100
electronic, 110, 132, 159, 164
energy of transition, 99
in isomeric transitions, 120–122, 168, 170–171
Gulf wars, 73, 279

H

Half-life, 25
branched decay, 32–34
in equilibria, 35–41
nuclide production, 218, 242
for radiopharmaceuticals, 193
Half-value layer (HVL), 149–150, 151, 286
Hanford, Washington, 83
Health effects, 13–14, 76, 78, 179–180, 279–281, 282–283
Helium, 57, 137, 183, 253, 258
Hiroshima, 88, 179, 262, 279, 283
Hybrid technologies, 197
Hydraulic fracturing, 85

I

Idaho National Laboratory, 84, 174
Inelastic collisions, 132
Inelastic scattering, see Electrons
Interference, 301
Internal conversions, see Decay, internal conversions
International Commission on Radiological Protection (ICRP), 276–277
International Thermonuclear Experimental Reactor (ITER), 270
Inverse square law, 284, 310, 313–314, 319, 321–322
Ionization, 2, 13, 131, 132–133, 187
air, 135
chamber, see Detectors, gas–filled, ionization chambers
in detectors, 155–158
specific, 135–139, 273
X–ray production, 292
Ion pairs, 132, 135, 137, 139–140, 273
in detectors, 155–157, 159
Irradiation facilities, 187–188

Isobars, 5, 114–119, 127, 189, 240, 242, 323
beta decay, 9, 114–119
fission products, 230–231
Isodose curves, 323–324, 326, 329
Isomeric transitions, see Decay, isomeric transitions
Isomers, 8, 119–120, 168, 191–192
Isotones, 5
Isotopes, 4–5, 72, 241–242, 251
average atomic mass, 23
chemical behavior, 15, 90, 94, 191–192, 196, 198
magic numbers, 10

J

Japanese government, 78, 267, 283–284
Jornada del Muerto, New Mexico, 89

K

Kerma, 275
Kinetic energies, 131–133, 142, 205–206
alpha decay, 56, 98–99
Auger electron, 112
compound nucleus (E_{KC}), 207–208, 234
conversion electron, 121
fission products, 227–229, 235–236
pair production, 146
positron, 107
projectile (E_{KX}), 205, 207
relativistic, 134, 139, 301–302
released in media, see Kerma
temperature, 239
thermal neutron, 234
X–ray production, 295
Kits, 191
Kitty litter, 263
Knolls Atomic Power Laboratory (KAPL), 19, 119
Klystrons, 300, 303
K-25, Oak Ridge, Tennessee, 72

L

Large Hadron Collider (LHC), 222
Ligands, 194–195
Linacs, see Accelerators, linear
Linear energy transfer (LET), 135–136, 139, 273–274
Linear interpolation, 318, 322
Linear non–threshold (LNT) model, 14–15, 282–283
Little Boy, 88, 90
Litvinenko, Alexander, 152
Los Alamos National Laboratory, 183

M

Magic numbers, 10–12, 54, 123, 227, 231–233, 242
Magnetic resonance imaging (MRI), 180, 197
Magnetrons, 300, 302–303
Mail irradiation, 187
Manhattan Project, 83, 87–89
Mass defect, 52, 54
Mass number (A), 5–8, 28, 99, 204, 206–207, 227
    trends with, 53, 55, 115–119, 229–233
Mattauch's rule, 119, 242
Mean, 41–45
Megatons to megawatts, 74, 86
Metalloids, 163
Metastable states, 8, 119–120, 122, 168
Microwaves, 1–2, 220, 280, 300, 302
Monitors, 278, 302–304, 311
    units (N_{mu}), 317
Monoenergetic, 97, 102, 120, 148, 150, 161
Multichannel analyzers (MCA), 161
Muons, 186, 223

N

Nagasaki, 83, 89, 179, 262, 279, 283
National Council on Radiation Protection and
        Measurement (NCRP), 17, 277
Naturally occurring decay series, see Decay, series
Neutrinos, 176, 238–239
    cosmic radiation, 223
    electron capture, 108, 110
    positron decay, 106
Neutron activation analysis, 217
Neutron numbers (N), 5, 7, 9–10, 16, 19,
        115–116, 117
Neutrons, 4–5, 7, 187, 223, 232
    bomb, 92
    capture, 8, 198, 204, 269
        in nuclear reactors, 70, 74, 86, 189,
            234–235, 247, 258
        in stars, 240–241
        in weapons, 90–92
    cold, 213
    delayed, 126, 237, 250
    economy, 251–253, 256–257
    emission, see Decay, proton/neutron emission
    fast, 69–70, 76, 88, 91, 213, 235, 258–259
    flux, 215
        in stars, 240–241
        reactor, 69–70, 89, 249, 259, 260
        weapons, 242, 262–263
    howitzer, 218
    incident, energy, 213–214, 232, 235, 249,
            254, 256
    initiator, 87–89
    mass, 4, 51, 54–55, 405
    prompt, 67–68, 88, 124, 227, 229, 236, 250

quality factor, 275
reflector, 88, 90
stability, 10–11, 103
thermal, see Projectiles
Neutron to proton ratios (N/Z), 9–10, 12, 115–118,
        241–242
    in alpha decay, 101–102
    extreme values, 124–126
    in fission, 230
    multiple decay modes, 114
Nevada Test Site, 82
Nonproliferation treaty (NPT), 86
Nuclear fuel cycles, 80–81, 86, 94
Nuclear medicine, 17, 188–201, 282
    gamma cameras, 199–201
        detector heads, 200–201
    hybrid/fusion, 197
    molecular imaging, 196–197, 221
    nuclide production, 188–192, 221–222
    radiopharmaceuticals, 188, 192–196
        biodistribution, 193
        diagnostic agents, 193–196
        myocardial imaging, 194
        synthesis, 194–196
        therapeutic agents, 186, 197–198
    typical dose, 280
Nuclear power reactors, 68–85, 247–270
    boiling water (BWR), 71–72, 253–254, 259,
            261–262, 266–268
    calandria, 252, 253, 260
    Canadian deuterium uranium (CANDU),
            251–252, 259, 260–261
    control rods, 69–70, 249, 252, 260, 265
    core, 70–71, 77–79, 142, 250, 254, 265
    coolant, 70–71, 76–77, 249–253, 258–259
    cost, 84–85, 255
    fission product decay heat, 260, 263, 267–268
    fuel bundles, 252–253
    fuel rods, 69, 76–78, 79, 249–250, 252, 260,
            262, 264
    fusion, 269–270
    gas–cooled (GCR/GFR), 253, 256, 258
    generation III, 258, 261
    generation IV, 257–259
    light water graphite, (RBMK), 251, 253–254,
            264–266
    liquid fluoride thorium (LFTR), 257
    metal–cooled fast, 258
    mixed oxide (MOX) fuel, 74
    moderator, 70, 249, 251–254, 258
    meltdown, 69, 77, 250, 262
        Chernobyl, 254, 264–266
        Fukushima, 66, 78–79, 266, 284
        Three–Mile Island, 77, 263–264
    natural, 81
    nuclide production, 74, 188–189
    pebble bed, 253

poison, 256
pressurized water (PWR), 70–71, 252, 259–262
pressurized heavy water (PHWR), 243–245, 248
radiation releases, 16, 77–78, 262–263, 278
reactions in, 247–250
safety, 75–79, 259–262
small modular (SMR), 254–255
supercritical water–cooled, 259
thorium–fueled, 74, 255–257
very high–temperature, 258
void coefficient, 253–254, 264–265, 270
Nuclear reactions, 6, 8, 203–219, 330
with cosmic radiation, 15, 222–224
cross section (σ), 69, 212–214, 233–234, 239
effective Coulomb barrier (E_{ecb}), 206–207, 210, 239
energy, 70–71, 204–212, 233–237, 238–239
fission, 67–68, 87–88, 188, 229–237
low–energy, 210–211
medical applications, 188–189, 328
shorthand, 203–204
thermonuclear, 239
threshold energy (E_{tr}), 206, 209
yield, 212, 214–219, 230–233
Nuclear Regulatory Commission (NRC), 277
Nuclear stability, 9–12, 101
toward beta decay, 114–119, 124–125, 230, 242
toward cluster decay, 123
toward fission and fusion, 53–55, 234, 237, 240
in fission products, 231
Nuclear wastes, 16, 79–84, 256–257, 258, 259
forms, 84
from fusion, 269
geologic repository, 80–82
high–level, 82–84, 263
low–level, 81, 84
Nuclear weapons, 16, 82, 87–93
bunker buster, 92–93
clean vs. dirty, 91–93
fallout from, 16, 262, 279
fission, 87–90
gun–type, 87–88
implosion–type, 88–90
neutron, 92
proliferation, 73, 86–87, 93, 189, 257
testing, 82
thermonuclear, 90–92, 242, 270
fission primary, 90–91
neutron reflector, 90–91
yield, 88–91
Nucleons, 5–7, 204, 222–223
and binding energy, 51–55
in isomeric transitions, 8, 120
magic numbers, 10–11
Nuclides, 4–5, 7, 34, 110
binding energy, 53–54

cosmogenic, see Cosmogenic nuclides
charts of, see Chart of the nuclides
positron vs. electron capture decay, 109
in reactions, 203
specific activity, 29
stability, 9–12
Nuclide symbols, 4–5, 8, 119
Nuclidic states, 57, 97, 100–101, 120

O

Oak Ridge National Laboratory, 84, 213, 257
Object scattering, 174
Oklo mine, Gabon, 81
Oppenheimer, J. Robert, 89
Output factor (O_f), 327

P

Pair production, 143, 145–148, 166
Pantex plant, 74
Parents, 34
in alpha decay, 97, 99, 102
in beta decay, 103
decay diagrams, 57, 104–105
in equilibria, 34–41, 94
in positron and electron capture decay, 109
in radionuclide generators, 190–191
Parity, 120–121
Partial half-lives, 34
Peak scatter factor, 316
Penumbra, 298–299, 304, 323, 328–329
Phantom, 309, 316, 320, 321, 324
Phases
state of matter, 77, 194–195, 254
electromagnetic wave, 220, 300–301
Photoelectric effect, 143, 145
in detectors, 157, 160, 165–166, 200
in gamma spectra, 165–166, 170, 173–175
relative probability, 146–147, 166–167, 175, 274
Photoelectrons, 145, 160–161
Photomultiplier tubes (PMT), 159–160, 161, 162, 164, 172, 199–201
Photons, 1–3; see also Electromagnetic, radiation; Gamma rays; X–rays
annihilation, 106, 146, 173, 194, 200
attenuation, 147–151, 285–286
in detectors, 157–158, 159, 160–161, 162, 163
energy and dose depth, 297–298
incident, 143–144, 146, 161
interactions with matter, 6, 157–158, 220; see also Compton scattering; Pair production; Photoelectric effect
in nuclear medicine, 196, 200
polyenergetic beams, 148, 151
torpedoes, 106, 129

Photosynthesis, 198
Pions, 186, 222–223
Planck's constant (*h*), 3, 405
Plasma, 91, 238, 239, 269
Plutonium-239, 67, 230, 232–234, 242
    in reactors, 73–74, 247–248, 251–252,
        256, 258
    in waste, 79–81
    in weapons, 86, 87–90, 91, 94
Poisson distributions, 45–46
Positrons, 7, 57–60, 106–107, 131, 146, 238
    decay, *see* Decay
    emission tomography (PET), 107, 194,
        196–197, 200, 210
    gamma spectra, 171–173
    interactions with matter, 137–140, 146
    maximum energy, 62, 107–108
Postal Service, U.S., 187
Products, 8, 203–204, 210–211, 213–214
    fission, 67, 189
    for nuclear medicine, 189, 198
Projectiles, 8, 187, 203–210, 243
    from accelerators, 189, 219–222
    cosmic, 222
    electrons, 138–142, 295
    energy and product(s), 213–214
    effective Coulomb barrier and *Z*, 210
    fission, 237
    flux, 215
    fusion, 239
    kinetic energy, *see* Kinetic energies,
        projectiles
    recoil, 98
    thermal neutrons, 67, 72, 74, 204–205,
        213–214, 230–236
Promethium, 241–242
Proton cycle, 238–239, 269
Protons, 4–7, 12, 57, 210, 221–222
    delayed, 125–126
    emission, *see* Decay, proton/neutron
        emission
    LET, 139
    mass, 4, 51, 54, 405
    nuclear stability, 9–11, 53–54
    radiation therapy, 185–186, 328–330
    scattering
        active, 329–330
        passive, 329
Pulse height analyzers (PHA), 200

Q

Quality factors (*Q*), 274–275
Quantum number, principal (*n*), 109–110
Quarks, 5, 7, 102
Quench, 158, 162

R

Radiation, 1–3, 12–13
    anthropogenic sources, 17, 242
    electromagnetic, *see* Electromagnetic,
        radiation
    fear, 1, 3, 13, 75, 82, 92, 93, 155, 179, 273,
        281–282
    hormesis, 14–15, 180, 282–283
    ionizing, 12–13, 131
    naturally occurring, 15–17, 223–224
    primary and secondary, 132, 275,
        311–312, 328
Radiation therapy (RT), 182–186, 197–198,
        282, 309; *see also* X–ray beams,
        high–energy
    electron beam, 185–186, 220, 306, 325–327
    gamma knife, 185
    history, 182–184, 296
    image–guided, 184–185
        cyberknife, 185
    implants, 183, 185, 186
        remote afterloading, 185
    intensity–modulated, 184, 324
    isocentric, 183–184, 299–300, 324–325
    megavoltage machines, 183–184, 296–297
    molecular, 186, 198
    proton beam, 185–186, 328–330
    respiratory–gated, 185
Radioactive, 9, 15
Radioactivity, 3–4
Radiological dispersal devices (RDD), *see* Dirty
        bombs
Radionuclide generators, 190–192
Radiotracers, 198
Radon, 16–17, 75, 179, 273, 279
Ramsar, Iran, 289
Random errors, *see* Uncertainties
Ranges, 5–6, 136, 138–140, 326, 328–329
Range shifters, 329
Recoil, 143, 228
    alpha decay, 56, 62, 98–99, 101, 198
    beta decay, 105–106
    beta plus neutron decay, 125–126
    cluster decay, 123–124
    positron decay, 108
    proton/neutron emission, 124
Reflectance, 294
Relative biological effectiveness (RBE), *see*
        Quality factors
Relative standard deviations (RSD), 43
Relative stopping powers (RSP), 138
Relativity, theory of, 134
Reprocessing, 79–81, 86, 93, 256–257
Recovery times, 158
Resolving times, 152, 164

Resonance, 300
Resonance capture, 213
Resonance energies, 213–214
Resonance integrals, 214, 256
Risks, 273, 280–283
    food irradiation, 187
    from ionizing radiation, 13–15, 179–180,
        281–283
    from nuclear reactors, 75–76, 78–79, 281–283
Rutherford, Ernest, 205, 306

S

Saturation, 216–219
Scattering angles (θ), 141, 143–144, 168
Scintillation cocktails, 162
Scintillators, 159–163, 165, 182, 199–201; see
        also Detectors
Secondary electrons, 132, 138, 275, 299, 328
Secondary ionizations, 132
Semi-empirical mass equation, 115, 117
Sestamibi-^{99m}Tc, 194–195
Shielding, 73, 138, 141–142, 165, 278,
        284–287, 330
    atmospheric, 15, 280
    in electron beam therapy, 325–327
    for high–energy photons, 13, 165, 173, 185
    in nuclear reactors, 72, 257, 262
    for nuclear waste, 79
Shroud of Turin, 31
Significant figures, 25, 31, 45, 52, 229
Single photon emission computed tomography
        (SPECT), 196
Skin sparing, 297–298
Smoke detectors, 136–137, 152
Society of Nuclear Medicine (SNM), 197
Source to axis distance (SAD), 320, 321–322, 324
Source to collimator distance (SDD), 299
Source to surface distance (SSD), 299, 310,
        312–315, 317–319, 320, 321, 324, 325
Specific activities, see Activities
Specific ionization (SI), 135–137, 139–140, 273
Spent fuel, see Nuclear wastes
Spin, 102, 120–121, 169–171
Standard deviation, 42–44
    of the mean, 43–44
Standing wave accelerator guides, 301
Stanford linear accelerator, 220
Star Trek, 7, 51, 106, 129, 177
Star Wars, 6, 54
Statistics, 41–46
Steady state, 40
Stellar nucleosynthesis, 240–241
    p–process, 240–241
    r–process, 241
    s–process, 240–241

Strong force, 5–6, 51
    and binding energy, 52–54
    in fission and reactions, 228, 233
    in large nuclides, 10, 242
Subcritical masses, 87, 88, 259
Sum of All Fears, The, 93
Supercritical masses, 68, 87–90, 248
Supernovae, 241
Swipe tests, 176, 278, 284
Systematic errors, 46

T

Tampers, 88–89, 91–92
Tampons, 187
Targets
    in accelerators, 219, 241–243
    organs, in nuclear medicine, 192–194,
        196, 198–199
    in radiopharmaceutical production,
        188–190, 222
    in reactions, 8, 203–204, 207, 210, 239
        and yield, 213–215
    in X–ray production (anodes), 185, 291–295,
        302–305
Technetium, 5, 194, 241
Technetium-99m, 8, 119, 145, 190–196, 285
Tenth-value layers (TVL), 149, 286–287
Therapeutic ratios, 182–183, 185–186
Thermoluminescence, 164
Thermoluminescent dosimeters (TLDs), 164–165
Thermonuclear reactions, 239
Thermonuclear weapons, 90, 242
Three Mile Island (TMI), 16, 66, 76–77, 85,
        262–264, 266, 280
Tissue-air ratio (TAR), 320–321
Total energy of decay (E), 19, 55–57, 62, 126
    in alpha decay, 56, 98–99
    in beta decay, 58, 104–105
    in electron capture, 60, 109–110, 112
    in gamma decay, 61–62, 121
    in positron decay, 59, 107–108, 110, 117
    in spontaneous fission, 228–229
Total skin irradiation (TSI), 327
Transition, energy of, 57, 97, 99–100, 107, 121, 126
Transmission, 294–295, 302
Transmission factors, 148, 287
Traveling wave accelerator guides, 301
Trinity, 89
Tunneling, 212–213, 233, 239–240, 269

U

Uncertainties, 42–45
United States government, 17, 66, 73–74, 80–81,
        85, 86, 279

Uranium, 93, 241, 256, 280
  consumer products, 179
  depleted, 73, 88, 94, 258, 279
  disposal, 80, 82
  enrichment, 72–73, 80, 86, 94, 251, 258
  major deposits, 72, 75
  mining, 75, 80
  weapons–grade, 73, 86, 88, 94, 189

V

Valley of beta stability, 114–119, 124–125, 242;
    see also Belt of stability
Vitrification, 84

W

Waste Isolation Pilot Plant (WIPP), 81–82, 263
Waveguides, 300–303
Wavelength, 2–3, 148
Wedge filter, 324
Weizsäcker, C.F. von, 115, 117
W-quantity, 135

X

X-ray beams, high energy, 183, 296–305
  flatness, 304–305
  symmetry, 184–185, 304–305
  why?, 296–299
X-rays, 2–3, 13
  attenuation, 148–151
  characteristic, 111–113
  continuous, 111, 141; *see also* Bremsstrahlung

detection of, 155, 160, 162, 164, 182
dosimetry, 309–325
in gamma spectra, 168–169, 170, 173
generation of, 110–111, 131, 140–141,
    291–306
  conventional tubes, 291–294
  efficiency, 295–296
  linear accelerators, 183, 300–305
interactions with matter, 143–148, 275–276,
    286, 298
medical applications, 151, 180–182,
    182–185
quality factor, 274–275
risk, 17, 279–281, 286
safety interlocks, 305
X-ray tubes, conventional, 181, 291–294
  filament, 291–292
    current, 291
  focal spot, 291–292
  inherent filtration, 292–293
  orthovoltage, 182, 296–299, 305
  target, 291–292, 294–295
    shape, 294
  tube current, 291
  tube voltage, 291–292, 294

Y

Yucca Mountain, 82–84
Y-12, 74

Z

Zirconium, 250, 253, 264–265, 268